Geomorphology of Rocky Coasts

Tsuguo Sunamura

*Institute of Geoscience, University of Tsukuba,
Ibaraki 305, Japan*

JOHN WILEY & SONS
Chichester · New York · Brisbane · Toronto · Singapore

Copyright © 1992 by John Wiley & Sons Ltd,
Baffins Lane, Chichester,
West Sussex PO19 1UD, England

All rights reserved.

No part of this book may be reproduced by any means, or transmitted, or translated into a machine language without the written permission of the publisher.

Other Wiley Editorial Offices

John Wiley & Sons, Inc., 605 Third Avenue,
New York, NY 10158-0012, USA

Jacaranda Wiley Ltd, G.P.O. Box 859, Brisbane,
Queensland 4001, Australia

John Wiley & Sons (Canada) Ltd, 22 Worcester Road,
Rexdale, Ontario M9W 1L1, Canada

John Wiley & Sons (SEA) Pte Ltd, 37 Jalan Pemimpin #05-04,
Block B, Union Industrial Building, Singapore 2057

Library of Congress Cataloging-in-Publication Data

Sunamura, Tsuguo, 1941–
  Geomorphology of rocky coasts / Tsuguo Sunamura.
    p. cm. — (Coastal morphology and research)
  Includes bibliographical references and index.
  ISBN 0-471-91775-3
  1. Coasts. I. Title. II. Series.
  GB451.2.S86 1992
  551.4′57—dc20   91–47199
                                                                  CIP

British Library Cataloguing in Publication Data

A catalogue record for this book is
available from the British Library

ISBN 0-471-91775-3

Typeset in 10/12 Times by Mathematical Composition Setters Ltd, Salisbury, Wiltshire
Printed and bound in Great Britain by Biddles Ltd, Guildford and King's Lynn

Staying at WG

Geomorphology of Rocky Coasts

Coastal Morphology and Research

Series Editor: Eric C. F. Bird

CORAL REEF GEOMORPHOLOGY
André Guilcher

COASTAL DUNES
Form and Process

Edited by
Karl Nordstrom, Norbert Psuty and Bill Carter

GEOMORPHOLOGY OF ROCKY COASTS
Tsuguo Sunamura

Contents

Foreword ix
Preface xi

Chapter One: Introduction 1

Chapter Two: Nearshore wave field 5
  Wave development, analysis, and forecasting 5
  Linear wave theory 8
    Water surface elevation, wavelength, velocity, and period 10
    Water particle motions and pressures 13
    Wave energy 16
    Wave height 17
    Wave refraction 18
  Breaking waves 20
  Broken waves and wave set-up 25
  Nearshore currents 28
  Types of waves at the cliff base and their pressure 29

Chapter Three: Water level factors: tides and storm surges ... 38
  Introduction 38
  Tides 38
  Storm surges 44

Chapter Four: Strength of landform materials 48
  Introduction 48
  Basic physical properties 48
  Mechanical strength 50
    Unconfined (or uniaxial) compressive strength 52
    Tensile strength 54
    Shear strength 56
    Impact strength 59
    The Schmidt hammer rebound number 60
    Penetration strength 63

Compressive wave velocity 63
Representative strength parameter 65
Influences of discontinuities on strength 65
Deterioration of strength 68
   Fatigue 68
   Weathering 70
   Biological influences 74

Chapter Five: Processes of cliff erosion 75
Factors affecting cliff-base erosion by waves 75
Assailing force of waves 75
Resisting force of cliff materials 79
Force–erosion relationships: a laboratory approach 80
Erosion rates and measuring techniques 86
Temporal variations in erosion rates 88
Spatial variations in erosion rates 96
Changes in cliff profiles 107

Chapter Six: Underwater bedrock erosion 117
Introduction 117
Factors affecting bedrock lowering 117
   Assailing force 117
   Resisting force 120
   Basic relationships for bedrock lowering 122
Rates of lowering 123
Processes of lowering 126
Wave-base problems 132

Chapter Seven: Shore platforms 139
Major contemporary morphologies 139
Demarcation of platform types 150
Type-A platforms 153
Type-B platforms 163
   Seaward drops and platform initiation 166
   Platform elevation 172
   Platform width 177
A rocky coast evolution model 180

Chapter Eight: Some characteristic erosional landforms 184
Notches 184
Sea caves 188
Sea arches and stacks 190
Ramparts 193

Contents vii

Ramps 196
Potholes 196
Solution pools 200
Tafoni and honeycombs 202

Chapter Nine: Effects of human activity on rocky coasts 209
Introduction 209
Effectiveness of engineering structures 209
Bedrock scouring in front of sea walls 217
Cliff erosion controlled by beach sediment volume 218
Human activities and mass movement 221
Future sea level rise and cliff erosion 225

References 229
Appendix One: Conversion factors 257
Appendix Two: Worldwide coastal cliff erosion rates 264
Author Index 279
Location Index 287
Subject Index 293

Foreword

The Coastal Morphology and Research Series, having dealt with coral reefs and coastal dunes, continues with this volume by a Japanese scientist, Tsuguo Sunamura, on rocky coasts.

Professor Sunamura is chairman of the Graduate School of Geosciences at the University of Tsukuba, in Ibaraki, not far from Tokyo. He has a distinguished record of research and publication in Japanese and English, and has specialized in the study of cliffs and shore platforms on the coasts of Japan. Over the past two decades I have discussed these topics with him in correspondence and during my visits to Japan on behalf of the United Nations University in Tokyo. I thus became aware of the diversity and interest of Japanese rocky coasts, and of Professor Sunamura's interpretations of them, using laboratory experiments and mathematical modelling.

Treatment of rocky coasts in geomorphological textbooks is generally disappointing, perhaps because cliffs and shore platforms are a deceptively simple landform association. We are told that the sea has taken a bite out of the land; that cliffs are receding; and that hard rock formations persist as protrusions when softer outcrops have been more quickly excavated. Yet there are many different kinds of rocky coast, some of them without either cliffs or shore platforms, and closer investigations have found an array of physical, chemical and biological processes at work, with weathering effects, both destructive and indurative, playing a varying rôle.

In contrast with their prolific work on beaches, barriers, lagoons, marshes, deltas and coastal plains, American geomorphologists have given little attention to rocky coasts. Doubtless this is because of the scarcity of cliffs and shore platforms along the Gulf and Atlantic seaboards of the United States. In Japan, as in Britain, France and New Zealand, there are many rocky coasts that have attracted the attention of geomorphologists, and it seemed appropriate to ask Tsuguo Sunamura to prepare this volume, giving the perspective of a Japanese scientist on rocky coast problems. As in his research papers, he has adopted the stance of an engineer as well as a geomorphologist, expressing relationships in symbolic terms, and developing equations to illustrate process and response on these solid structures.

My editorial task included some minor modification of the English text for the sake of clarity, but I have endeavoured to conserve Professor Sunamura's style and approach. The result, we hope, will provide students with an introduction to the study of rocky coast geomorphology and an insight into research problems. It also provides a timely discussion of the effects on rocky coasts of the forecast global sea level rise during the coming century, one of the problems that geomorphologists and engineers must now address.

Eric Bird

Preface

Deeper understanding of contemporary processes, the resulting landforms, and changes in them is of vital importance in furthering the study of evolutionary development of landscapes with which the discipline of geomorphology has been primarily concerned. This understanding also facilitates the solution of the engineering, environmental, and planning problems associated with changes of the earth's surface, and furthermore, it provides a sound foundation for studying the prediction of landform changes that might occur in the near future. To emphasize the importance of the study of present-day processes and landforms I have been teaching undergraduate and graduate courses in coastal geomorphology at the University of Tsukuba. It was in 1987 that Professor Eric Bird of the University of Melbourne, the editor of the Coastal Morphology and Research Series, asked me to write a book on rocky coasts.

This book is based primarily on my lecture notes, incorporating up-to-date research results selected on a global basis. The book first describes the fundamentals of wave dynamics, tides, and rock mechanics, to provide a physical basis for studying rocky coasts. Emphasis is given to illustrating contemporary processes of cliff recession and sea-floor erosion. Shore platforms are discussed and contrasted with plunging cliffs on the basis of process–form relationships, and a new model for rocky coast evolution is presented. Other characteristic morphological units, such as notches, sea caves, arches, potholes, and tafoni are described. Finally, various impacts of human activities on rocky coasts, including the effects of a global sea level rise which could occur during the next century, are discussed. The descriptions and discussions are made as quantitatively as possible. Although the SI system of units are used throughout the book, the conventional metric or English units are adopted in some places without converting to the SI system. Conversion factors are tabulated in the appendix for the convenience of readers.

I would like to express my appreciation to Professor Bird for affording me the opportunity to present my research approach, for providing fruitful comments, and for correcting my English. I am indebted to many people for assistance during the manuscript preparation. Dr Masahiko Isobe of the University of Tokyo reviewed a preliminary version of the manuscript of wave mechanics in Chapter 2. Dr Yukinori Matsukura of the University of Tsukuba discussed the topics of weathering (Chapter 4) and tafoni (Chapter 8). Professor Shintaro Hotta of Nihon University, Dr Tesuzo Yasunari of the University of Tsukuba, and Dr Alex Zeman of the National Water Research Institute, Canada Centre for Inland Waters, provided useful information on recent research. Dr Ichirou Takeda of Kyoto Kyoiku University, Mr Koichi Watase of the Institute of Geoscience Library at the University of Tsukuba, and the staff of the Interlibrary Loan Department at the University of Tsukuba Library assisted me in literature searches. The British Library and the United States Library of Congress lent me invaluable literature. Mr Shiro Ozaki of the Institute of Geoscience, University of Tsukuba, drew and redrew many figures, and Ms Hisami Mizuno typed the manuscript.

I would like to express my thanks to Mrs Helen Bailey, Mrs Louise Portsmouth, Dr Lewis Derrick, and other staff of John Wiley & Sons for their editorial work. I wish to thank all authors, institutions, learned societies, and publishers who have kindly given permission for reprinting of figures from their publications.

Just thirty years have passed since I began to study the landform of rocky coasts, having been enchanted by a wide variety of their morphological features. Most of my original research before 1979 was performed when I was at the Coastal Engineering Laboratory (CEL), University of Tokyo. I should like to acknowledge Professor Emeritus Kiyoshi Horikawa, formerly Director of CEL, now Professor of Saitama University, who afforded me the opportunity to study rocky coasts from the point of view of dynamics, and also Professor Yuichi Nishimatsu of the Department of Mineral Development Engineering at the University of Tokyo from whom I learned the fundamentals of rock mechanics.

I owe a special debt to those who have provided much input through discussions on science and/or geomorphology. They are: Dr Nicholas C. Kraus of the Coastal Engineering Research Center, Waterways Experiment Station, USA; Professor Eiju Yatsu of Takaoka College of Law, formerly Professor of the University of Tsukuba; and Professor Takasuke Suzuki and Dr Ken'ichi Takahashi, both of Chuo University.

Finally, I acknowledge the Ministry of Education, Science and of Culture, Japan, who have often funded my research.

Tsuguo Sunamura
February 1992

Chapter One

Introduction

Geomorphology, simply defined as the study of landforms (Ritter, 1986, p. 2), should be a four-dimensional science which makes it possible firstly to explain the origin and development of landforms on a physical and/or chemical basis, and secondly to predict future landform changes on a quantitative basis. To accomplish these, a thorough understanding of contemporary geomorphological changes is essential, and causal relationships must be formulated for predictions of the time and scale of landform changes which could occur in the near future. Such predictions will contribute to the solution of environmental, planning, and engineering problems. This is an important role which modern geomorphology should play (Sunamura, 1982b). In this context, *the present is the key to the future*. This metaphor does not imply that the Huttonian doctrine of uniformitarianism can extend to the future, but simply that the research of present-day processes and landform changes provides the key to the solution of future problems.

This book deals with geomorphology of rocky coasts focusing on contemporary processes and resultant landform changes. As shown in Figure 1.1, many sea-level fluctuations of glacio-eustatic origin have occurred during the last 250 000 years, the most recent being a rapid sea-level rise, called the Holocene transgression, which occurred after the last glacial maximum about 18 000 years ago. Coastal features have been etched at the margins of land and sea under the influence not only of the eustatic sea-level changes but also of local crustal uplift or subsidence due to tectonic movements and glacial isostasy. Where the latter influences have not occurred, it is thought that the Holocene transgression brought the sea close to its present level about 6000 years ago, since when sea level has remained almost stationary, although some minor relative fluctuations have been recorded. Six thousand years is a considerably longer period of fairly stable sea level compared with Pleistocene inter-glacial stillstands (Figure 1.1), although the accuracy decreases with increasing time. 'The present' provides us with an opportunity to study the processes and resulting landforms under a stationary sea level condition. Such

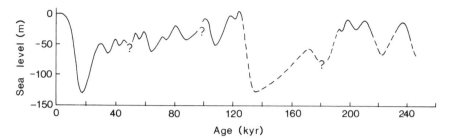

Figure 1.1 Quaternary sea-level fluctuations from 250 000 years BP to present. From Chappell and Shackleton (1986). Reprinted by permission from *Nature*, **324**, 137–40, copyright © 1986, Macmillan Magazines Ltd

investigations will improve our understanding of rocky coast geomorphology and may facilitate (1) studies of evolutionary processes under varying sea-level conditions, (2) modelling of landform changes in the near future, and (3) more accurate reconstruction of the relative sea levels of the past from coastal features such as marine terraces.

A definition should be given to the term 'rocky coast'. According to Webster's *Third New International Dictionary*, the first meaning of 'rock' is 'a usually bare cliff, promontory, peak, or hill that is one mass', and the etymology of 'rocky' is 'cliffy'. A rocky coast is defined here as 'a coast that is cliffed and yet composed of consolidated material irrespective of its hardness'. Rocky coasts treated in this book include shores composed of materials ranging from hard rocks such as granite and basalt to soft cohesive material such as glacial deposits.

Rocky coasts, occupying approximately 80% of the world's coastline (Emery and Kuhn, 1982), have tended to receive relatively little attention in scientific literature, compared with beaches and coasts where urban, industrial, or recreational developments have been concentrated. Due to increased use of the coast, however, considerable attention has recently been focused on rocky coasts. Rocky coast geomorphology has been described in parts of several coastal books (Guilcher, 1958; Zenkovich, 1967; Davies, 1972; King, 1972; Bird, 1976; Pethick, 1984; Carter, 1988), and a text on this subject has been published by Trenhaile (1987).

Rocky coasts suffer erosion, which is an irreversible process (Philpott, 1984). There is no way to restore a rocky coast once it has been eroded. On the other hand, beaches which are composed of unconsolidated material such as sand or gravel, or a mixture of these, undergo reversible change: erosion and accretion. Even if nature fails to restore the beach, recovery is possible by means of artificial nourishment.

Factors to be considered for the study of erosion of rocky coasts are waves and rocks: erosion depends on the relative magnitude of the two. The intensity

Introduction

of the wave factor is greatly influenced by the volume of beach sediment overlying the profile of rocky coasts. Figure 1.2 shows a control section or a compartment which is set up to consider a sediment budget on a rocky coast. In this diagram, L denotes the longshore sediment transport rate, O is the onshore/offshore sand transport rate, R is the rate of sediment discharge from rivers, and R_c is the rate of sediment supply from eroding rocky coasts. Open arrows indicate an inflow of sediment to this compartment, and solid arrows an outflow. If the sum of the inflow rates is smaller than that of the outflow rates, the sediment volume will decrease. If the sum of inflow rates is larger than that of the outflows, sediment will accumulate. If beach mining or dumping takes place within the compartment, these man-made factors affect the sediment budget. Thus, many variables control the volume of sediment on a rocky coast.

Where there is abundant sediment it may prevent wave action from reaching the bedrock. Such a coast should be treated as a beach, even if there is a cliff behind. In contrast to this, there is a coast where no sediment overlies the sea floor in front of a cliff, although this is rare. There are many transitions between the two extremes (Philpott, 1984). Beach sediment can either accentuate bedrock erosion by acting as an abrasive, or hinder it by working as a protective layer. These two contrasting effects characterize the coastal process of rocky shores. A basic knowledge of the sediment movement under wave action and resultant beach changes (e.g. Komar, 1976) is needed for better understanding of rocky coast erosional processes.

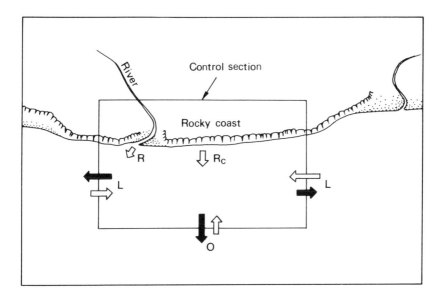

Figure 1.2   Sediment compartment on a rocky coast

This book first explains the fundamental characteristics of waves, which drive the process of rocky coasts (Chapter 2) and then describes water level factors, which are tides and storm surges (Chapter 3). In Chapter 4 are illustrated mechanical properties of consolidated landform materials. Chapter 5 deals with the present-day processes of cliff recession, emphasizing the importance and effectiveness of a laboratory approach. Chapter 6 considers vertical erosion of the sea floor including discussion of the wave base. In Chapter 7 major contemporary morphologies—shore platforms—are discussed in contrast with plunging cliffs, and the mechanisms of platform formation are examined. A model for rocky coast evolution is presented. Other characteristic landforms such as notches, sea caves, arches, and potholes are discussed in Chapter 8. Finally, Chapter 9 discusses various impacts of human activities on rocky coasts including the future influence of a possible sea-level rise induced by the intensified greenhouse effect.

Chapter Two
Nearshore wave field

WAVE DEVELOPMENT, ANALYSIS AND FORECASTING

Waves grow as they acquire energy from winds blowing over a water surface. The amount of energy transferred from winds to waves is a function of the wind speed, the wind duration, and the fetch (Figure 2.1); the more energy transferred the greater the waves in the fetch area. The wind-generated waves, called wind waves, are short-crested in form and highly irregular and steep in cross section, with a wide spectrum range of wave periods and directions. Once the wind waves travel out of the generation area, wave components with shorter periods die out more quickly, and the range of wave propagation direction becomes narrower, so that the waves become more regular with longer periods and longer crests. Such waves are known as swell (Figure 2.1), and have a narrow range of wave periods and directions.

The actual form of waves, even if they are swell, is not a simple sinusoidal curve. Figure 2.2a is a sketch of a wave record, illustrating considerable irregularity in wave form. Statistical analysis can be applied to wave records. The zero-upcrossing method, now generally accepted, has been introduced to define the wave height and period (e.g. Goda, 1985, pp. 12–14). Wave height is the vertical distance between the maximum and minimum levels between the two adjacent zero-upcrossing points, and wave period is the interval of these two points. From a data set of wave height and period, obtained from a complete wave record taken over a selected time, usually 20 minutes, the significant wave height $H_{1/3}$ and period $T_{1/3}$ can be calculated statistically: $H_{1/3}$ is the average height among the waves of which wave heights are within the highest one-third of the data set, and $T_{1/3}$ is the average period of these waves. The significant wave height, a widely used measure, is approximately equal to visually observed height. There are other wave statistics such as (1) H_{max} and T_{max} which are the height of the maximum wave in the data set and its period, (2) $H_{1/10}$ and $T_{1/10}$ which are the average height of the highest one-tenth waves and the average period of these waves, and (3) $\bar{H}$ and $\bar{T}$, which are the mean height and the mean period of all waves. Height parameters such as H_{max},

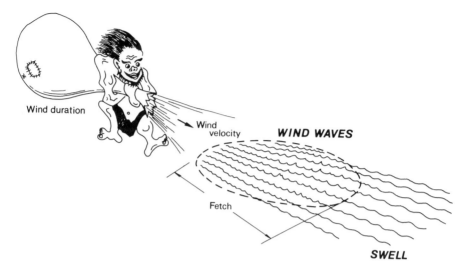

Figure 2.1 Schematic diagram illustrating wave generation and two types of waves: wind waves and swell

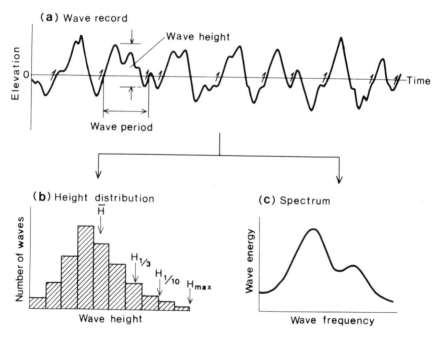

Figure 2.2 Sketch of wave record (a) and two ways of wave analysis: statistical analysis of wave height (b) and spectral analysis (c). After Komar (1983)

$H_{1/10}$, and $\bar{H}$ are illustrated in Figure 2.2b, with each being related to $H_{1/3}$. Period parameters, T_{max}, $T_{1/10}$, and $\bar{T}$, are also related to $T_{1/3}$ (e.g. Wiegel, 1964, pp. 198–204; Horikawa, 1978, pp. 58–61; Goda, 1985, pp. 19–22).

Another way of processing irregular wave data, shown in Figure 2.2a, is spectral analysis. It is known that any complex wave form can be expressed as the sum of many simple sinusoidal waves with different heights and periods. In spectral analysis, the amount of energy of each component (sinusoidal) wave is plotted against the frequency of the wave, the reciprocal of wave period (Figure 2.2c). Because the wave height is proportional to the square root of wave energy as described later, the height–period characteristics of irregular waves can be visualized from the spectrum as shown in Figure 2.2c.

Although full understanding of wave generating mechanism has not been achieved, studies of wave generation relationships in the 1950s have enabled us to forecast (or hindcast) wave height and period from wind parameters, i.e. wind speed, duration, and length of fetch. These relationships evolved from theoretical considerations, but various constants and coefficients in the relationships were determined with actually measured wave data. The best-known forecasting technique is the S–M–B method, originally introduced by Sverdrup and Munk (1947), and subsequently revised by Bretschneider (1958). This method is used to predict significant wave height and period when the wind speed and duration are assumed constant over the fetch. Significant wave characteristics can also be obtained with the wave spectrum method, commonly called the P–N–J method, which Pierson *et al.* (1955) developed to predict development of the wave spectrum. Wilson (1955) established predictive relationships for significant waves which can be applied in moving fetches and variable wind speeds.

These methods described above are all used for prediction of wind waves in deep water, and cannot be applied to shallow water environments where the development of waves is suppressed because of the energy lost by bottom friction and wave breaking. Bretschneider (1954) provided relationships for shallow-water wave prediction in constant wind speed and fetch length. For moving fetch conditions, Sakamoto and Ijima's (1963) method is available.

Swell, i.e. wind waves that have travelled out of the fetch area, gradually decays if no further wind energy is supplied to the waves. Bretschneider's (1968) relationships are useful for forecasting significant wave characteristics of swell.

Wave data are requisite for the study of rocky coasts from the standpoint of dynamic geomorphology. One of these predictive techniques should be employed to express quantitatively the wave climate of the coast to be studied if no recorded wave data are available. A description of procedures for wave prediction is beyond the scope of this book. Readers are recommended to consult appropriate textbooks (e.g. Bretschneider, 1966a; Silvester, 1974a, pp. 65–141; Komar, 1976, pp. 81–95; Horikawa, 1978, pp. 57–88; Coastal Engineering Research Center, 1984, pp. 3:39–3:66).

LINEAR WAVE THEORY

The irregular and complex fluid motions within real waves make the mathematical treatment very difficult. Idealized, regular waves are usually employed in existing wave theories, of which there are basically two: small-amplitude wave theory, and finite-amplitude wave theory. The former, also called Airy wave theory, named after Airy (1845), is developed under the assumption that waves have small height; it is alternatively known as linear wave theory because of the use of linear equations in the derivation of the theory. The latter, the finite-amplitude wave theory, describing waves of large height, includes Stokes-wave, cnoidal-wave, and solitary-wave theories (e.g. Wiegel, 1964, pp. 11–76; Dean and Eagleson, 1966; Le Méhauté, 1976, pp. 239–272). Waves described by these theories are much closer to reality, especially in a shallow-water region, compared with those dealt with in linear wave theory; but these theories are difficult to handle mathematically. A thorough understanding of these theories, including linear wave theory, requires a considerable knowledge of mathematics. Linear wave theory represents the simplest formulation, and although its range of application is fairly narrow, it gives us the essentials, and seems sufficient for the present purpose of understanding the rudimentary wave dynamics in the nearshore zone.

Basic parameters shown in Figure 2.3 are necessary for description of periodic water waves. The wave height, H, is defined as the vertical distance from trough to crest; the wavelength, L, as the horizontal distance between

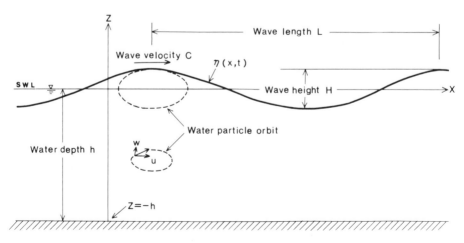

Figure 2.3 Notation necessary for describing linear wave theory

successive crests (or troughs); the wave (or phase) velocity, C, as the speed of propagation of wave form; and the water depth, h, as the vertical distance from still-water level (SWL) to the sea floor. Another characteristic parameter is the wave period, T, which is defined as the time interval required for successive crests (or troughs) to pass a given fixed point. From the geometrical relationship,

$$C = \frac{L}{T} \tag{2.1}$$

This holds for any kind of periodic waves.

In the theoretical treatment of water waves, the fluid can be considered to be incompressible and inviscid, and the wave motion is regarded as irrotational. This physical setting leads to the following equation for water mass continuity, known as the Laplace equation:

$$\frac{\partial^2 \Phi}{\partial x^2} + \frac{\partial^2 \Phi}{\partial z^2} = 0 \tag{2.2}$$

where Φ is called the velocity potential, a scalar quantity from which the horizontal velocity u and the vertical velocity w of the water particles can be easily derived as $u = -\partial \Phi/\partial x$ and $w = -\partial \Phi/\partial z$, x is the horizontal axis taken in the direction of wave propagation, and z is the vertical axis which is positive upwards from SWL (Figure 2.3). In order to solve this equation, two appropriate boundary conditions are needed. One of them is that the vertical velocity on the sea bottom is zero:

$$w = -\frac{\partial \Phi}{\partial z} = 0 \quad \text{on} \quad z = -h \tag{2.3}$$

The other condition is that the pressure p on the water surface should be equal to the atmospheric pressure: $p = 0$ on $z = \eta$, where η is the elevation of water surface (Figure 2.3). Substituting this requirement into the Bernoulli equation for unsteady fluid motion:

$$-\frac{\partial \Phi}{\partial t} + \tfrac{1}{2}(u^2 + w^2) + \frac{p}{\rho} + gz = 0 \tag{2.4}$$

we have

$$-\frac{\partial \Phi}{\partial t} + \tfrac{1}{2}(u^2 + w^2) + gz = 0 \quad \text{on} \quad z = \eta \tag{2.5}$$

where t is the time, ρ is the density of fluid, and g is the acceleration due to gravity. Linear wave theory is based on the assumption that waves have very small height: $\eta \approx 0$. This suggests that the water particle velocities u and w are also small and then the squares of the velocities (u^2 and w^2) are negligible in comparison with the remaining terms. Under this assumption, Eq. (2.5) reduces to

$$\eta = \frac{1}{g}\left(\frac{\partial \Phi}{\partial t}\right) \quad \text{on} \quad z = 0 \qquad (2.6)$$

Under these boundary conditions [Eqs (2.3) and (2.6)], the Laplace equation [Eq. (2.2)] can be solved holding that the solutions should be for periodic waves changing both in time t and in space x. One of the solutions, the velocity potential for a progressive wave moving in the positive x-direction in uniform-water depth, is given by

$$\Phi = \frac{ag \cosh k(h+z)}{\sigma_* \cosh kh} \cos(kx - \sigma_* t) \qquad (2.7)$$

where $a = H/2$ (amplitude), $k = 2\pi/L$ (wave number), and $\sigma_* = 2\pi/T$ (angular frequency).

Water surface elevation, wavelength, velocity, and period

From Eqs (2.6) and (2.7) we have the water surface elevation, i.e. wave profile:

$$\eta = a \sin(kz - \sigma_* t) \qquad (2.8)$$

The assumption of a small-amplitude wave implies that the slope of the wave surface is also small and therefore that the vertical velocity of the water particles on the surface is approximately equal to the rising speed of the water surface. Hence, we can write

$$w = \frac{\partial \eta}{\partial t} \quad \text{on} \quad z = 0 \qquad (2.9)$$

From Eqs (2.6), (2.7), and (2.9), and $w = -\partial \Phi/\partial z$,

$$\sigma_*^2 = gk \tanh kh \qquad (2.10a)$$

which is called the dispersion relation. Because $\sigma_* = 2\pi/T$, $k = 2\pi/L$, and $C = L/T$, we can rewrite this as

$$L = \frac{gT^2}{2\pi} \tanh \frac{2\pi h}{L} \qquad (2.10b)$$

This is a fundamental relationship among wavelength, wave period, and water depth. Some difficulty is encountered in solving for wavelength L, because this equation contains L on both sides. A trial-and-error procedure is required to calculate L. Otherwise, special tables, originally developed using a computer by Wiegel (1964, pp. 514–24), are available (Coastal Engineering Research Center, 1984, Appendix, pp. C:2–C:29). Using Eqs (2.1) and (2.10b),

$$C = \frac{gT}{2\pi} \tanh \frac{2\pi h}{L} \qquad (2.11)$$

Nearshore Wave Field

The right side of Eqs (2.10) and (2.11) contains the hyperbolic tangent, which is illustrated in Figure 2.4 together with other hyperbolic functions. It is found that the hyperbolic tangent approaches unity as the value of kh $(=2\pi h/L)$ increases:

$$\tanh \frac{2\pi h}{L} \approx 1 \tag{2.12}$$

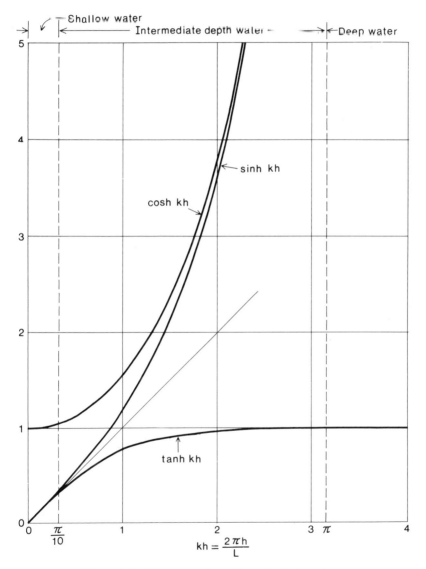

Figure 2.4 Characteristics of hyperbolic functions

This approximation holds for $kh \geqq \pi$, i.e. $h/L \geqq 1/2$, which is commonly called the 'deep water' condition. By use of Eq. (2.12), Eq. (2.10b) reduces to

$$L_0 = \frac{gT^2}{2\pi} \tag{2.13}$$

where L_0 is the wavelength in deep water. This is an important relationship in which L_0 is uniquely determined once the wave period T is fixed. The subscript 0 denotes hereafter deep-water conditions. Because T is independent of the water depth and remains constant (e.g. Eagleson and Dean, 1966), the subscript is omitted. Similarly, the deep-water wave velocity is obtained from Eq. (2.11):

$$C_0 = \frac{gT}{2\pi} = \frac{L_0}{T} \tag{2.14}$$

Figure 2.4 also shows that

$$\tanh \frac{2\pi h}{L} \approx \frac{2\pi h}{L} \tag{2.15}$$

for $kh \leqq \pi/10$, i.e. $h/L \leqq \frac{1}{20}$, which is the 'shallow water' conditions. Substitution of this approximation into Eqs (2.10b) and (2.11) yields

$$L = T\sqrt{(gh)} \tag{2.16}$$

for shallow-water wavelength; and

$$C = \sqrt{(gh)} \tag{2.17}$$

for shallow-water wave velocity.

For the region of $\frac{1}{20} < h/L < \frac{1}{2}$, i.e. for 'intermediate-water' conditions, the general equations, Eqs (2.10b) and (2.11), should be used for the calculation of wavelength and wave velocity, respectively. Equation (2.10b) may be rewritten, using Eq. (2.13), as

$$\frac{h}{L_0} = \frac{h}{L} \tanh \frac{2\pi h}{L} \tag{2.18}$$

The relationship between h/L_0 and h/L has been tabulated by Wiegel (1964, pp. 514–524) (Coastal Engineering Research Center, 1984, Appendix, pp. C:2–C:29). By use of this, we can easily determine L at an arbitrary water depth of h when the wave period is given, because L_0 is known through Eq. (2.13). The wave velocity C at a water depth of h can be expressed, from Eqs (2.11) and (2.14), as

$$\frac{C}{C_0} = \frac{L}{L_0} = \tanh \frac{2\pi h}{L} \tag{2.19}$$

Variations in C/C_0 or L/L_0 with h/L_0 are plotted in Figure 2.5.

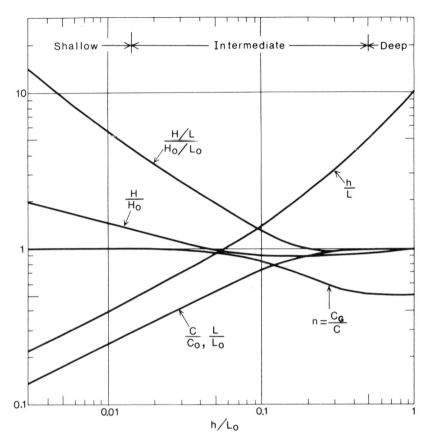

Figure 2.5 Theoretical results of shoaling transformation for linear waves

Water particle motions and pressures

Horizontal and vertical components of water-particle velocity, u and w, are given from Eq. (2.7) with the aid of Eq. (2.10a)

$$u = -\frac{\partial \Phi}{\partial x} = \frac{a\sigma_* \cosh k(h+z)}{\sinh kh} \sin(kx - \sigma_* t) \tag{2.20a}$$

$$w = -\frac{\partial \Phi}{\partial z} = -\frac{a\sigma_* \sinh k(h+z)}{\sinh kh} \cos(kx - \sigma_* t) \tag{2.20b}$$

It can be mathematically shown that the water particle with u and w expressed by the above equations follows an elliptical path (e.g. Eagleson and Dean,

1966), of which the major (horizontal) axis, M, and the minor (vertical) axis, N, are given by

$$M = \frac{2a \cosh k(h+z)}{\sinh kh} = 2a \frac{e^{kh}e^{kz} + e^{-kh}e^{-kz}}{e^{kh} - e^{-kh}} \quad (2.21a)$$

$$N = \frac{2a \sinh k(h+z)}{\sinh kh} = 2a \frac{e^{kh}e^{kz} + e^{-kh}e^{-kz}}{e^{kh} - e^{-kh}} \quad (2.21b)$$

For deep-water conditions, i.e. for $kh \geqq \pi$ ($h/L \geqq \frac{1}{2}$), e^{-kh} becomes very small and can approximate zero. Then we have

$$\left. \begin{array}{l} M = 2ae^{kz} \\ N = 2ae^{kz} \end{array} \right\} \quad (2.22a)$$

The result shows that the major and minor axes are equal: the water particle moves in a circular orbit with the diameter being equal to the wave height ($2a = H$) at the surface ($z = 0$) and decreasing exponentially with water depth (Figure 2.6a). For shallow-water conditions, i.e. for $kh \leqq \pi/10$ ($h/L \leqq \frac{1}{20}$), Figure 2.4 shows that $\cosh kh \approx 1$ and $\sinh kh \approx kh$. Because $z < 0$, $k(h+z) < kh$; therefore

$$\cosh k(h+z) \approx 1$$
$$\sinh k(h+z) \approx k(h+z)$$

Then, Eqs (2.21a) and (2.21b) reduce to

$$\left. \begin{array}{l} M = \dfrac{2a}{kh} \\ \\ N = \dfrac{2ak(h+z)}{kh} \end{array} \right\} \quad (2.22b)$$

It is found that the major axis M, independent of water depth, is constant from the surface to the bottom, whereas the minor axis N decreases from wave height ($2a = H$) at the surface ($z = 0$) to zero at the bottom ($z = -h$). This situation is shown schematically in Figure 2.6c. For intermediate-depth water conditions, $\frac{1}{20} < h/L < \frac{1}{2}$, the water particle moves in elliptical orbits with both the major and the minor axes decreasing with water depth [Eqs (2.21a) and (2.21b)] as illustrated in Figure 2.6b. The minor axis N becomes zero at the bottom ($z = -h$), similar to the case of shallow water waves. This is also evident from the bottom boundary condition, Eq. (2.3). It should be noted that the to-and-fro water particle motion takes place on the bottom for intermediate or shallow water waves. This to-and-fro motion is directly related to the erosion of underwater bedrocks, which will be discussed in detail in Chapter 6.

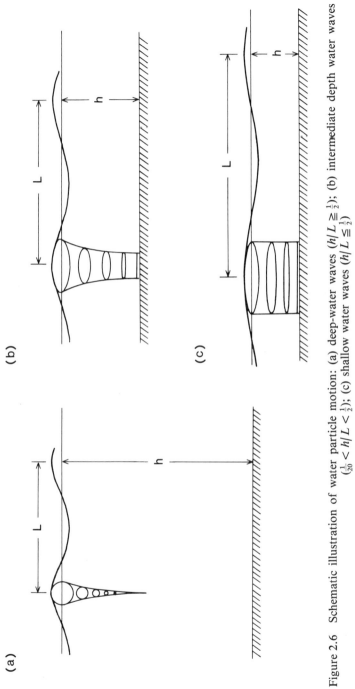

Figure 2.6 Schematic illustration of water particle motion: (a) deep-water waves ($h/L \geqq \frac{1}{2}$); (b) intermediate depth water waves ($\frac{1}{20} < h/L < \frac{1}{2}$); (c) shallow water waves ($h/L \leqq \frac{1}{2}$)

Neglecting high-order velocity terms, u^2 and w^2, in the Bernoulli equation [Eq. (2.4)] gives

$$\frac{p}{\rho} = \frac{\partial \Phi}{\partial t} - gz \qquad (2.23)$$

Using Eqs (2.7) and (2.8), and (2.23), we obtain the underwater pressure:

$$p = \rho g \eta \frac{\cosh k(h+z)}{\cosh kh} - \rho g z \qquad (2.24)$$

The first- and second-terms on the right-hand side of the equation are known as the dynamic and the static pressure, respectively.

Wave energy

The energy of waves exists in two forms: potential and kinetic energies. The former is due to the displacement of the wave surface from SWL. Because the potential energy, E_p, is given by (mass) × (gravitational acceleration) × (elevation), E_p in a wave per unit width of crest can be expressed as

$$E_p = \int_0^L \int_0^\eta \rho g z \, dz \, dx$$

Integrating the equation using Eq. (2.8), we have

$$E_p = \tfrac{1}{16} \rho g H^2 L \qquad (2.25)$$

The kinetic energy is due to the orbital motion of water particles under the waves. Because the kinetic energy, E_k, is given by $\tfrac{1}{2} \times$ (mass) × (velocity)^2, E_k in a wave per unit crest width is

$$E_k = \int_0^L \int_{-h}^\eta \frac{\rho}{2} (u^2 + w^2) \, dz \, dx$$

Integration of the equation using Eqs (2.20a) and (2.20b) and an approximation of $\eta \approx 0$ yields

$$E_k = \tfrac{1}{16} \rho g H^2 L \qquad (2.26)$$

The total energy in a wave per unit width of crest is

$$E_T = E_p + E_k = \tfrac{1}{8} \rho g H^2 L \qquad (2.27)$$

The total wave energy per unit area is

$$E = \tfrac{1}{8} \rho g H^2 \qquad (2.28)$$

The wave energy is transmitted in the direction of wave propagation. The rate of energy transmission, usually called the energy flux, is equal to the work done by pressure per unit time, which is given by the product of (1) the force acting on a vertical plane under the waves, (pressure) × (area), and (2) the velocity of the water particle across this plane. Then, the energy flux per unit width of wave crest, averaged over a period, can be described as

$$P = \frac{1}{T}\int_0^T \int_{-h}^{\eta} pu \, dz \, dt$$

Again using $\eta \approx 0$ and applying Eqs (2.20a) and (2.24), we have

$$P = \tfrac{1}{8}\rho g H^2 \frac{C}{2}\left[1 + \frac{2kh}{\sinh 2kh}\right] = EC_G \tag{2.29}$$

where

$$C_G = \frac{C}{2}\left[1 + \frac{2kh}{\sinh 2kh}\right] = nC \tag{2.30}$$

and

$$n = \frac{1}{2}\left[1 + \frac{2kh}{\sinh 2kh}\right] \tag{2.31}$$

The quantity P is often referred to as the wave power, and C_G is called the group velocity, the propagating speed of a group of waves (e.g. Eagleson and Dean, 1966). Equation (2.29) indicates that the wave energy travels with the group velocity.

As seen from Eq. (2.31), $n = \tfrac{1}{2}$ in deep water ($kh \to \infty$) and $n = 1$ in shallow water ($kh \to 0$); this variation is shown in Figure 2.5. In deep water, individual waves advance twice as fast as the wave group [Eq. (2.30)], so that a wave travelling through the group finally leads the group and gradually disappears, while a new wave begins to emerge at the rear of the group (Figure 2.7). As the waves travel into shallow water, individual waves tend to advance with the same speed as the group.

Wave height

Let us consider the change in the height of waves approaching perpendicular to a straight coast with gradually decreasing water depths expressed by parallel bottom contours. Assuming that there is no energy dissipation, reflection, and generation during wave travelling, the wave-energy flux is maintained constant from deep to shallow regions. From Eq. (2.29), we have

$$E_0(C_G)_0 = EC_G = \text{constant} \tag{2.32}$$

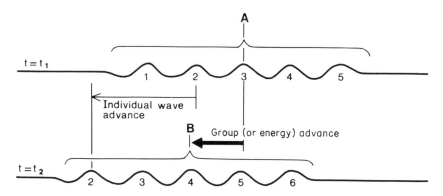

Figure 2.7 Schematic presentation of advance of wave group with a half velocity of individual waves. While the centre of the wave group advances from A to B during the time from t_1 to t_2 wave number 1 fades out and a new wave, 6, emerges. After Bascom (1980, Figure 23)

Because $E = \rho g H^2/8$, $E_0 = \rho g H_0^2/8$, $C_G = nC$, and $(C_G)_0 = C_0/2$, Eq. (2.32) gives

$$\frac{H}{H_0} = \sqrt{\left(\frac{1}{2n}\frac{C_0}{C}\right)} = K_s \qquad (2.33)$$

where K_s is called the shoaling coefficient. From Eqs (2.18), (2.19) and (2.31), we find that Eq. (2.33) is a function of only h/L_0; the H/H_0–h/L_0 relationship is illustrated in Figure 2.5. This figure shows that a slight reduction occurs in the height of waves when they move into intermediate water depths, but after this, the wave height increases with further decreasing water depths. On the other hand, the wavelength steadily decreases as the waves enter shallower water (Figure 2.5). This means that the wave steepness, H/L, drastically increases (Figure 2.5), so that the waves become more unstable as they advance.

Wave refraction

The waves travelling into shallow water begin to 'feel bottom' when the water depth is less than half the wavelength in deep water ($h/L_0 \leq \frac{1}{2}$), and the wave velocity (moving speed of wave crest) decreases with decreasing water depth (See the C/C_0 vs. h/L_0 relationship in Figure 2.5). Imagine long-crested waves obliquely approaching a straight coast with parallel contours. The inshore part of each wave crest, compared with the offshore part, is moving more slowly because of shallower depth. The result is that the offshore part swings forwards and the wave crests tend to become parallel to the depth contours. This process—wave refraction—significantly occurs to waves approaching a

Nearshore Wave Field

headland–bay complex where the nearshore contours approximately follow the coastline configuration (Figure 2.8). The orthogonals, the lines drawn normal to the wave crests, converge at the headlands to concentrate wave energy, and diverge at the bays to spread out energy.

If no energy losses occur during wave advance and no lateral flow of energy takes place along the wave crest, the energy flux between two orthogonals is constant. Then,

$$b_0 E_0 (C_G)_0 = bEC_G = \text{constant} \tag{2.34}$$

where b_0 is the spacing of the orthogonals in deep water and b is their spacing at arbitrary depths in shallow water. Applying the same equations as used to obtain Eq. (2.33), Eq. (2.34) yields

$$\frac{H}{H_0} = \sqrt{\left(\frac{1}{2n}\frac{C_0}{C}\right)} \sqrt{\left(\frac{b_0}{b}\right)} \tag{2.35}$$

The term $\sqrt{(b_0/b)}$ represents the wave refraction effect, and is called the refraction coefficient, K_r:

$$K_r = \sqrt{\left(\frac{b_0}{b}\right)} \tag{2.36}$$

Using Eqs (2.33) and (2.36), Eq. (2.35) reduces to

$$H = K_s K_r H_0 \tag{2.37}$$

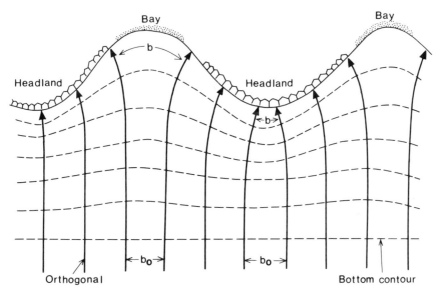

Figure 2.8 Wave refraction on a schematized headland–bay complex

The height of waves which advance shallow water undergoing refraction should be calculated using this equation. The refraction coefficient may generally be obtained by drawing a wave refraction diagram. For the procedures in the refraction diagram construction, see an appropriate reference (e.g., Coastal Engineering Research Center, 1984, pp. 2:66–2:74).

BREAKING WAVES

When waves advance into shallow water with the depth approximately equal to the wave height, they become unstable and finally break. No theory is available at present to successfully describe the phenomenon of wave breaking (e.g. Longuet-Higgins, 1980). Most breaking wave studies depend on laboratory experiments.

Breaking waves are commonly classified into three types according to their shape: spilling, plunging, and surging (Iversen, 1952). Galvin (1968) defined a

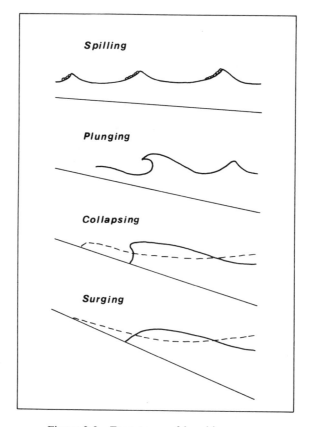

Figure 2.9 Four types of breaking waves

fourth, using the term 'collapsing' for a type intermediate between plunging and surging. These breaker types are shown schematically in Figure 2.9. Spilling breakers are characterized by the appearance of foam cascading down from the top of the peaking wave crest. In plunging breakers the waves curl over and a mass of water collapses against the sea surface. In collapsing breakers the crest of waves peaks up as if they plunge, but then the base of the deforming wave rushes up the shore as a thin layer of foaming water, so that the crest collapses and disappears. In surging breakers the form of waves remains smooth with no marked crest and they slide up a slope with minor air entrainment.

The occurrence of these breaker types is found to depend on deep water wave steepness, H_0/L_0, and the bottom slope, $\tan \beta$ (Iversen, 1952; Patrick and Wiegel, 1955; Hayami, 1958; Galvin, 1968; Weggel, 1972; Battjes, 1974). Most of the existing studies have attempted to classify, using these two quantities, the type of breaking waves occurring on the bottom slope gentler than $\frac{1}{10}$, i.e. $\tan \beta \leq 0.1$. Rocky coasts have the shore-zone slope which is generally much steeper than that of sandy beaches; it becomes almost vertical on the coast of plunging cliffs. A recent laboratory study by Okazaki and Sunamura (1991) conducted with the purpose of elucidating the breaker type on a steep slope, presented a breaker-type classification (Figure 2.10) which can be applied to the bottom slope with any gradient. Figure 2.10 indicates that if the wave steepness remains constant, spilling, plunging, collapsing, and surging breakers occur in this order as the bottom slope increases. Application of the breaker-type classification always involves subjective judgements, because transitional types appear when breakers change from one type to another. The curves demarcating breaker types in Figure 2.10 are given by

$$\left.\begin{array}{l} H_0/L_0 = 2.4 \, (\tan \beta)^{1.8} \text{ for spilling–plunging boundary} \\ H_0/L_0 = 0.19 \, (\tan \beta)^{2.5} \text{ for plunging–collapsing boundary} \\ H_0/L_0 = 0.074 \, (\tan \beta)^{2.4} \text{ for collapsing–surging boundary} \end{array}\right\} \quad (2.38)$$

These three boundaries can be well approximated by Battjes' (1974) surf similarity parameter, $\xi_0 \, (= \tan \beta / \sqrt{(H_0/L_0)})$, as $\xi_0 = 0.5$, 2.4, and 3.7, respectively (Okazaki and Sunamura, 1991).

The height of waves just starting to break is called the breaker height, denoted by H_b. Munk (1949) derived from solitary wave theory the relationship between H_b/H_0 and H_0/L_0. A poor agreement of the theoretically obtained relationship with available data in the range of high H_0/L_0-value was found by Komar and Gaughan (1972). They, using the energy flux relationship in the linear wave theory, obtained the semi-empirical equation:

$$\frac{H_b}{H_0} = 0.563 \left(\frac{H_0}{L_0}\right)^{-1/5} \qquad (2.39)$$

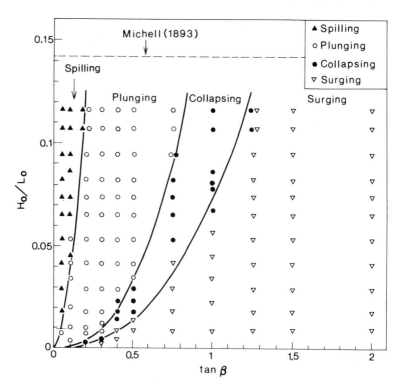

Figure 2.10 Demarcation of breaker types. From Okazaki and Sunamura (1991), by permission of Coastal Education and Research Foundation. The dashed line indicates Michell's (1893) limiting wave steepness for breaking, $H_0/L_0 = 0.142$

where the coefficient, 0.563, was determined by a best fit to laboratory and field data. This equation has been reconfirmed by Weishar and Byrne (1978) from field data obtained at Virginia Beach, with a bottom slope ranging from $\frac{1}{50}$ (seaward) to $\frac{1}{20}$ (shoreward), on the Atlantic coast of the United States.

Laboratory studies by Iversen (1952), Galvin (1969), Goda (1970) and others indicate that the breaker height is also dependent on the bottom slope. Figure 2.11 is a plot of Goda's empirical relationship between H_b/H_0 and H_0/L_0 for some selected bottom slopes, tan β. This figure shows that the Komar–Gaughan relationship [Eq. (2.39)], depicted by the dashed curve, corresponds closely to the curve for tan $\beta \leq \frac{1}{50}$ in the mid range of H_0/L_0 values.

Some attempts have been made to formulate the H_b/H_0–H_0/L_0 relationship incorporating the effect of tan β by Le Méhauté and Koh (1967):

$$\frac{H_b}{H_0} = 0.76(\tan \beta)^{1/7} \left(\frac{H_0}{L_0}\right)^{-1/4} \qquad (2.40)$$

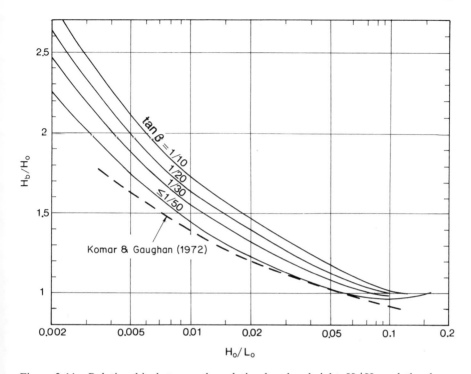

Figure 2.11 Relationship between the relative breaker height H_b/H_0 and the deep-water wave steepness H_0/L_0. From Goda (1970), by permission

and by Sunamura and Horikawa (1974):

$$\frac{H_b}{H_0} = (\tan \beta)^{0.2} \left(\frac{H_0}{L_0}\right)^{-0.25} \quad (2.41)$$

with the latter being reconfirmed using data obtained from prototype-scale wave tank experiments, although the number of data is limited (Sunamura, 1982c). These two empirical equations are very similar to each other and also to Goda's curves in Figure 2.11 (Mizuguchi et al., 1988).

Work by McCowan (1894), based on solitary wave theory demonstrated that

$$\gamma_b = \frac{H_b}{h_b} = 0.78 \quad (2.42)$$

where h_b is the water depth for wave breaking. Subsequent theoretical studies indicated the γ_b-value lying in a range of 0.73 to 1.03 (Galvin, 1972; Komar, 1976, p. 56). Many experimental studies conducted since the 1950s have demonstrated that γ_b depends on the bottom slope (Ippen and Kulin, 1954;

Camfield and Street, 1969; Galvin, 1969; Goda, 1970; Weggel, 1972). Figure 2.12 shows Goda's (1970) relationship between h_b/H_0 and H_0/L_0 for several bottom slopes, $\tan \beta$.

If H_0, L_0 (or T), and $\tan \beta$ are given, we can easily predict H_b using Figure 2.11 or Eq. (2.39) or (2.40), and h_b using Figure 2.12. Using H_b thus obtained, h_b can also be obtained through calculation of

$$\frac{h_b}{H_b} = \frac{1}{C_1 - (C_2 H_b / gT^2)} \qquad (2.43)$$

where $C_1 = 1.56/(1 + e^{-19.5 \tan \beta})$ and $C_2 = 43.75(1 - e^{-19 \tan \beta})$. This is a normalized relation (Coastal Engineering Research Center, 1984, p. 2:130) of the empirical equation originally proposed by Weggel (1972). A nomograph for the equation is also provided by Coastal Engineering Research Center (1984, p. 2:132). If wave refraction occurs, H_0 in Figures 2.11 and 2.12, and also in Eqs (2.39), (2.40) and (2.41), should be replaced by the corrected wave height, $H_0' = K_r H_0$, where K_r is the refraction coefficient [Eq. (2.36)].

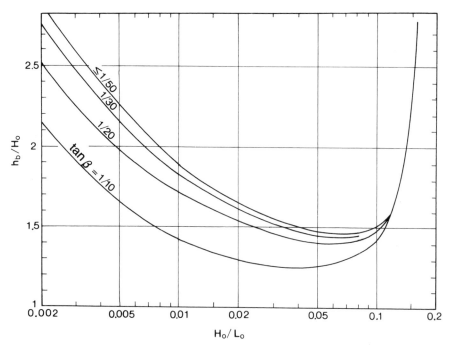

Figure 2.12 Relationship between the relative breaker depth h_b/H_0 and the deep-water wave steepness H_0/L_0. From Goda (1970), by permission

BROKEN WAVES AND WAVE SET-UP

After they break, broken waves advance through the surf zone toward the shore, losing energy because of turbulence and bottom friction, so that the height of broken waves decreases with increasing distance from the breaking point. Denoting H as the wave height in the surf zone, H_b as the breaker height, h as the water depth measured from SWL, h_b as the breaking depth, and $\tan \beta$ as the gradient of uniformly sloping bottom, Figure 2.13, synthesized by Sunamura (1984) from the existing laboratory data, shows wave-height attenuation curves. It is seen that the degree of wave-height reduction increases with decreasing bottom slope. The figure also shows that the curves are not linear but concave upward, suggesting an exponential decay in wave height (e.g. Ijima et al., 1956). Figure 2.13 is useful for easy prediction of approximate height of broken waves. The following relationship (Sunamura, 1985) is also useful: it was empirically obtained considering the influence of wave period and assuming the exponential wave decay, and calibrated using

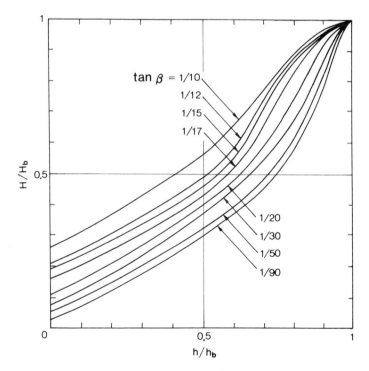

Figure 2.13 Laboratory relationship of wave height attenuation after breaking, for various bottom slopes. Data from Sasaki and Saeki (1974); Mizuguchi et al. (1978)

available field data and prototype-scale wave-tank experiment data:

$$\frac{H}{H_b} = \exp\left(-\frac{35.2\, l \tan \beta}{T\sqrt{(gh_b)}}\right) \quad (2.44)$$

where l is the distance from the breaking point (Figure 2.14). Broken waves propagating on a gentle slope tend to re-form their shape and again break at a shallower depth than the first breaking point, producing multi-breakers. Equation (2.44) was derived from ignoring the wave re-forming and subsequent breaking process, i.e. local wave-height increase associated with multi-breaking, so that this relation can only describe the general trend of wave attenuation in the surf zone.

There have been several physical models to predict the surf-zone wave height, beginning with the work of Ijima *et al.* (1956) who considered internal energy dissipation due to eddy viscosity as a dominant wave-decay process. These studies differ in modelling of energy-dissipation mechanisms. See for details Dally *et al.* (1984), Svendsen (1984), and Mizuguchi *et al.* (1988), and the papers cited therein.

A significant phenomenon occurring in the surf zone is wave set-up: a shoreward rise in the mean water level (MWL) from the wave breaking point (Figure 2.14). This is due to landward decrease in the radiation stress (Longuet-Higgins and Stewart, 1963, 1964) produced in broken waves with their height decreasing toward the shore. Longuet-Higgins and Stewart have theoretically shown

$$\frac{d\eta_*}{dx} = -K' \frac{dh}{dx} \quad (2.45)$$

where η_* is the change in MWL, h is the still-water depth, x is the offshore coordinate (Figure 2.14), and K' is a coefficient, which is a ratio of set-up slope, $d\eta_*/dx$, to the bottom slope, $\tan \beta = -dh/dx$ (Figure 2.14). Bowen

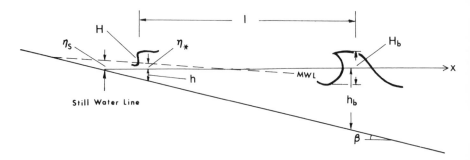

Figure 2.14 Definition sketch for waves and wave set-up in the surf zone

et al. (1968) theoretically obtained

$$K' = \frac{1}{1 + (8/3\gamma'^2)}, \qquad \gamma' = \frac{H}{\eta_* + h}$$

and experimentally demonstrated the validity of the relation. Integration of Eq. (2.45) yields

$$\eta_* = -K'h + \eta_s \qquad (2.46)$$

where η_s is the rise in MWL at the still-water line. Battjes (1974) theoretically obtained $\eta_s = 0.3\gamma'H_b$. The nondimensional quantity γ' and hence K' also are considered to be a function of $\tan \beta$, respectively, as anticipated from the previously mentioned bottom-slope dependency of γ_b applied to the breaking point. Sasaki and Saeki's (1974) laboratory work indicated $\eta_s = (1.63 \tan \beta + 0.048)H_b$ and $K' = 3.85 \tan \beta + 0.015$, so that Eq. (2.46) becomes

$$\eta_* = -(3.85 \tan \beta + 0.015)h + (1.63 \tan \beta + 0.048)H_b \qquad (2.47)$$

After their theoretical development to explain set-up induced by random waves, Battjes and Janssen (1978) performed laboratory experiments in which irregular waves were broken on a uniform bottom slope ($\tan \beta = \frac{1}{20}$). Theory and experiment were in good agreement, and they proposed

$$0.15 < \frac{\eta_s}{(H_{\text{rms}})_0} < 0.21 \qquad (2.48)$$

where $(H_{\text{rms}})_0$ is the root-mean-square deep-water wave height. The range of the value in this relation depends on the wave steepness. Because $(H_{\text{rms}})_0$ is approximately related to the significant deep-water wave height as $(H_{\text{rms}})_0 = 0.706(H_{1/3})_0$ (Longuet-Higgins, 1952), Eq. (2.48) can be rewritten as

$$0.11 < \frac{\eta_s}{(H_{1/3})_0} < 0.15 \qquad (2.49)$$

An attempt made in the real world was a study by Guza and Thornton (1981). They measured wave set-up at Torrey Pine Beach, San Diego, California ($\tan \beta \approx \frac{1}{50}$) and obtained a simple relationship:

$$\eta_s = 0.17(H_{1/3})_0 \qquad (2.50)$$

which is close to the upper limit value of Eq. (2.49). Equations (2.49) and (2.50) suggest that the amount of sea-level rise at the shoreline would be between 10 and 20% of the significant wave height in deep water.

NEARSHORE CURRENTS

Currents occurring in the nearshore zone are different in origin from ocean currents, tidal currents, and wind-induced currents. Nearshore currents are generated by the action of waves; they include (1) longshore currents, (2) rip currents, and (3) onshore currents. It is well known that an oblique wave approach is responsible for the generation of longshore currents that dominated in the surf zone, flowing parallel to the shoreline. Rip currents or rips, which are fed by the longshore current, develop with a certain alongshore spacing; they are strong currents flowing offshore through a narrow zone across the breaker zone. Onshore currents are characterized by a shoreward slow water movement (mass transport) passing through the breaker zone in the area between two adjacent rips. Waves arriving parallel to the shoreline form a closed current system (Figure 2.15), i.e. 'nearshore circulation' (Shepard and Inman, 1950). In this system, there are two rip currents which are fed respectively by two longshore currents flowing in opposite directions from midway between the two rips, and onshore currents develop in a wide area between the rips, returning water to the surf zone to compensate for the volume of water carried offshore by the rip currents.

Nearshore currents seem to have no direct influence on the evolution of rocky coasts. Indirectly, however, they are of great importance, because they can transport clastic sediment which can accelerate wave erosion of coastal

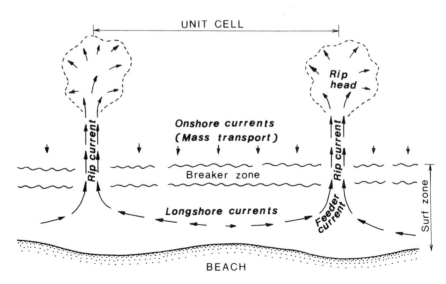

Figure 2.15   Nearshore current system consisting longshore currents, rip currents, and shoreward slow water movement (mass transport). After Shepard and Inman (1950)

Nearshore Wave Field

rocks working as an abrasive tool, decelerate it working as a wave-energy dissipator, or hinder it working as a protective layer (Chapters 5 and 6).

Many theories or models have been proposed to account for the generation of nearshore currents (Komar, 1976, pp. 168–202; Horikawa, 1978, pp. 185–229; Pethick, 1984, pp. 33–45). Empirical relationships that predict the rate of sediment transport by longshore currents have been presented, for example, by Kraus *et al.* (1988).

TYPES OF WAVES AT THE CLIFF BASE AND THEIR PRESSURE

There are three types of waves occurring at the foot of a cliff when waves arrive parallel to the coast: standing, breaking, and broken waves. The occurrence of these wave types depends on the relative magnitude of the breaking depth of incoming waves, h_b, and the water depth in front of the cliff, h: (1) if $h > h_b$, then standing waves are created by reflection from the cliff, (2) if

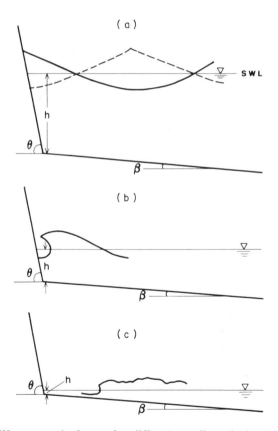

Figure 2.16 Wave types in front of a cliff: (a) standing; (b) breaking; (c) broken

$h = h_b$, then breaking waves are produced, and (3) if $h < h_b$, then broken waves are formed (Figure 2.16). As shown in Figure 2.12, h_b can be expressed by a function of the deep-water wave characteristics and the nearshore bottom slope. Therefore, these two factors plus the nearshore water depth control the type of waves. As the frontal water depth decreases (Figure 2.16), standing, breaking, and broken waves appear in this sequence if other controlling factors are kept constant. The same sequential occurrence of wave types is found at a cliff located in water of a certain depth, as incoming waves with a constant period increase their height. When incident waves are of small height, standing waves are generated in front of a cliff. Their pressure acting on the cliff is characterized by a smooth temporal variation with a single or double humps as illustrated in Figure 2.17. Somewhat larger waves produce the breaking of standing waves which results in an asymmetrical pressure–time curve with the first hump (arrow in Figure 2.17a) slightly peaking up. When the height of incoming waves attains a critical condition, breaking waves form, producing impulsive pressure with the highest intensity (Figure 2.17b). Beyond this condition, the waves break before arriving at the cliff, on which broken waves produce the pressure with a reduced peak (Figure 2.17c).

Distinctive variations in pressure intensity with the wave type are well illustrated in Figure 2.18. This diagram is depicted based on laboratory data of Hom-ma *et al.* (1962), in which a vertical wall situated at a water depth of 5 cm on a uniformly sloping bottom with slope $\frac{1}{15}$ was exposed to input waves with a constant period but with different heights. A transition exists from standing to breaking waves, and from breaking to broken waves.

Standing waves generated at a vertical wall are characterized by up-and-down movement of water surface with no marked disturbance. When the wave crest reaches the wall, the largest dynamic pressure is produced with its

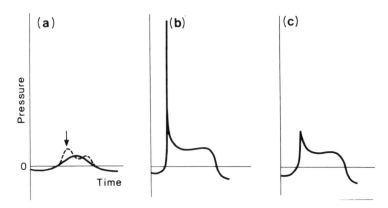

Figure 2.17 Schematic illustration of pressure–time curves which are assumed to be recorded at SWL: (a) standing; (b) breaking; (c) broken

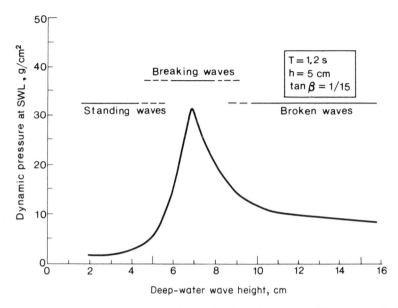

Figure 2.18 Wave pressure on a vertical wall plotted against offshore wave height. The curve is drawn through averaging highly scattered data of Hom-ma *et al.* (1962)

maximum intensity at SWL as illustrated in Figure 2.19. Theoretical studies indicate a similar pressure distribution (e.g. Sainflou, 1928; Miche, 1944; Biésel, 1952; Rundgren, 1958). It is the elevation of the maximum dynamic pressure that is of geomorphological significance. The dynamic pressure can be estimated using one of the existing theories, but the difference among the calculated values is small. Using the modified Sainflou theory (Hudson, 1952), the maximum pressure intensity, p_m, (Figure 2.19) is given by

$$p_m = (p_1 + \rho g h)\left(\frac{H + h_0}{h + H + h_0}\right) \quad (2.51)$$

where $p_1 = \rho g H / \cosh(2\pi h/L)$, $h_0 = (\pi H^2/L)\coth(2\pi h/L)$, H is one half of standing wave height, i.e. the incoming wave height, and L is the wavelength at a water depth of h. The static pressure, p_s, is distributed in a triangular shape, with $p_s = 0$ at the elevation of wave crest at the wall and $p_s = \rho g(H_0 + h_0 + h)$ at the base of the wall.

Impulsive dynamic pressure on a vertical wall by breaking waves occurs when the water depth at the wall, h, is equal to the water depth, h_M, calculated from the following empirical relation (Mitsuyasu, 1962):

$$\frac{h_M}{H_0} = C_M \left(\frac{H_0}{L_0}\right)^{-1/4} \quad (2.52)$$

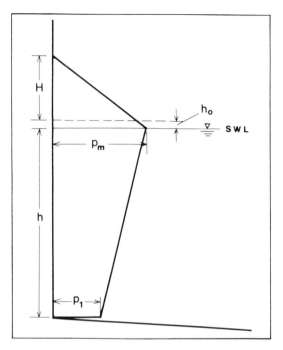

Figure 2.19 Vertical distribution of dynamic pressure caused by standing waves

where C_M is a function of the bottom slope, $\tan \beta$, and is expressed as $C_M = 0.59 - 3.2 \tan \beta$. A more appropriate expression for C_M is given by

$$C_M = 0.034/(10.038 + \tan \beta) \qquad (2.53)$$

This is based on more data with a wider range in the value of $\tan \beta$ than those used in Mitsuyasu's study (Hom-ma *et al.*, 1962, Figure 4; Horikawa, 1978, Figure 2.8.11). The height of deep-water waves which will break in front of the wall, $(H_0)_b$, can be obtained by substituting $h_M = h$ into Eq. (2.52) and by solving this for H_0:

$$(H_0)_b = (h/C_M)^{1.33} L_0^{-0.33} \qquad (2.54)$$

With knowledge of the bottom slope, the wall-front water depth, and the deep-water wavelength (wave period), it is possible to calculate the height of deep-water waves which will produce the shock pressure on the wall by use of Eqs (2.53) and (2.54).

Theoretical studies have not succeeded in estimation of breaker-induced dynamic pressures because of the complicated phenomenon of wave breaking against the wall. The well-known empirical relationship used to calculate the pressure distribution on a vertical wall is Minikin's (1963, pp. 36–48) model, which was originally proposed in 1950 (Figure 2.20). This model is based on field measurements made in Europe and on the laboratory work of Bagnold (1939), who examined the effect of air enclosed in breaking waves on the intensity of impulsive pressure. Laboratory experiments by Nagai (1961) and Kirkgoz (1982) indicated that the Minikin model provides considerable underestimates. According to prototype-scale experiments with a wave flume 320 m long (Partenscky, 1988), a measured value of the maximum pressure produced by 1.5 m breakers was four times greater than a value calculated by the Minikin formula. With these findings in mind, the following simple expression seems sufficient to provide approximate estimates for the maximum pressure intensity:

$$p_m = 35 \rho g (H_0)_b \tag{2.55}$$

This is base on curves (Figure 2.20) drawn from the experiments of Denny (1951), Ross (1955), and Mitsuyasu (1963). Blackmore and Hewson (1984) presented a summary of existing field data of impact pressures measured at various places in the world.

Dynamic pressures generated by breaking waves show a maximum value at or slightly above SWL, decreasing abruptly above and below this level as

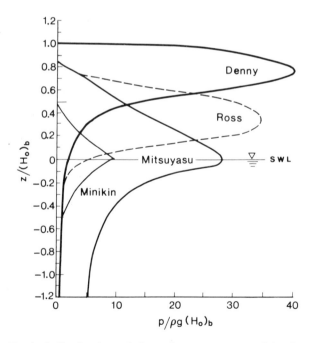

Figure 2.20 Vertical distribution of dynamic pressure caused by breaking waves. After Mitsuyasu (1963), by permission of Japan Society of Civil Engineers

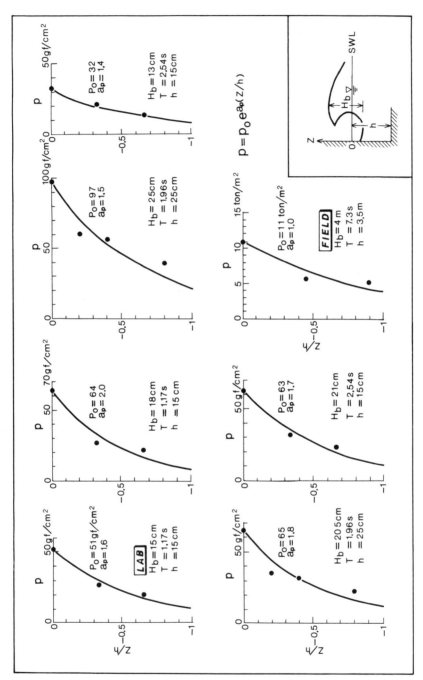

Figure 2.21 Exponential reduction in the intensity of breaking wave pressure. After Sunamura (1990b)

shown in Figure 2.20. The pressure distribution with the maximum being located at SWL is also described in Nagai's (1961) and Plakida's (1970) models. Occurrence of a maximum value above SWL is found in the laboratory experiments of Kirkgoz (1982). A recent prototype-scale test by Partenscky (1988) indicates that a maximum value occurred at a considerably higher level located at $0.7H_b$ (H_b = breaker height at the wall) above SWL. Based on field measurements at sea walls on the southern and western coasts of England, Blackmore and Hewson (1984) constructed a model in which impact pressures are uniformly distributed in a zone around SWL.

It is clear from these studies that dynamic pressure intensity decreases with increasing depth of water below SWL either in the laboratory or in the field situations. Figure 2.21 shows some selected laboratory data of Nagai and Otsubo (1968) and field data obtained at Haboro Harbour, Japan by Hokkaido Development Bureau (cited in Nagai and Otsubo, 1968). The reduction in pressure, p, with increase in water depth, z, can be expressed by an exponential function (Sunamura, 1990b):

$$p = p_0 e^{a_p(z/h)} \qquad (2.56)$$

where p_0 is the pressure at SWL, and a_p is a reduction coefficient ($1 \lesseqgtr a_p \lesseqgtr 2$). The static pressure is usually assumed to be triangular-shaped with $p_s = 0$ at the elevation of wave crest at the wall, H_b', and $p_s = \rho g(H_b' + h)$ at the sea floor.

Broken waves are characterized by shoreward rushing of highly turbulent water mass in the surf zone. Hom-ma and Horikawa (1964) have presented an experiment-based model in which the distribution of dynamic pressure resulting from broken waves on a vertical wall is described as a triangular pattern with a peak value located at SWL (Figure 2.22a). The maximum pressure is given by

$$p_m = 1.6\rho gh \qquad (2.57)$$

The static pressure is assumed to have a triangular distribution with $p_s = 0$ at an elevation of $1.2h$, above SWL and $p_s = 2.2\rho gh$ at the base.

Another predictive formula, originally developed in 1961 by the US Army Beach Erosion Board, shows that the dynamic pressure of broken waves is distributed uniformly above SWL up to a height of $0.78H_b$ (H_b = breaking wave height) (Figure 2.22b) and the maximum pressure is expressed by (Coastal Engineering Research Center, 1984, pp. 7:192–3):

$$p_m = 0.5\rho g h_b \qquad (2.58)$$

where h_b is the water depth at the wave breaking point. The static pressure distribution is triangular-shaped with $p_s = 0$ at a height of $0.78H_b$ and $p_s = \rho g(h + 0.78H_b)$ at the sea bottom.

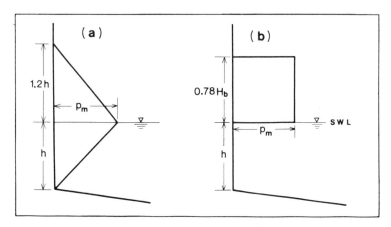

Figure 2.22 Vertical distributions of dynamic pressure caused by broken waves: (a) after Hom-ma and Horikawa (1964), and (b) after Coastal Engineering Research Center (1984, Figure 7:104)

For estimating wave pressure acting on a vertical wall located shoreward of the waterline, the only available model was that of CERC (Coastal Engineering Research Center, 1984, pp. 7:194–7). It is recently found that this model gives considerable overestimates (Camfield, 1991). Camfield revised the model to provide the total pressure, p_t:

$$p_t = 4.5\rho gh^2 = 0.18H_b^2\left(1 - \frac{x_1}{x_2}\right)^2 \qquad (2.59)$$

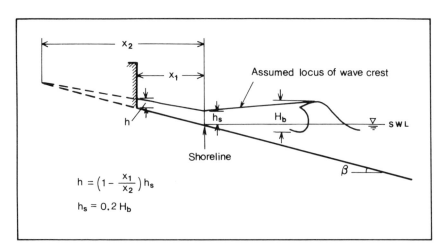

Figure 2.23 Definition sketch for the estimation of the height of broken waves at a wall located landward from the shoreline. From Camfield (1991), by permission of American Society of Civil Engineers

where h is the height of broken waves at the wall, x_1 is the horizontal distance from the shoreline to the wall, and x_2 is the distance from the shoreline to the limit of wave run-up when the wall is not present (Figure 2.23). This equation holds for a range of slopes defined as $0.01 < \tan \beta < 0.1$. Further research is required to obtain pressure distributions (Camfield, 1991).

The effect of cliff inclination on dynamic pressure may be given by (Homma and Horikawa, 1964; Coastal Engineering Research Center, 1984, p. 7:200):

$$p_m^* = p_m \sin \theta \tag{2.60}$$

where p_m^* is the pressure intensity on an inclined wall with an angle of θ, and p_m is the pressure on a vertical wall. Estimation of dynamic pressure produced by waves obliquely incident to a vertical wall is in terms of the following relationship (Coastal Engineering Research Center, 1984, p. 7:198):

$$p_m' = p_m \sin^2 \theta_s \tag{2.61}$$

where p_m' is the pressure exerted by oblique waves, p_m is the pressure that would occur if waves act normal to the wall, and θ_s is the angle between the incident wave direction and the axis of the wall.

Chapter Three
Water level factors: tides and storm surges

INTRODUCTION

Water level fluctuations (other than waves) have no direct effect on the morphology of rocky coasts. Indirectly, however, they control the types of waves approaching rocky coasts (Chapter 2), and determine the elevation of wave attack, so that they are one of the significant controlling factors in the magnitude and position of the assailing waves. Changes in water level may influence weathering and biological activities, both of which affect the mechanical properties (i.e. resisting force) of coast-forming materials.

Water level fluctuations are primarily due to astronomical tides, which are quasi-regular rises and falls of sea level caused by tide-generating forces associated with the motion of the Earth–Moon–Sun system (Dean, 1966; Wood, 1982, Pugh, 1987, pp. 59–88). The tidal variations are frequently disturbed by meteorological factors such as gales and abrupt changes in atmospheric pressure, which result in episodic water level changes, usually known as storm surges.

TIDES

The Earth–Moon–Sun system (Figure 3.1) has a characteristic motion in which the Earth rotates about the Sun and the Moon rotates about the Earth, spinning about its own axis. A centrifugal force due to the Earth's spinning causes no tidal variations. The gravitational forces of the Sun and Moon produce tides. According to Newton's law of gravitation (every particle of matter in the universe attracts every other particle with a force proportional directly to the product of their masses and, inversely to the square of the distance between them), the Sun is found to have only half the tide-producing effect of the Moon, so that the Moon is the major tide-producing body.

Water Level Factors: Tides and Storm Surges 39

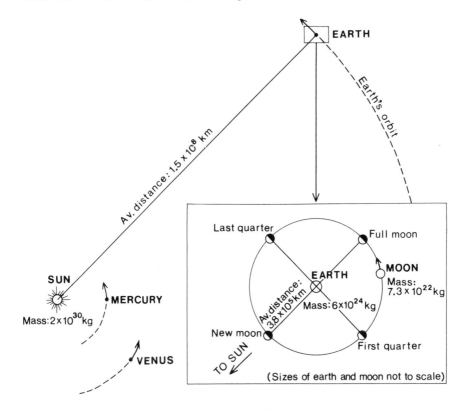

Figure 3.1 Schematic diagram showing the Earth–Moon–Sun system

Let us first consider the Earth–Moon system, and further assume that the Earth does not rotate about its own axis. This assumption indicates that an irrotational Earth and the Moon revolve about a common axis. It should be noted here that no eccentric rotation of the Earth does occur about the common axis. Every point on the Earth revolves through a circle with the same radius which is approximately three-quarters of the Earth's radius (e.g. Dean, 1966). As shown in Figure 3.2, the centrifugal force due to the rotating system, f_c, has the same magnitude at any point on the Earth (because of the same radius of revolution) and is always directed away from the Moon parallel with the line connecting the centres of the Earth and Moon. The attractive force of the Moon, f_a, which is always directed toward the centre of the Moon, increases its magnitude with decreasing distance from a point on the Earth to the Moon centre. The vector subtraction of these forces is the net force, denoted by the bold arrow in Figure 3.2, which is responsible for tide generation. The bulge of sea water occurs not only on the Moon side but also on the opposite side of the Earth: a rugby ball-shaped tidal bulge forms.

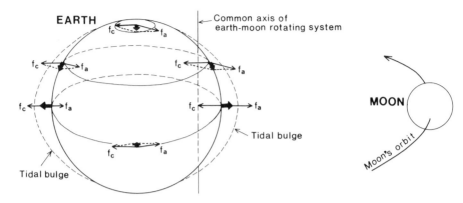

Figure 3.2 Diagram illustrating tide generating force (bold arrow) and tidal bulge; f_a is attractive force of the moon and f_c is centrifugal force due to the Earth–Moon rotating system

Because the Earth rotates daily on its own axis, a fixed point on the Earth experiences two high tides and two low tides in a lunar day or tidal day (the time of the rotation of the Earth with respect to the Moon—24 hr 50 min). The tides with a period of one-half tidal day (12 hr 25 min) are called semidiurnal tides (Figure 3.3). The height of tides is maximum on the equator and it decreases toward higher latitudes.

The Earth's axis is actually inclined at 66.5° to the ecliptic plane, the plane in which the Earth revolves about the Sun: the Earth's equatorial plane makes an angle of 23.5° with the ecliptic plane. It can be considered that the Moon's orbit about the Earth almost coincides with the ecliptic plane, because the difference in angle between the ecliptic plane and the Moon's orbital plane is small, about 5°. The inclination of the Earth's axis does not much affect tidal variation in low latitudes, where semidiurnal tides predominate, but in higher latitudes, the characteristics of the diurnal tides (tides with one high and one low water in a tidal day; see Figure 3.3) appears. Figure 3.4 shows an example of tidal curves from Los Angeles, California and St Michael, Alaska: the semidiurnal tides dominate at Los Angeles, and diurnal features at St Michael, the latter being at a higher latitude. Tides in most places show characteristics of the mixed tides, i.e. tides with two highs and two lows usually occurring each tidal day with recurring marked difference in the heights of successive tides (Figure 3.3).

The actual occurrence of semidiurnal, diurnal, or mixed tides depends not only on the latitude of location but also on the resonance characteristics of a local water basin (e.g. Komar, 1976, p. 139). The tidal type is of significance in determining the length of the time for exposure to air between tides, which

Water Level Factors: Tides and Storm Surges

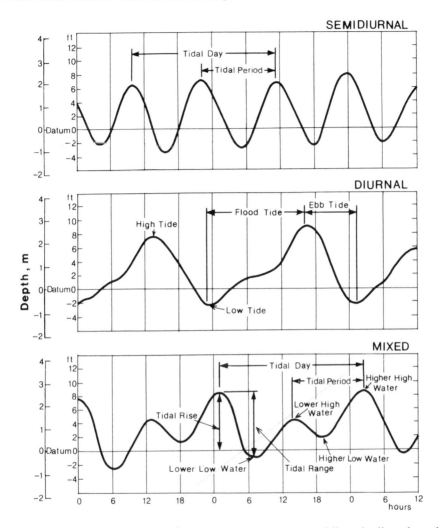

Figure 3.3 Schematic presentation of three tidal types: semidiurnal, diurnal, and mixed. From Coastal Engineering Research Center (1984, Figure A:10), by permission of Coastal Engineering Research Center, US Army Engineers Waterways Experiment Station

is closely related to biological zonation and weathering of rocks. It is evident that the exposure time in a tidal day is longest for diurnal tides.

The Sun also affects tidal variations. As already mentioned, the tide-producing effect of the Sun is half that of the Moon. The Moon and Sun increase the tide-generating force during new and full Moon (Figure 3.1b), causing the greatest tides in a lunar month, which are called the 'spring tides'. During the first and last quarters, on the other hand, the tide-producing effect

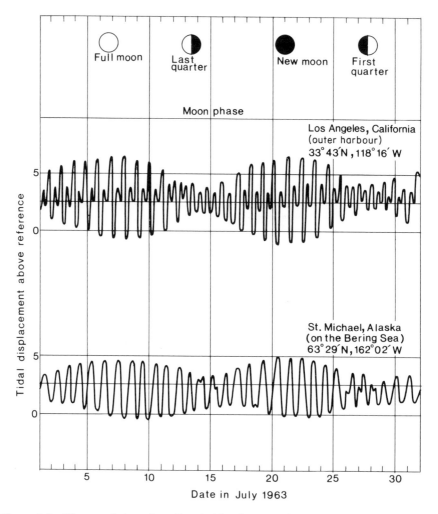

Figure 3.4 Characteristics of predicted tides for two sites located at different latitudes. Semidiurnal tides dominate at Los Angeles and diurnal tides at St Michael. From Dean (1966), by permission

is offset to produce the minimum tides, called the 'neap tides'. Actually the spring and neap tides occur a few days behind the corresponding Moon (Figure 3.4) due to (1) the inertia of large bodies of water and (2) the friction between the water body and the sea bottom.

The difference in height between consecutive high and low tides is the tidal range (Figure 3.3). According to spring tidal ranges, tides are categorized into three (Davies, 1972, p. 51): microtides with a tidal range of less than 2 m,

Water Level Factors: Tides and Storm Surges 43

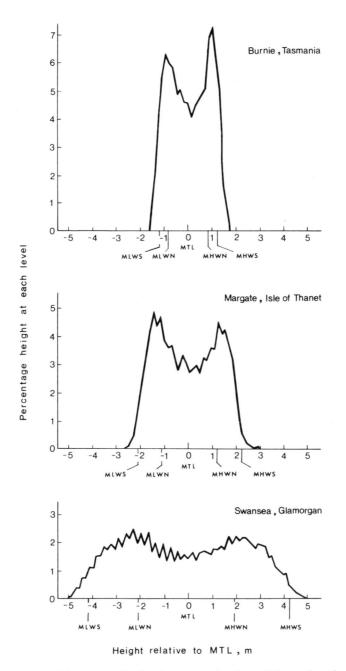

Figure 3.5 Sea-level frequency distribution curves for three different locations. After Carr and Graff (1982), by permission of the Institute of British Geographers

mesotides with a range between 2 and 6 m, and macrotides with a range of greater than 6 m. Most of the world's ocean coasts are in microtidal or mesotidal environments. But there are some highly localized areas exposed to macrotides, such as the Bay of Fundy, Canada; the south shore of the North Sea; English Channel coasts; the Pacific continental coast of Alaska; the Yellow Sea coasts; and the northwestern shore of Australia, with the first having a tidal range of approximately 15 m at the bay head, the largest tidal range in the world. Such a strong regional difference in tidal ranges is due to the effects of the coastline configuration and the underwater topography on the behaviour of the water mass (tidal waves) entering shallow regions.

Some commonly used terms in describing tidal levels are listed here (mainly from Coastal Engineering Research Center, 1984):

Mean higher high (or *lower low*) *water* (abbreviated as MHHW or MLLW): the average height of the higher high (or lower low) waters (Figure 3.3) over a 19-year period.

Mean high (or *low*) *water* (MHW or MLW): The average height of the high (or low) waters over a 19-year period.

Mean high (or *low*) *water neap* (MHWN or MLWN): The average height of the high (or low) waters occurring at the time of neap tide.

Mean high (or *low*) *water spring* (MHWS or MLWS): The average height of the high (or low) waters occurring at the time of spring tide.

Mean sea level (MSL): The average height of the sea surface for all stages of the tide over a 19-year period. It may slightly vary in different years. Not necessarily equal to mean tide level.

Mean tide level (MTL): A plane midway between mean high water and mean low water.

Figure 3.5 illustrates sea-level frequency distribution curves originally plotted by Carr and Graff (1982) using digitized hourly sea level records for three sites with different tidal ranges. These curves demonstrate a double peak distribution, with maxima at about MHWN and MLWN, respectively, indicating that maximum tidal duration is associated with high and low water zones rather than with mid-tide level.

STORM SURGES

The other kind of water level fluctuation is due to meteorological variations. 'Storm surges' are water level fluctuations produced by extreme meteorological disturbances such as gales and atmospheric pressure changes associated with typhoons (or hurricanes, or cyclones, the preferred term depending on geographical location). The most significant phenomenon occurring on the coast is an abnormal rise in sea level, caused by strong onshore winds and abrupt atmospheric pressure reduction.

Figure 3.6 shows observed sea-level fluctuations at Atlantic City, New Jersey, during the hurricane of 14–15 September 1944 (Harris, 1963). In this figure the predicted astronomical tide is plotted by a smooth curve. The difference between the actual water level and the predicted tide level is called the surge. The surge at Atlantic City due to this hurricane is illustrated in Figure 3.7, which was constructed from Figure 3.6. Figure 3.7 shows three water level undulations after the surge peak (arrowed). Such undulations are termed resurgences, water level oscillations occurring after the passage of hurricanes or typhoons (e.g. Groen and Groves, 1962).

Winds blowing over a body of water exert a tangential stress on the water surface, and raise water level downwind: such a water level rise is called the wind set-up. The amount of wind set-up on the coast depends on many factors: wind speed and direction, fetch length, water depth, coastline configuration, and underwater topography (Bretschneider, 1966b; Silvester, 1974b, pp. 178–201). Extremely low-pressure systems results in a raised sea level, as

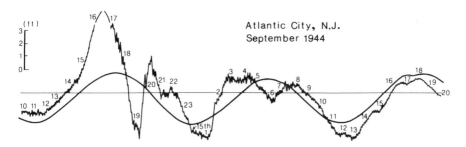

Figure 3.6 Observed and predicted tides at Atlantic City, New Jersey during 14 to 15 September 1944. From Harris (1963), by permission of US Weather Service

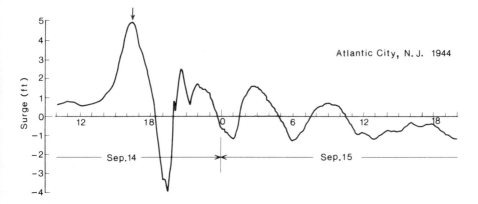

Figure 3.7 Surge at Atlantic City, New Jersey during 14 to 15 September 1944

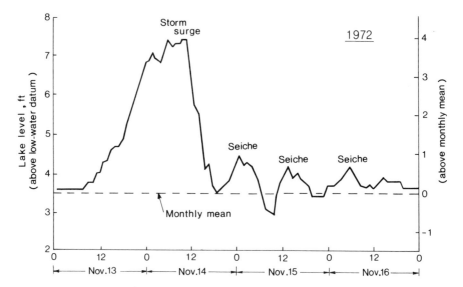

Figure 3.8 Variation of Lake Erie water level at Toledo, Ohio. From Carter et al. (1987, Figure 2.5), *Living with the Lake Erie Shore*, by permission of Duke University Press

an inverted barometer. The static pressure relationship shows that the amount of water level rise in centimetres, ξ, is given by $\xi = 0.99\,\Delta p$, where Δp is the pressure difference in millibars between the centre of the low-pressure system and the surrounding area. It is empirically known that increase in Δp gives rise to an increase in wind velocity. A combined effect of increasing pressure difference and resultant wind velocity augmentation causes further rise in water level, and the latter factor does also generate large waves. The larger storm surges generally accompany the larger waves.

Actual storm surges on the coast vary as the low-pressure system migrates accompanying changes in (1) the atmospheric pressure, (2) the wind speed and direction, and (3) the length of fetch. Mathematical models are widely used for predicting-storm surges. See reviews in Groen and Groves (1962), Horikawa (1978, pp. 153–67), Coastal Engineering Research Center (1984, p. 3:115), and references cited therein.

Enclosed waters, such as the Great Lakes, have no significant tidal fluctuations, but are subject to changes in water level due to hydrological and meteorological factors. The hydrological factors include precipitation, evaporation, inflows, and outflow. An imbalance in these variables determines changes in mean water level. Episodic water level rises are induced by meteorological factors such as strong wind stress and atmospheric pressure reduction, which have been explained before. These meteorological disturbances may cause long-period oscillations on a closed body of water, termed seiches (e.g. Raichlen, 1966). Figure 3.8 shows an example of the variation of Lake Erie water level observed at Toledo, Ohio (Carter *et al.*, 1987). This figure well illustrates the remarkable surge, caused by the storm of 13–14 November 1972, and three subsequent seiches.

Chapter Four

Strength of landform materials

INTRODUCTION

Landform materials have been conventionally classified into rocks and soils, but the basis of classification has not always been sound. Rocks are usually defined as a consolidated assemblage of mineral grains, and soils are defined as an unconsolidated or poorly consolidated assemblage of rock-derived substance which may or may not contain organic matter. Although a transition exists between the two, and no definite boundary can be drawn, this book follows this distinction.

Rocky coasts are composed of a wide variety of materials ranging from hard basalt and granite to soft soils such as clays and tills. For studies of processes and evolution of rocky coasts, it is important to treat coastal rocks or soils as *materials* with their quantified 'hardness' always in mind. In this context, the geological classification of landform materials is of less significance.

Before discussing the resistance of coast-forming materials to wave erosion (Chapter 5), a description will be made of the various kinds of mechanical strength, and then strength reduction due to (1) space-dependent factors such as discontinuities and (2) time-dependent factors such as fatigue, weathering, and biological activity. This chapter is written with an intention of providing the fundamental knowledge necessary not only for a dynamic approach to rocky coast geomorphology but also to further research in other fields of process geomorphology.

BASIC PHYSICAL PROPERTIES

Physical properties of rocks and soils are their intrinsic characteristics which can be obtained through physical testing. Basic physical properties necessary for understanding the mechanical strength of these materials include density, porosity, and water content. Testing procedures for these properties have been developed by the American Society of Testing and Materials and the International Society for Rock Mechanics.

Strength of Landform Materials

Rocks and soils consist of solids (usually mineral grains) and voids filled with air and/or water. Figure 4.1 is a diagram illustrating the three constituents (solids, water, and air) in an imaginary sample. Density, ρ, is defined as mass per unit volume:

$$\rho = \frac{m}{V} \qquad (4.1)$$

where m is the mass and V is the volume. Unit weight, γ, is expressed as the weight (force) of a unit volume:

$$\gamma = \frac{W}{V} = \frac{mg}{V} = \rho g \qquad (4.2)$$

where W is the weight and g is the acceleration due to gravity. Bulk density (frequently simply called 'density') of rocks or soils, ρ_*, and their unit weight, γ_*, can be written as

$$\rho_* = \frac{m_w + m_s}{V_a + V_w + V_s} \qquad (4.3)$$

and

$$\gamma_* = \frac{W_w + M_s}{V_a + V_w + V_s} \qquad (4.4)$$

where the suffixes, a, w, and s, denote air, water, and solids, respectively. If

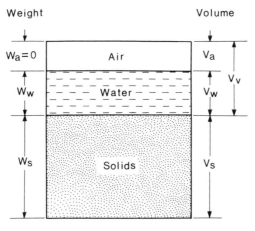

Figure 4.1 Diagram illustrating the three components, solids, water, and air, in a rock or soil sample

the material is completely dry, then its dry density, ρ_{*d} and dry unit weight, γ_{*d}, are expressed by

$$\rho_{*d} = \frac{m_s}{V_v + V_s} \tag{4.5}$$

and

$$\gamma_{*d} = \frac{W_s}{V_v + V_s} \tag{4.6}$$

where V_v is the volume of voids which is equal to $V_a + V_w$. Either ρ_{*d} or γ_{*d} is usually determined and quoted as an index of basic physical properties; γ_{*d} is preferably used in engineering because it is expressed in force units.

Porosity, n_*, is defined as the ratio of the volume of voids, pores, or interstices in the material to its bulk volume:

$$n_* = \frac{V_v}{V_v + V_s} \tag{4.7}$$

Void ratio, e, is given by

$$e = \frac{V_v}{V_s} \tag{4.8}$$

Therefore, these two parameters are interrelated:

$$n_* = \frac{e}{1+e} \quad \text{and} \quad e = \frac{n_*}{1+n_*} \tag{4.9}$$

Water (or moisture) content, w, can be obtained from the relationship

$$w = \frac{W_w}{W_s} \times 100(\%) \tag{4.10}$$

The degree of saturation, s, is expressed by

$$s = \frac{V_w}{V_v} \times 100(\%) \tag{4.11}$$

MECHANICAL STRENGTH

Prior to describing various strengths, some rudimentary knowledge on stress and strain is necessary. When an external force is applied to a solid body, internal stresses arise in the solid material. Figure 4.2a illustrates a solid body on which the axial force, σ_1, and the lateral force, σ_3, are acting (assume $\sigma_1 > \sigma_3$). Two kinds of stresses result in this solid body: (1) stress normal to a plane inclined at an angle, θ, normal stress, σ, and (2) stress tangential to

Strength of Landform Materials

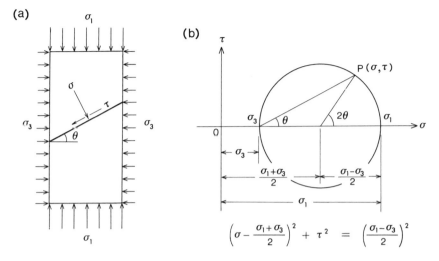

Figure 4.2 Representation of stresses on Mohr's circle. See text for explanation

this plane, shear stress, τ. They are expressed as follows (e.g. Jaeger, 1972, pp. 151–3; Kézdi, 1974, pp. 184–5; Cernia, 1982, pp. 244–8)

$$\left.\begin{aligned}\sigma &= \frac{\sigma_1 + \sigma_3}{2} + \frac{\sigma_1 - \sigma_3}{2} \cos 2\theta \\ \tau &= \frac{\sigma_1 - \sigma_3}{2} \sin 2\theta\end{aligned}\right\} \quad (4.12)$$

Squaring both sides of these equations respectively and then adding them yields

$$\left(\sigma - \frac{\sigma_1 + \sigma_3}{2}\right)^2 + \tau^2 = \left(\frac{\sigma_1 - \sigma_3}{2}\right)^2 \quad (4.13)$$

Graphical presentation of Eq. (4.13) is the well-known Mohr's circle (Figure 4.2b); a point on the circumference of the circle, P, represents the normal and shear stresses on an arbitrary plane inclined θ by in the material. The stress has the dimension of force per unit area.

When stresses arise in a solid body, it suffers deformation, i.e. strain, which is expressed as a ratio of deformation to the original dimension. Suppose that a cylindrical specimen is loaded axially in compression. Deformation occurs laterally and vertically. The ratio of lateral to vertical strains is called Poisson's ratio.

Figure 4.3 illustrates typical axial stress–strain (σ–ε) curves. Strong materials tend to show a linear relation which characterizes elasticity (Figure 4.3a). For

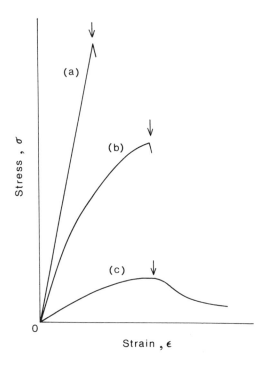

Figure 4.3   Three curves illustrating stress–strain relationships

weaker materials more notable nonlinearity, representing less elastic characteristics, appears in the σ–ε relation (Figure 4.3b and c). Considering that the material ruptures when stress exceeds a certain critical value, i.e. the peak stress (arrows in Figure 4.3), the strength of materials can be expressed in terms of the peak stress, σ_{peak}. The modulus of elasticity or deformation modulus, $\hat{E}$, is given by $\hat{E} = \sigma/\varepsilon$ when the stress–strain curve is linear. Otherwise, it is usually expressed by the slope of a tangent to the curve at a stress equal to $0.5\sigma_{peak}$.

Units of strength are given by the applied force at rupture per unit area; they are expressed by N/m^2(Pa), kg/cm^2, ton/m^2, lb/in^2 (psi), or ton/ft^2. A conversion table from SI units (N/m^2) to the conventional metric or English units, or vice versa, is given in Appendix 1.

Unconfined (or uniaxial) compressive strength

This is the most widely used strength measure in the field of rock mechanics. In soil mechanics field this measure is often used for assessing strength of cohesive soils. A regular-shaped specimen, usually of cylindrical shape, is installed with its longitudinal axis vertical in a testing machine, and an axial force (load) is applied between the platens of the machine (Figure 4.4a). The compressive strength, S_c, is given by

$$S_c = \frac{P_c}{A_c} \qquad (4.14)$$

where P_c is the applied force at failure of the specimen and A_c is the cross-sectional area in the direction normal to the axial force.

The compressive strength value thus determined is affected by (1) geometry of specimens such as height/diameter ratio and size, (2) water (moisture) content in the specimens, (3) the rate of loading. The compressive strength decreases as the height/diameter ratio of the specimen increases up to a ratio of 2 to 3, after which the strength reduction becomes very small or nil (e.g. Grosvenor, 1963; Hobbs, 1964; Mogi, 1966; Protodyakonov, 1969; Hawkes and Mellor 1970). The decrease in strength usually occurs as the size of specimens becomes larger (e.g. Evans and Pomeroy, 1958; Mogi, 1962; Bernaix, 1967; Lundborg, 1967, 1968). The strength reduction also takes place with increasing water content in the specimen (e.g. Colback and Wiid, 1965; Feda, 1966; Burshtein, 1969). The increase in the rate of loading gives rise to an increase in the strength value (e.g. Houpert, 1970; Kobayashi, 1970).

Thus, compressive strength values vary with different testing conditions. The testing method suggested by the American Society of Testing and Materials (ASTM) is recommended for a standard (ASTM D2938-86 for rocks and D2166-85 for soils).

It is easy to prepare cylinder specimens from intact materials, i.e. the materials which are free from joints or fissures. However, preparation of such

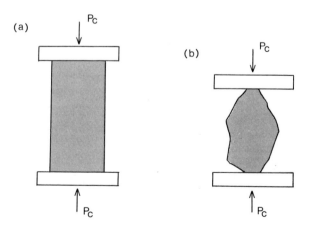

Figure 4.4 Unconfined compression tests using a regular-shaped specimen (a) and an irregular specimen (b)

regular specimens is much more difficult from densely jointed or fragile rocks. For these cases it is better to use the compressive strength testing originally proposed by Protodyakonov (1960) and recommended by the International Bureau for Rock Mechanics (1961). In this test (Figure 4.4b), irregular specimens (spindle-shaped) with a volume of about 100 cm^3 are used and the number of specimens to be tested is 15–20. The compressive strength obtained through this test, S_{ci}, can be described by

$$S_{ci} = \frac{P_c}{A_i} \qquad (4.15)$$

where P_c is the applied force at failure and A_i is the maximum cross-sectional area of the specimen. The strength S_{ci} is related by Protodyakonov (1960) to S_c:

$$S_{ci} = 0.19 S_c \qquad (4.16)$$

Tensile strength

The uniaxial tensile test for rocks (ASTM D2936-84), frequently called the direct test, makes use of a cylindrical specimen which is installed vertically in a testing machine. An axial force is applied to the grips mounted at the ends of the specimen to pull it down (Figure 4.5a). The tensile strength, S_t, is given by

$$S_t = \frac{P_t}{A_c} \qquad (4.17)$$

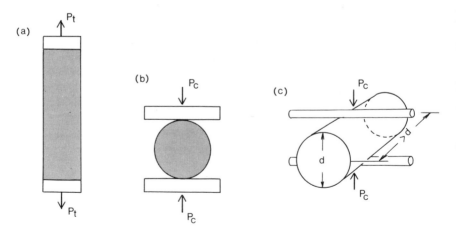

Figure 4.5 Three types of tensile tests: direct test (a), Brazilian test (b), and point-load test (c)

where P_t is the applied force at rupture of the specimen and A_c is the cross-sectional area. Because there are difficulties in gripping the specimen and applying a load parallel to the specimen axis, this direct test is not easy to conduct. Instead, a number of indirect methods for assessing the tensile strength have been proposed (e.g. Vutukuri et al., 1974, pp. 95–131).

Among these, the Brazilian test (Hondros, 1959) is most popular (ASTM D3967-86). In this test a disc specimen is loaded diametrically between the platens of a compression testing machine (Figure 4.5b). The tensile strength S_t can be calculated by

$$S_t = \frac{2P_c}{\pi dl} \qquad (4.18)$$

where P_c is the applied force at failure, d is the diameter of the disc, and l is the thickness of the disc. The above relationship is valid only when fracturing originates from the centre of the disc, with a vertical crack propagated upward and downward. The specimen geometry and also loading rates influence the strength (Vutukuri et al., 1974, pp. 110–14). Mellor and Hawkes' (1971) suggestions are that d should be greater than NX core size (54 mm in diameter) and l should be greater than d.

The point-load testing (e.g. Broch and Franklin, 1972; Guidicini et al., 1973) is also available for estimating the tensile strength: a cylindrical specimen placed horizontally, is compressed by the load applied vertically through two point contacts (Figure 4.5c). The tensile strength S_t can be obtained by

$$S_t = \frac{0.96 P_c}{d^2} \approx \frac{P_c}{d^2} \qquad (4.19)$$

where P_c is the applied force and d is the distance between the loading points, i.e. the diameter of the specimen.

A brittleness index, B_t, is used to designate the degree in brittleness of materials (Aida and Okamoto, 1960). This is expressed by a ratio of compressive to tensile strengths:

$$B_t = \frac{S_c}{S_t} \qquad (4.20)$$

Figure 4.6 shows the tensile strength plotted against the unconfined compressive strength for saturated rock samples. The tensile strength data employed here were obtained from the point load test (Irfan and Dearman, 1978; Bell, 1983, Tables 8.2, 8.3, and 8.6) and from the Brazilian test (Sunamura, 1973; Takahashi, 1976; Tsujimoto, 1987). Although considerable scatter of the data exists in Figure 4.6, there is a clear tendency for tensile strength to increase with rising compressive strength. It is also found that the brittleness index ranges from about 5 to 25. Incidentally, Bieniawski (1975)

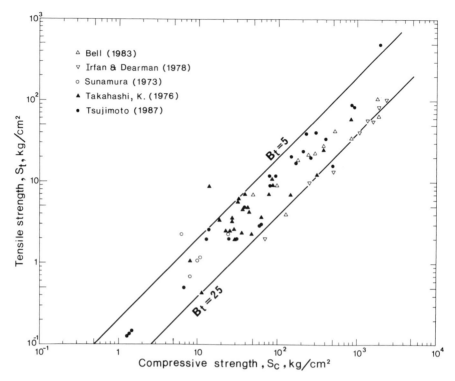

Figure 4.6 Relationship between tensile strength and unconfined compressive strength for saturated rock samples

and Irfan and Dearman (1978) have already reported a linear correlation between the point load tensile strength and the compressive strength.

Shear strength

The best way to estimate shear strength is to apply the triaxial compression test in which uniform and known stress conditions can be produced in the interior of the specimen. The triaxial test (ASTM D2664-86 for rock and D2850-82 for soils) is performed by applying an axial force vertically and a uniform confining pressure laterally on a cylindrical specimen, the former being greater than the latter. The vertical stresses at the failure, σ_1, vary for different values of lateral stress applied σ_3: σ_1 becomes larger as σ_3 increases. Through a series of tests conducted, the σ_1-values corresponding to specific values of σ_3 are obtained and these data are plotted on a τ–σ plane to construct a family of Mohr's circles (Figure 4.7), the tangent to which is called the Mohr envelope. The intercepting point of this envelope and the τ-axis, labelled S_s on this

Strength of Landform Materials 57

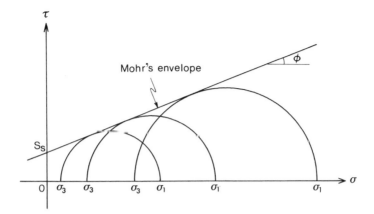

Figure 4.7 Mohr's circles and Mohr's envelope

diagram, is defined as the cohesive strength in the field of soil mechanics, and is frequently referred to as the shear strength in the field of rock mechanics, although the shear strength should be exactly expressed in terms of the Mohr envelope. For some rocks and soils the Mohr envelope is of slightly convex-upward curves, but for most materials the envelope is assumed to be linear. The straight line in Figure 4.7 is given by

$$\tau = S_s + \sigma \tan \phi \quad (4.21)$$

where τ and σ are shear and normal stresses acting on the failure plane, respectively, S_s is the cohesive or shear strength, and ϕ is the angle of internal friction. The triaxial test requires expensive equipment, especially for testing hard rocks, and the procedure accompanies a time- and labour-consuming job.

Other tests for determining the shear strength are available (e.g. Vutukuri *et al.*, 1974, pp. 142–65); they include (1) the torsion test, (2) the direct shear tests with single or double shear planes, (3) the punch test, (4) the constricted oblique test, and (5) the vane shear test, the last being applicable to *in situ* measurement of soils (ASTM D2573-72). The most popular method is the direct shear test with a single shear plane (ASTM D3080-72 for soils). A shear box in which a specimen is placed consists of an upper and a lower section. The upper section moves horizontally relative to the lower section, which is fixed, when a shearing force is applied laterally to the upper section. The normal force is acting on the specimen. Because the shear stress at rupture, τ, depends on the magnitude of the normal stress applied, σ, several sets of data on τ and σ should be collected by varying σ. These data sets are plotted on a plane to obtain a failure envelope similar to Eq. (4.21). Then the shear strength is determined.

A geometrical relationship in Figure 4.8 leads to:

$$S_c = \frac{2S_s \cos \phi}{1 - \sin \phi} \quad (4.22)$$

Derivation of this equation is based on the assumption that the Mohr envelope obtained from the triaxial test or the failure envelope from the direct shear test can be expressed by a straight line which is tangent to the Mohr circle of the unconfined compressive strength. No unconfined compression tests are applicable to very weak materials because preparation of the test specimen is not possible. Equation (4.22) is useful for estimating the compressive strength of such materials from the results of the triaxial or the direct shear tests.

Another method is available for determining shear strength. This does not depend on shear tests but on a calculation using the following relation (Kobayashi and Okumura, 1971):

$$S_s = \frac{S_c S_t}{2\sqrt{[S_t(S_c - 3S_t)]}} \quad (4.23)$$

This was obtained by considering two Mohr's circles plotted, respectively, for (1) tensile strength determined from the Brazilian test and (2) unconfined compressive strength, and then by assuming that the shear strength is represented by the intercept of the common tangent of the two circles with τ-axis.

Yamaguchi and Nishimatsu (1977, pp. 135–6) reported that, for the shear strength of rocks, a good agreement exists among results obtained through these three methods: (1) the triaxial test, (2) the single shear test, and (3)

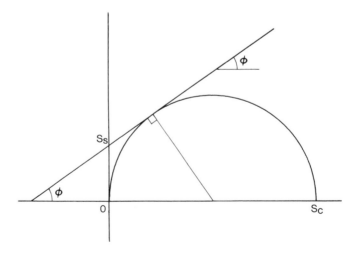

Figure 4.8 Mohr's circle for unconfined compressive strength, with a straight envelope

Strength of Landform Materials 59

calculation using Eq. (4.23). Therefore, any of these is suitable for estimating the shear strength of rocks. For soils, especially the less cohesive materials, either the triaxial or the single shear tests should be employed.

Figure 4.9 shows the relationship between the shear strength calculated from Eq. (4.23) and the unconfined compressive strength for saturated rock specimens. A good correlation between the two is found.

Impact strength

For measurements of the resistance of rocks against impact force, Protodyakonov's (1962) test is of interest. This test, widely used in the former USSR, was originally developed to determine the drillability of rocks. An irregular-shaped rock specimen, placed on the bottom of a cylinder, is pounded by a heavy weight falling from a certain height. The Protodyakonov index is assessed from (1) the number of pounding and (2) the amount of finer fraction of the crushed specimen. A good correlation between this index and the unconfined compressive strength has been reported (Singh, 1981).

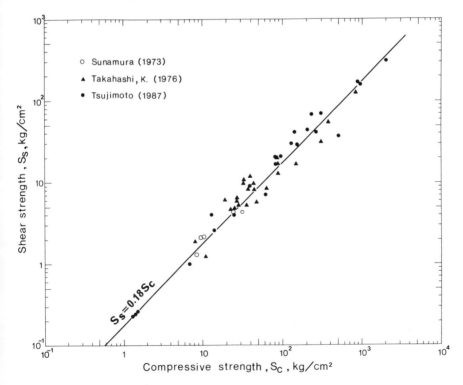

Figure 4.9 Relationship between shear strength calculated using Eq. (4.23) and unconfined compressive strength for saturated rock samples

A method basically similar to that of Protodyakonov has been proposed in Britain by Evans and Pomeroy (1966, pp. 132–7). The strength measure obtained through this method is called an 'impact strength index' (ISI), which is different from Protodyakonov's coefficient. It is found that ISI is strongly related to the unconfined compressive strength (Fig. 4.10).

The Schmidt hammer rebound number

The height of rebound of a small steel ball after its collision with a rock surface depends on elasticity of the surface, which in turn reflects to a certain extent

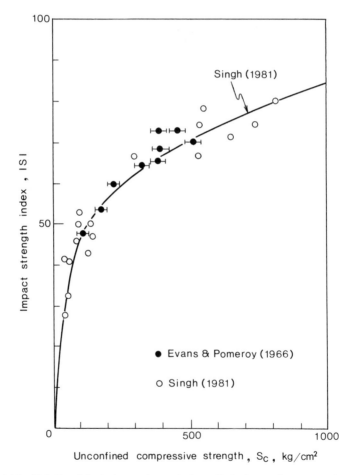

Figure 4.10 Relationship between impact strength index and compressive strength. Data from Evans and Pomeroy (1966, p. 135), obtained from coals, and Singh (1981) from mostly sandstone and shale

mechanical strength of rocks. Applying this, E. Schmidt in 1948 designed a light, portable hammer to conduct non-destructive tests on concrete. The hammer, now used for *in situ* rock tests, measures the distance of rebound of a plunger which is released by a spring towards the surface of rock. This measure is called the Schmidt hammer rebound number, R_s. The R_s-values vary with (1) the orientation of the hammer to the rock surface, (2) water content in rocks, (3) the presence of discontinuities such as cracks, joints, and bedding planes, and (4) irregularities of the surface. The rebound number R_s has been correlated with the unconfined compressive strength (Duncan, 1967; Irfan and Dearman, 1978). Figure 4.11 shows a correlation diagram plotted using data of Irfan and Dearman (1978) and Bell (1983). Figure 4.12 is a diagram originally constructed by Deere and Miller (1966), which enables us to relate R_s to the unconfined compressive strength if the hammer orientation and the unit weight of rocks are known.

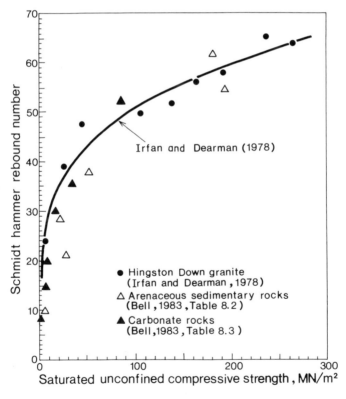

Figure 4.11 Relationship between Schmidt hammer rebound number and compressive strength

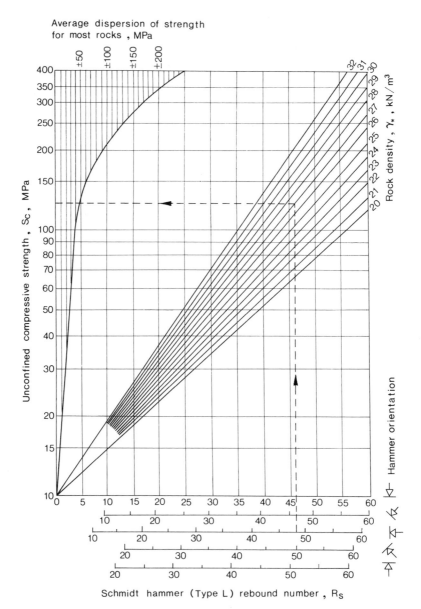

Figure 4.12 Graph for estimating compressive strength of rocks from Schmidt hammer (Type L) reading. From Hoek and Bray (1981, Figure 37), by permission of the Institute of Mining and Metallurgy

Penetration strength

The Schmidt hammer test cannot be applied to assess the strength of soils and weakly consolidated rocks. For an easy, rapid, crude index of the strength of such materials, a cone-type penetrometer test is useful (Sanglerat, 1972). Two kinds of strength parameters are used. One is a parameter obtained from reading the force required to push a cone-shaped probe a certain distance into the material, and is expressed in terms of the force divided by the area of the cone base. The other is a parameter, used for some pocket-size penetrometers, based on the measurement of volume of a probe intruded under a certain constant force. The two strength parameters are inversely related to each other. Figure 4.13 exemplifies that the former parameter can be linearly related to the unconfined compressive strength. The difference in gradient of the lines in this figure is due to the shape and size of cones used.

Compressive wave velocity

Propagation velocity of compressional (or longitudinal, or primary) waves in rocks (e.g. Lama and Vutukuri, 1978, pp. 195–6)—usually denoted as

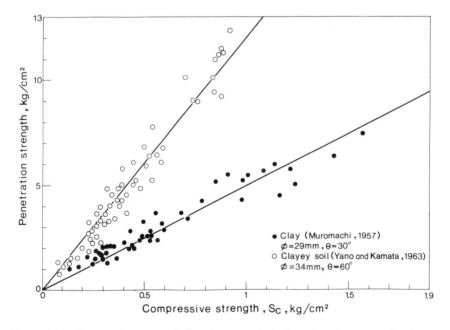

Figure 4.13   Penetration strength (bearing capacity) plotted against unconfined compressive strength. In this figure ϕ and θ denote diameter and vertex angle of the cone, respectively

V_p—is expressed as

$$V_p = \sqrt{\frac{(1-\nu)\hat{E}}{\rho(1+\nu)(1-2\nu)}} \qquad (4.24)$$

where $\hat{E}$ is the deformation modulus, ρ is the density, and ν is Poisson's ratio. Because (1) $\hat{E}$ is positively correlated with the unconfined compressive strength S_c and (2) ρ- and ν-values are respectively in a small range for rocks, it is generally said that V_p increases with increasing S_c. This relation has been confirmed. Figure 4.14 exemplifies S_c vs. V_p relationship, plotted using data from Inoue and Omi (1971), who conducted the V_p test and the compression test

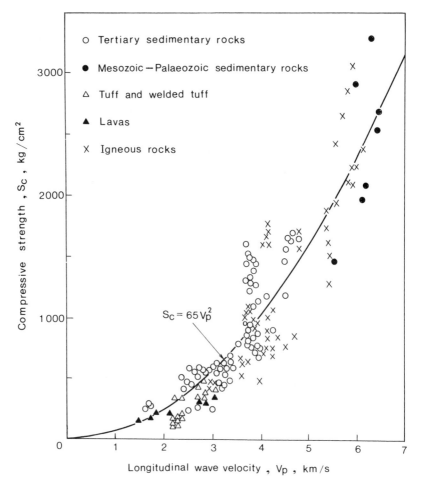

Figure 4.14 Relationship between compressive strength and longitudinal wave velocity. Data from Inoue and Omi (1971)

using air-dry, regular-shaped specimens taken from various types of rocks in Japan. The result (Figure 4.14) shows that a general trend can be expressed by $S_c \propto V_p^2$. Aside from ρ and ν, other factors affecting V_p are porosity, water content, texture, and anisotropy (Lama and Vutukuri, 1978, pp. 236–98). Duncan (1967) found a positive, linear relation between laboratory determined V_p-values and the Schmidt hammer rebound number.

Representative strength parameter

Because unconfined compressive strength is a widely used parameter with testing criteria having been well established, attempts have been made to correlate this with other strength measures. Some of the results are illustrated in Figures 4.6, 4.9, 4.10, 4.11, 4.13, and 4.14. Although the relations shown in these figures are not always linear in a mathematical sense, the compressive strength is found to be positively related to other measures. This suggests that unconfined compressive strength can serve as a representative strength parameter for intact materials. For the case of very weak materials, for which compression tests are not possible, the strength can be evaluated through Eq. (4.22). To solve actual geomorphological problems, however, the most appropriate strength parameter should be employed depending on (1) the type of forces acting on a site of interest and (2) the resultant of processes occurring. If this parameter is not easy to quantify, we can substitute the unconfined compressive strength.

INFLUENCES OF DISCONTINUITIES ON STRENGTH

Structural discontinuities, which are any visible fractures within a rock or soil masses, include cracks (or fissures), cleavage planes, joints, faults, folds, and stratifications. Cracks usually refers to discontinuities of weathering origin, which are often observed both on a rock and soil surfaces. The remaining five terms are used for description of interruptions inherent in a rock body. Joints, i.e. fractures with little or no displacement, are of major importance because of their existence within almost all rock types.

Discontinuities reduce the overall strength of rock or soil masses. The strength reduction is controlled by the opening, interval, extent (continuity), inclination and orientation of discontinuities. Figure 4.15 illustrates experimental results from Hayashi (1966), who performed uniaxial compression tests on a quadrangular prism-shaped, jointed specimen made of plaster. The joints were simulated by a wax paper inserted during casting. This figure shows that uniaxial compressive strength decreases with an increasing number of joints, even if the total area of the intermittent joints is equal, and also that the joint inclination affects the strength. The effect of cracks on the shear

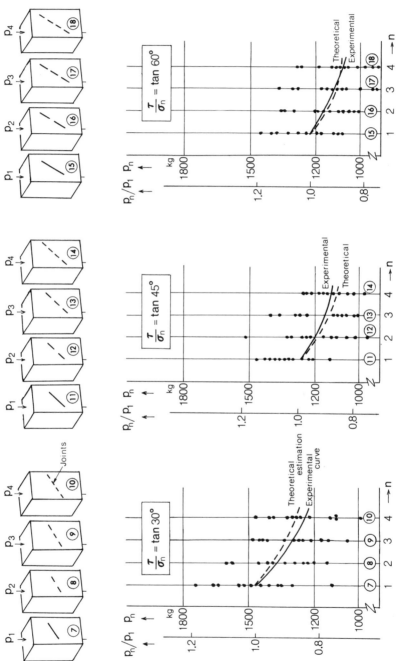

Figure 4.15 Experimental results showing dependence of compressive strength on the number of joints and joint inclination. From Hayashi (1966), by permission

strength of fissured clays was studied by Marsland (1972). Figure 4.16 shows that the laboratory shear strength (obtained through triaxial tests), normalized by the *in situ* strength, drops abruptly as the spacing of cracks in the specimens decreases.

Several attempts have been made in connection with engineering design to objectively and quantitatively assess the quality of a rock mass containing fractures. Onodera (1963) proposed a parameter which can be expressed by a ratio, $\hat{E}_f/\hat{E}_l$, where $\hat{E}_f$ and $\hat{E}_l$ are respectively the deformation moduli for a rock mass *in situ*, and for an intact laboratory specimen taken from the same rock mass. Hobbs (1974) examined the relationship between $\hat{E}_f/\hat{E}_l$, now known as the rock mass factor, and the fracture frequency, the mean number of fractures per metre. He found that the rock mass factor diminishes with increasing fracture frequency following a near-exponential decay curve. Because the deformation modulus is related to the square of the compressional wave velocity, or sonic velocity [Eq. (4.24)], $\hat{E}_f/\hat{E}_l$ is essentially the same as $(V_{pf}/V_{pl})^2$, called the velocity index (e.g. Farmer, 1983, p. 30), where V_{pf} is the sonic velocity measured in the field on a rock mass and V_{pl} is the velocity measured in the laboratory on an intact specimen. When a rock mass has no discontinuities, the velocity index should be unity. With increasing degree of fracturing, the velocity index decreases towards zero.

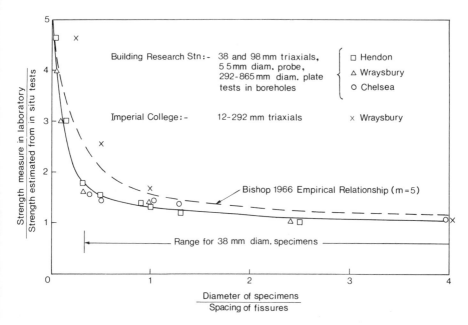

Figure 4.16 Shear strength reduction with increasing number of cracks in fissured clays. From Marsland (1972), by permission of G. T. Foulis & Co., Ltd

The concept of rock quality designation (RQD) was introduced by Deere (1964). The determination of RQD-value requires the data of a drill core, say, of NX size (54 mm in diameter). The RQD is expressed as

$$\text{RQD} = 100 \frac{\Sigma l}{l_\text{T}} \qquad (4.25)$$

where Σl is the collective length of intact core sticks longer than 10 cm and l_T is the total length of core required for the RQD-determination (about 1.5 m). The RQD decreases as the degree of fracturing increases (Deere et al., 1966). A good correlation between the RQD and the velocity index was also suggested by Deere et al. (1966).

An engineering-oriented work by Ikeda (1979), who attempted to estimate the strength of fractured rock masses, shows that

$$S_\text{c}^* = \left(\frac{V_\text{pf}}{V_\text{pl}}\right)^2 S_\text{c} \qquad (4.26)$$

where S_c^* is the compressive strength of an overall rock mass in the field and S_c is the compressive strength of the intact specimen in the laboratory. This relation was derived from Eq. (4.24) on the assumption that (1) the modulus of elasticity is linearly related to the unconfined compressive strength and (2) the density and Poisson's ratio of rocks, both affecting sonic velocity, are respectively constants, regardless of fracturing. Instead of the velocity index in Eq. (4.26), Suzuki (1982) employed a simpler parameter, V_pf/V_pl:

$$S_\text{c}^* = \left(\frac{V_\text{pf}}{V_\text{pl}}\right) S_\text{c} \qquad (4.27)$$

This relationship was applied to express the resistance of river channel bedrock to fluvial erosive forces in his study on lateral erosion. His work was the first geomorphological research in which the influence of fracturing within a rock mass was quantified. Recently Sunamura (1988a) discussed the possibility of applying existing parameters on visibly measured fracture frequency to strength-assessment of fractured rocks.

DETERIORATION OF STRENGTH

Fatigue

Repeated stress generation within a rock specimen due to cyclic loading brings about fracture occurring at stress level considerably lower than the normally determined strength, i.e. ultimate strength. Such a strength reduction process is closely associated with crack initiation and propagation facilitated by the cyclic stress generated. This process is known as fatigue, and the reduced strength level is called the fatigue strength, σ_f. The ratio of fatigue strength to ultimate strength, σ_f/S_c, is plotted against the number of loading cycles, N, to illustrate the fatigue characteristics; the diagram is usually known as the $S-N$ curve. Figure 4.17 shows $S-N$ curves of unconfined compression tests on Emochi andesite specimens under dry and wet conditions. Dry specimens generally show a fatigue limit (i.e. lower critical value of σ_f/S_c) of several tenths of ultimate strength (e.g. Farmer, 1983 p. 137; Yatsu, 1988, pp. 43–7). As shown in this diagram, a decrease in fatigue limit occurs when specimens are water-saturated (Horibe et al., 1970; Brighenti, 1979).

Strength reduction due to repeated (dynamic) loading also occurs on cohesive soils. Cyclic loading tests conducted in the field of soil mechanics indicate that the reduction in the static unconfined compressive strength rises with increasing (1) magnitude of repeated stresses and (2) number of cycles (e.g. Seed and Chan, 1966; Andersen, 1976; Lee and Focht, 1976; Houston and Herrmann, 1980).

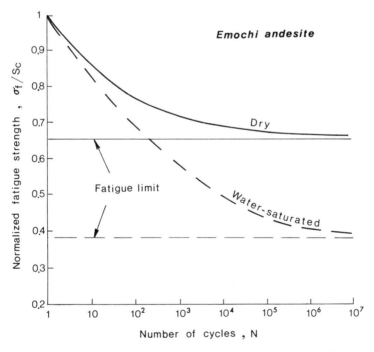

Figure 4.17 An example of fatigue ($S-N$ curve) showing the relationship between normalized fatigue strength and the number of loading cycles. Curves are drawn from data given by Horibe et al. (1970)

Weathering

Weathering, i.e. *in situ* alteration of rocks and minerals on exposure to atmospheric agents, plays an important role in reducing the strength of rocks. Weathering is generally either chemical and/or mechanical. Although these processes actually proceed almost spontaneously and interrelatedly, a separate description will be given as a matter of convenience.

Chemical weathering

This is the process by which rocks and minerals at or near the Earth's surface are transformed into chemically stable substances through reactions such as hydrolysis, oxidation–reduction, and solution. Water, a carrier of dissolved oxygen, carbon dioxide, and various acids, plays a crucial role in chemical weathering processes. This book will provide only a basic account of chemical weathering. Readers who wish to study this topic further are advised to read such texts as Curtis (1976), Brunsden (1976), Ollier (1984, pp. 30–51), and Yatsu (1988, pp. 167–283), the last including a critical review with an extensive literature.

Hydrolysis is a chemical process through which rock-forming minerals are altered or decomposed by water, i.e. the reaction between mineral ions and the H^+ and OH^- ions of water. In the silicate minerals (e.g. feldspar and mica), the metal cations such as K^+ and Na^+ are replaced by H^+ ions to form clay minerals, and the OH^- ions combine with these cations to form soluble (and thus mobile) products. The chemical decomposition of a substance takes place in the presence of water, and the water itself is altered. Feldspar-rich rocks such as granite and granodiorite are particularly vulnerable to hydrolysis.

Marked weathering is frequently seen on granite. Figure 4.18 exemplifies the influence of weathering on the reduction in strength of coarse-grained granite of the Neo-Ukrainian type in the former USSR. In this figure, both compressive strength and porosity of the samples taken from various depths are plotted against the vertical distance from the Earth's surface. Considering that porosity is an effective index for expressing the degree of rock weathering (e.g. Belikov *et al.*, 1967), Figure 4.18 shows that weathering of the granite takes place close to the surface, resulting in considerable strength reduction. In the case of granite which is too highly weathered to prepare regular-shaped specimens for strength tests, Suzuki *et al.* (1977) employed point-load testing by use of irregular specimens, and found a similar strength reduction on the Rokko granites in Japan. Weathering-induced vertical variations in physical and mechanical properties have been investigated in UK on the Lias clay (a Jurassic shale) (Chandler, 1972) and on London Clay (Chandler and Apted, 1988). These studies indicated that, with decreasing depth towards the ground

Strength of Landform Materials 71

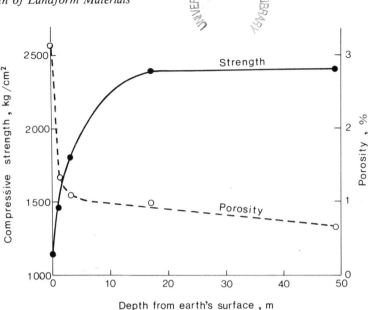

Figure 4.18 Variations with depth in the compressive strength and porosity of Neo-Ukrainian granite. Data from Belikov *et al.* (1967)

surface, water content (a measure of degree of weathering) increases, and shear strength decreases.

Oxidation is one of the modes of chemical weathering in which oxygen dissolved in water reacts with certain minerals to form oxides or hydroxides. Oxidation dominantly occurs in the aerated zone just above the level of permanent water saturation. One of the most easily oxidized minerals is iron, and oxidation produces ferric iron (Fe_2O_3). An assemblage of ferric iron forms a reddish–brown crust on the rock surface or similar-coloured stains along fractures; these features are evidence of chemical decomposition. Reduction, the reverse process of oxidation, removes oxygen from a compound, and usually occurs by the action of stagnant deoxgenated water. In the case of iron, ferric iron (Fe_2O_3) is changed to ferrous iron (FeO) which is more soluble and therefore may be more easily removed in the solution. The reduced products show greyish or bluish colours.

Solution is a process of chemical weathering by which rock-forming minerals are dissolved, resulting in rock decomposition. Solution of limestone, most of which is composed of calcium carbonate ($CaCO_3$), is well known. Through the reaction with water containing carbon dioxide (CO_2), $CaCO_3$ is dissociated into two ions, Ca^{2++} and HCO_3^-, which will diffuse away from a rock surface. The rock surface is always kept fresh. Calcium carbonate solubility is affected by water temperature, pH, and carbon dioxide concentration. Solution differs from hydrolysis and oxidation/reduction in that no residues are formed on the rock surface. In a broad sense, solution is itself a form of erosion—chemical erosion.

Because surface seawater is generally saturated or supersaturated with calcium carbonate, it is difficult to envisage dissolution of coastal limestone by seawater. However, nocturnal activity of some intertidal organisms produces a high concentration of carbon dioxide and lowers the pH in seawater inshore, creating conditions favourable for calcium carbonate dissolution. Thus, limestone is soluble in seawater under certain conditions. For a fuller discussion of solution mechanisms of coastal limestone by still sea water, see Trudgill (1985, pp. 127–37). In the spray zone on rocky shores of calcareous sandstone or limestone, solution processes produce intricate pitting on the rock surface, known as muricate weathering (Bird, 1974). At Jubilee Point, Victoria, Australia, solution is caused by the action of rain water, aerated surf and spray which may be richer in carbon dioxide, and corrosive exudations from marine organisms living in the spray zone (Bird, 1974).

Mechanical weathering

This comprises the processes of disintegration due to (1) frost action, (2) thermal stress, (3) pressure effects associated with salt-crystal growth, (4) swelling due to wetting and shrinking due to drying, and (5) unloading.

Frost action, which is characterized by volume expansion, upon freezing, of water occupying pores and/or interstices within a rock body, is responsible for the physical disintegration of rocks (Brunsden, 1976; Washburn, 1979, pp. 76–8; Yatsu, 1988, pp. 73–81). This is called frost weathering, which is dominant in cold environments where alternating freeze–thaw action frequently occurs. It is obvious that the fatigue effect facilitates this type of weathering.

When a rock mass is heated, the temperature of its outer few millimetres, as compared with the inner part, is much higher because of the low thermal conductivity of rocks. As a result, the outer part tends to expand more, and thermal stresses are generated, leading to rock disintegration (e.g. Ollier, 1984, pp. 18–22; Yatsu, 1988, pp. 120–40). Alternate heating and cooling bring about a fatigue effect, which will accelerate this thermal weathering.

In coastal and arid regions, crystal growth from saline solutions frequently takes place due to evaporation. Salt crystallization occurring in interstices within a rock produces stresses, leading to their widening, and resulting in granular disintegration. This is known as salt weathering (e.g. Wellman and Wilson, 1965). When salt crystals that have formed within pores are heated, or saturated with water, they expand and exert pressure against the constraining pore wall. This also produces thermal stress (e.g. Cooke and Smalley,

1968) or hydration stress (e.g. Winker and Wilhelm, 1970), both of which are also involved in salt weathering (Yatsu, 1988, pp. 81–7).

Clay minerals such as smectite or vermiculite have the inherent trait of swelling upon wetting, and shrinking when they dry out. Materials containing these clay minerals, such as mudstone and shale, show a considerable expansion on wetting (e.g. Matsukura and Yatsu, 1982; Yatsu, 1988, pp. 90–118), which sometimes initiates microcrack formation, widening of existing cracks, or disintegration of the rock mass. In Figure 4.19, the maximum expansion strain, ε_{max}, is plotted against the strength reduction index, I_s, which is defined as, $I_s = (1 - S_{cw}/S_{cd}) \times 100$ (%), where S_{cd} and S_{cw} are the unconfined compressive strength of dry and wet cylindrical specimens, respectively. $I_s = 100\%$ means that the specimens fragmented when submerged into water, i.e. $S_{cw} = 0$. It is found that the rock samples exhibiting larger (expansion) strain are prone

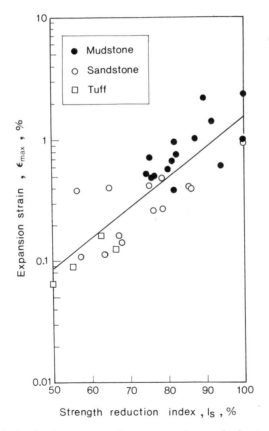

Figure 4.19 Relationship between maximum expansion strain due to water absorption (ε_{max}) and strength reduction index (I_s). From Takahashi (1976), by permission

to undergo greater strength reduction. Upon drying, evaporation of the absorbed water causes the expanded clays to shrink, and shrinkage cracks frequently form. Alternate swelling and shrinking associated with cyclic episodes of wetting and drying, plus the fatigue effect, culminate in the physical disintegration of rocks known as wet/dry weathering, or slaking.

Unloading (e.g. Ollier, 1984, pp. 4–10; Yatsu, 1988 pp. 140–50) is the removal by erosion of overlying materials. This may allow the pressure confined within rock masses to release. The pressure release is considered to be a major responsible factor for the development and widening of fractures approximately parallel to the exposed landform surface.

Biological influences

Biological activity may reduce the strength of rocks. It has been recently recognized that micro-organisms, including bacteria, algae, fungi, and lichens, play a significant role in chemical alteration of rock-forming minerals under certain conditions. These micro-organisms can cause the direct removal of lithic components, achieving the biological erosion of rock surfaces. One of the biomechanical effects is rock fracturing, due to the increasing pressure of plant roots which grow into the existing discontinuities such as joints and bedding planes, and then enlarge. Another effect is boring and grazing of coastal rocks by marine organisms, especially on limestones in the tropics, mainly through a combination of chemical and mechanical processes. Typical boring organisms are bivalve molluscs and clinoid sponges, while grazing organisms such as echinoids, chitons, and gastropods, dislodge lithic material from the rock surface.

For a full discussion of biological processes in general, refer to Yatsu (1988, pp. 285–396); for detailed descriptions of other effects of marine organisms, see Trudgill, (1985, pp. 137–55), Trenhaile (1987, pp. 64–82), and Spencer (1988). Unfortunately there have been few quantitative studies of decrease in mechanical strength due to marine biota, except by Healy (1968b). He measured compressive strength on sandstone and siltstone riddled by boring molluscs, and found that these rocks have only 20–35% of strength of solid (intact) rocks. This marked strength reduction was considered to be brought about by (1) the large void ratio due to many drilled holes and (2) chemical decay arising from metabolic activities of the boring organisms (respiration, excretion, and exudations). Further quantitative investigations of bio-induced strength deterioration are needed for studies on landform evolution in which biological action may play an important role.

Chapter Five
Processes of cliff erosion

FACTORS AFFECTING CLIFF-BASE EROSION BY WAVES

When waves erode the base of a cliff, the cliff becomes unstable, due to the increase in slope angle or in slope stress caused by basal erosion. This instability induces mass movement of various types such as falls, topples, slides, and flows. The occurrence of mass movement is greatly influenced by the lithology, geological structure, and geotechnical properties of the cliff-forming material, as well as by the magnitude of basal erosion. There may be a time lag between basal erosion and mass movement.

Mass movement supplies debris to the foot of a cliff. No basal erosion can occur while the debris obstructs wave attack at the cliff base. Continual wave action diminishes and disperses the debris-forming materials; they are transported alongshore and/or offshore by waves and associated nearshore currents, until the cliff base is again exposed to erosion. Figure 5.1 shows a descriptive diagram of cliff recession. Basal erosion by waves is obviously necessary for a cliff to continue to recede. Therefore, a thorough understanding of wave-induced basal erosion is of great importance in cliff recession studies.

Many factors are involved in wave erosion at the base of a cliff, as illustrated in Figure 5.2 showing possible interrelationships. Ultimately, the factors governing basal erosion are (1) the assailing force of waves at the cliff base, and (2) the resisting force of the material forming the lower cliff. The relative intensity of the two determines whether erosion occurs or not.

ASSAILING FORCE OF WAVES

Wave energy in deep wate is directly related to the assailing force exerted by waves at the cliff base, F_W, whereas the following three factors indirectly influence F_W: (1) water level (Chapter 3), (2) beach and shallow-water bottom topography in front of the cliff, and (3) beach sediment. A combination of

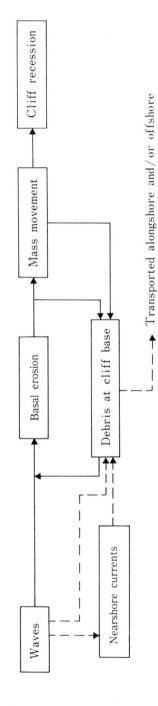

Figure 5.1 Coastal cliff recession system. Basal erosion by waves is essential for continued cliff retreat. After Sunamura (1983). Reprinted by permission from *CRC Handbook of Coastal Processes and Erosion*, copyright © CRC Press, Inc., Boca Raton, FL

Processes of Cliff Erosion

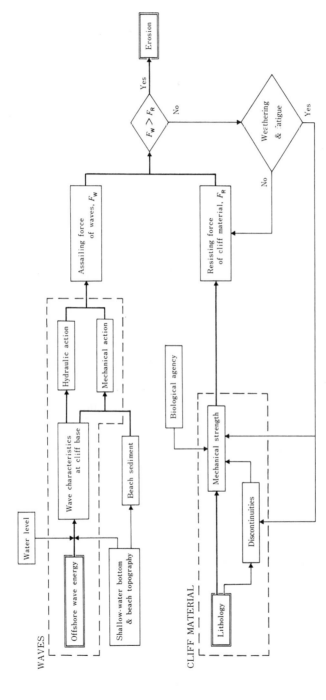

Figure 5.2 Factors affecting cliff base erosion by waves. After Sunamura (1983). Reprinted by permission from *CRC Handbook of Coastal Processes and Erosion*, copyright © CRC Press, Inc., Boca Raton, FL

water level, nearshore bottom morphology, and deep-water wave characteristics determines the type of waves immediately in front of the cliff, and controls their height. Waves always exert hydraulic action on a cliff face, and also mechanical action whenever they are armed with beach sediment.

Both hydraulic and mechanical action characterizes the wave assailing force (Figure 5.2). The former consists mainly of compression, tension, and shearing. When waves hit the cliff face, a compressive force (pressure—see Chapter 2) acts perpendicular to the face. If the cliff has joint- or fault-associated openings, the air in the interstices is suddenly compressed; as the wave recedes, the compressed air expands with explosive force to exert an outward stress. These forces may cause the interstices to grow by the process of wedge action (e.g. Thornbury, 1960, p. 432; Longwell et al., 1969, p. 333; King, 1972, p. 451; Holmes and Holmes, 1978, p. 509). Jointed, loose blocks may be easily removed from the cliff face by wave action, a process often referred to as quarrying or plucking. Tension which acts as waves recede may facilitate this removal process. Immediately after a wave hits a cliff, the water mass ascends the cliff face, exerting a shearing force. Large waves breaking against a steep cliff throw up huge masses of water (Baker, 1943, 1958). Cavitation (e.g., Barnes, 1956) probably occurs when the up-rush is of high velocity, although there is little information available on the effect and occurrence of cavitation in the coastal environment.

Waves armed with beach sediment exert mechanical action (Figure 5.2), which comprises abrasive and impact forces produced on a cliff face by mobilized solid particles. Abrasion (or corrasion) is the erosion of a bedrock surface caused by the wearing, grinding, or scraping action of sediment particles moved to and fro by waves. When rock fragments are hurled against the cliff, impact stresses are set up on the rock surface, the stress increasing as the mass and/or velocity of the impacting particles are increased. It is clear that these mechanical processes intensify the assailing force of waves. An extensive literature has stressed the importance of mechanical action accompanying wave erosion of rocky coasts (Emmons et al., 1955, p. 331 ; Gilluly et al., 1959, pp. 306–7; Cotton, 1960, pp. 408–9; Kuenen, 1960, pp. 222–3; Thornbury, 1960, p. 433; Strahler, 1963, p. 544; Guilcher, 1958, p. 19; Zenkovich, 1967, p. 141; Fairbridge, 1968, pp. 859–65; Longwell et al., 1969, p. 333; Davies, 1972, pp. 81–3; Bird, 1976, pp. 71–3; Komar, 1976, p. 15; Birkeland and Larson, 1978, p. 442; Sparks, 1986, p. 225; Ritter, 1986, p. 515).

The wave assailing force F_W is inclusive of these various actions or forces produced on a cliff face, which occur almost simultaneously. The magnitude of F_W changes cyclically, basically with wave period; this is a strong contrast with other geomorphological agents such as river flow and glaciers. No direct field measurements of F_W have been made, because of the complexity of the phenomena involved and the difficulty of measurement. Therefore, the most suitable physical and quantitative index for F_W has not yet been determined in actual field situations. This is one of the important problems to be solved in further studies of wave-induced rocky coast erosion.

RESISTING FORCE OF CLIFF MATERIALS

The lithology of cliff-forming materials reflects their mechanical strength, which principally determines the resisting force of cliff materials against waves, F_R (Figure 5.2). The deterioration of F_R is brought about by the occurrence of discontinuities in a rock mass which include bedding planes, cracks, cleavages, joints, and faults, some being inherent in lithology while others are of tectonic origin. In some areas, joints or faults are a dominant controlling factor for cliff recession and resultant coastline development (e.g. Wilson, 1952; Byrne, 1964).

Various mechanical strengths have been tested in cliff erosion studies: compressive strength (e.g. Sunamura, 1973; Ohshima, 1974; Aramaki, 1978; Carter and Guy, 1988; Allison 1989), tensile strength (e.g. Sunamura, 1973; Ohshima, 1974), cohesive strength (McGreal, 1979b), shear strength (Zeman, 1986; Kamphuis, 1987), penetration strength (Yamanouchi, 1964, 1977), and Schmidt hammer rebound value (Jones and Williams, 1991). Resistance of the cliff material against the wave assailing force F_W is considered to be different according to types of action involved in F_W. This varies, depending on temporal and spatial situations in the field. The most appropriate parameter for expressing the resisting force F_R has not yet been fully determined. Strength measurements as described above, are closely related (Chapter 4), and it is possible to use one of them as representative of F_R. Compressive strength is probably the most useful index.

Compressive strength has been used to represent erosion resistance by Flaxman (1963), Inozemtsev et al. (1965), and Kamphuis and Hall (1983), although these studies were not conducted with the intention of applying the results to the coastal environment. Flaxman (1963) indicated that resistance to erosion by running water of natural channels made of cohesive soils can be expressed in terms of their compressive strength. Kamphuis and Hall (1983) also found that critical shear stress for cohesive material erosion by unidirectional flow increased with compressive strength. A study by Inozemtsev et al. (1965) on cavitational erosion of concretes used for hydraulic structures suggested that compressive strength may serve as a criterion for evaluating cavitation resistance.

Another problem is how to relate the effects of discontinuities to the resisting force F_R. This problem becomes serious when cliffs are composed of densely jointed rocks. In his shore platform study, Tsujimoto (1987) attempted to evaluate multiplying compressive strength by V_{pf}/V_{pl} [Eq. (4.27)] where V_{pf} and V_{pl} are the sonic velocities measured, respectively, in the field on a rock mass with discontinuities and in the laboratory on an intact specimen. A decision on the most appropriate parameter to quantify the discontinuity effect on F_R is still awaited.

If cliff materials are subjected to weathering, their mechanical strength decreases, and discontinuities such as weathering cracks or joints develop (e.g. Carter and Guy, 1988), both giving rise to a reduction in F_R. Rocky coasts, which are always exposed to spray, splash, waves, and tides, are one of the most suitable environments for the multiple weathering processes described in Chapter 4. There have been many studies of weathering on rocky coasts, including the solution of limestone in tropical environments (e.g. Hodgkin, 1964; Trudgill, 1976; Spencer, 1985a), salt crystallization (e.g. Coleman *et al.*, 1966; Johannessen *et al.*, 1982), the wetting and drying process in the intertidal zone (e.g. Emery, 1960 p. 15; Suzuki *et al.*, 1972; Takahashi, 1975), and frost effects in cold or middle latitudes (e.g. Zenkovich, 1967, p. 173–8; Trenhaile and Mercan, 1984). Unfortunately, no attempt has been made to quantitatively evaluate weathering-induced deterioration of coastal rocks. The absence of such an attempt can be ascribed to (1) lack of development of *in situ* measuring techniques for the strength of weathered materials and (2) lack of a relevant index for the degree of weathering.

Cyclic wave action generates repeated stresses in cliff materials: fatigue (Chapter 4) probably occurs to deteriorate the mechanical strength. However, no studies on this deterioration process have been reported from actual rocky coasts.

Marine biological activities also affect mechanical strength. Rock-boring or snapping organisms (e.g. Neumann, 1966; Evans, 1968; Vita-Finzi and Cornelius, 1973) reduce overall rock resistance against wave erosion (Healy, 1968b). A striking contrast to this destructive action is a protective effect by encrusting organisms (e.g. Adey and MacIntyre, 1973; Trudgill, 1976; Wray, 1977; Focke, 1978a,b; Bosence, 1983; Bird, 1988). Recent publications by Trudgill (1985, pp. 137–60), Trenhaile (1987, pp. 64–82), and Spencer (1988) provide a useful summary of studies concerning biological agency on rocky coasts, each containing an extensive number of data on the rate of bioerosion.

FORCE–EROSION RELATIONSHIPS: A LABORATORY APPROACH

As shown in Figure 5.2, erosion occurs when the assailing force of waves F_W is greater than the resisting force of cliffs, F_R, i.e. when $F_W > F_R$. When $F_W \leq F_R$, however, no erosion takes place. If cliff materials are not vulnerable to weathering and fatigue, no reduction in F_R occurs, and there is no erosion as far as F_W does not exceed F_R. If a cliff is made of weathering- and/or fatigue-susceptible material, on the other hand, F_R-reduction will occur, so that erosion will take place even if F_W does not increase.

Processes of Cliff Erosion

A fundamental relation of basal erosion by waves can be expressed as

$$x = f(F_W, F_R, t) \tag{5.1}$$

where x is the eroded distance and t is the time. No field studies have been performed to determine the functional form of this equation, because complicated and severe natural environments have discouraged simultaneous measurements at the cliff base of the wave assailing force, the cliff resisting force, and resulting erosional events. To overcome this difficulty, the problem has been approached through laboratory experiments.

Such experiments are carried out under conditions determined by the experimenter in order to create a better understanding of the complex natural system. Prior to the experiment, it is essential to select major factors which would control the particular geomorphological process of interest, because the number of factors introduced is limited in the experiment. A deeper insight into the nature of actual processes leads to a more appropriate selection of controlling factors and a more relevant determination of test conditions.

Laboratory tests are largely classified into (1) small-scale experiments and (2) prototype- (or full-) scale experiments, with the latter having not yet been fully conducted in the research field of geomorphology mainly because the equipment is very expensive. It should be noted that the prototype experiment differs from the field experiment: the latter is an observation and/or measurement of phenomena occurring in the actual field situation in which one (or two) of major governing factors is kept controlled by the experimenter in order to unravel the natural system. Incidentally, some confusion seems to exist in nomenclature of 'field experiments' and 'field measurements'. The term 'field measurements' should be applied to denote the actual measurement of phenomena, as accurately as possible, in the field without disturbing nature's behaviour, and without controlling any factor.

The small-scale experiment, referred to hereafter as the laboratory experiment, can provide much information on unknown relationships among major factors which have been veiled in nature. A systematic and well-controlled study approach is required for this purpose. No laboratory experiments on geomorphological change on rocky coasts in response to wave action had been conducted before the late 1960s; this contrasts with the many model studies of beach changes (e.g. Sunamura, 1989). One reason for the absence of such studies was the difficulty in choosing a material which has mechanical properties similar to those of natural cliffs. The first laboratory study was performed simultaneously, but independently, at the University of Tokyo (Horikawa and Sunamura, 1968) and at the University of Tasmania (Sanders, 1968b). A wave tank was used for these experiments; at one end of the tank a model cliff was installed and exposed to the action of laboratory waves. In Horikawa and Sunamura's (1968) experiment, a cliff made of a mixture of Portland cement, sand, and water was constructed, whereas in Sanders' (1968b) test a mixture of plaster, sand, and water was used for model cliffs.

Model cliffs should have properties of being (1) sufficiently low in strength to be eroded by laboratory-scale waves in a reasonable time, but (2) capable of standing unsupported in a wave tank. After trial and error, Sanders (1968b) found the material made of plaster (15%) and fine-grained quartz sand (85%) met the above requirements. For modelling of glacial till coasts, Nairn (1986) discussed the selection of clays to be used as a cementing material, and finally developed a bentonite-silt mixture with a weight ratio of 3.5:96.5. To study longshore sediment transport on a laboratory cliffed coast, Chien (1956) had already used bentonite to build the cliff, but the detailed composition of his cliff material is unknown.

Because plaster, bentonite, diatomite, lime, and liquid glass are soluble in water, the strength of model cliffs cemented by one of these substances decreases with time due to solution after the cliffs are exposed to waves. These substances seem to be appropriate for modelling coastal cliffs made of calcareous rocks and clayey soils. The temporal strength-reduction is considered to depend on the weight ratio of the cementing material, but the dependency has not yet been quantitatively investigated.

Laboratory modelling for non-calcareous rocks or non-clayey materials should fulfil the condition of using insoluble substances as well as the other two requirements. Horikawa and Sunamura (1968) found that a cement–sand–water mixture is suitable for this modelling, and this material was used in subsequent laboratory experiments (e.g. Sunamura and Horikawa, 1971;

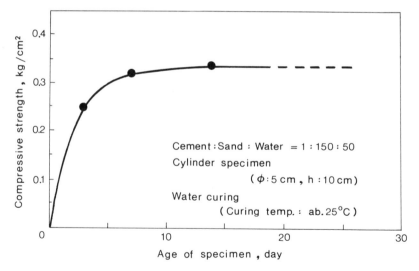

Figure 5.3 Relationship between compressive strength and age of specimen made of cement-sand-water mixture. From Sunamura (1973)

Processes of Cliff Erosion

Sunamura, 1975, 1976, 1982a, 1991). Because of use of Portland cement, this kind of cliff material takes a long time to attain stable strength. In their studies, High Early-Strength Cement which was used to abbreviate the curing time, and well-sorted find quartz sand (median diameter, 0.2 mm; Trask's sorting coefficient, 1.1; specific gravity, 2.65) were applied. Figure 5.3 shows the relationship between the compressive strength of this material, measured on a cylindrical specimen 5 cm in diameter and 10 cm in height, and time after water curing of the specimens. It was found that the strength became almost steady after two weeks. Compressive strength decreases as ratio of cement and sand increases, as shown in Figure 5.4. Construction of a model cliff with the desired strength is possible by changing the mix ratio. The cliff material is found to have a brittleness index (a ratio of compressive strength to tensile strength) of 6 to 7, a porosity of about 40%, and a dry bulk weight of 1.4 to 1.6 g/cm^3. The strength of this kind of material depends slightly on (1) the

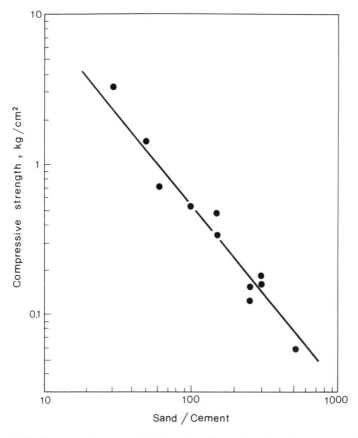

Figure 5.4 Compressive strength plotted against mix ratio of cement and sand

kind of cement, (2) curing conditions such as water temperature and quality, and (3) the grain size and sorting degree of sand used. These aspects have not yet been systematically examined.

In spite of many field studies emphasizing that discontinuities such as joints, faults, and bedding planes play a significant role in rocky coast erosion, laboratory modelling for discontinuities has not been attempted. Success in such modelling would facilitate a deeper understanding of the influence of discontinuities in the natural environment.

In an attempt to investigate erosional mechanisms, Sunamura (1976) conducted two kinds of laboratory experiments: one in a wave tank, and the other in a wave basin. In the former test, a model cliff made of a cement–sand–water mixture in a 1:150:50 weight ratio, with a compressive strength of 340 g/cm^2, was placed at the end of the tank and was exposed to broken waves. In the latter test, an identical cliff was installed at an angle of 60° to the input wave direction, so that longshore currents developed. Figure 5.5 is a plot of the result seen in the wave tank experiment, which shows the occurrence of cliff erosion and the formation of a sand beach in front of the cliff. The waves first eroded the cliff, resulting in a supply of sand to the base of the cliff. As erosion continued, sand was increasingly deposited at the cliff base: the input waves used the sand as an abrasive, so that the assailing force of the waves was intensified and the erosion was accelerated as clearly shown by the abrupt increase in eroded volume in Figure 5.6. In time, however, the sand deposit became thicker and a wider beach developed to bring about a dissipation of wave energy. Waves gradually lost their assailing force and the erosion rate was reduced. Although the assailing force remained time-independent when no sand was deposited, the force became automatically time-dependent, receiving feedback from the erosional process due to the

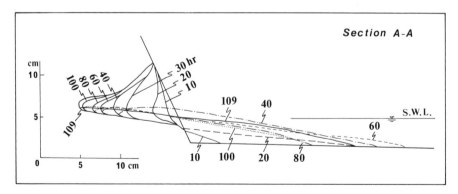

Figure 5.5 Laboratory result of erosion of cement-sand cliff and development of a beach in front of the cliff. The notch develops obliquely upward. From Sunamura (1976). With permission from *J. Geol.*, copyright © 1976 the University of Chicago

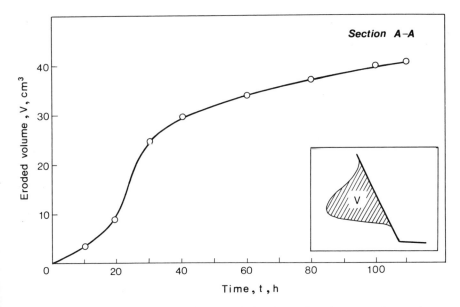

Figure 5.6 Temporal variation in cliff-eroded volume for the case of experiment shown in Figure 5.5. From Sunamura (1976). With permission from *J. Geol.*, copyright © 1976 the University of Chicago

action of sand produced by cliff erosion itself. In the wave basin experiment, on the other hand, no feedback system developed because no sand deposition took place at the cliff base: it was removed by longshore currents.

Applying the linear automatic control theory to erosional process in the wave tank experiment (Sunamura, 1976), Sunamura (1977) obtained the following basic relation:

$$\frac{dx}{dt} \propto \mathbf{F} \qquad (5.2)$$

where dx/dt is the rate of horizontal cliff erosion, and $\mathbf{F}$ is the erosive force of waves. It should be mentioned that the erosive force is different from the assailing force. Waves always exert their assailing force on the cliff face whenever they act on the cliff. Once the assailing force exceeds the resisting force of cliff material, then waves possess the erosive force $\mathbf{F}$ to accomplish erosion. Therefore the *erosive force* is defined as the force to produce *actual* erosion.

Equation (5.2) indicates that no erosion occurs if $\mathbf{F} \leq 0$, whereas erosion takes place if $\mathbf{F} > 0$ and the erosion rate is proportional to $\mathbf{F}$. The following relation was assumed for $\mathbf{F}$:

$$\mathbf{F} \propto \ln\left(\frac{F_W}{F_R}\right) \qquad (5.3)$$

This relation was chosen simply because $\mathbf{F} > 0$ when $F_W > F_R$, while $\mathbf{F} \leq 0$ when $F_W \leq F_R$ (Sunamura, 1977). From Eqs (5.2) and (5.3) one obtains

$$\frac{dx}{dt} = \varkappa \ln\left(\frac{F_W}{F_R}\right) \quad (5.4)$$

where $\varkappa$ is a constant. The functional form of Eq. (5.1) was thus determined with the aid of laboratory studies.

Two assumptions that (1) F_W can be expressed in terms of wave height at the cliff base and (2) F_R can be represented by the compressive strength of cliff material, together with dimensional considerations for both expressions, led to the following equations:

$$F_W = A\rho g H \quad (5.5a)$$

$$F_R = B S_c \quad (5.5b)$$

where H is the wave height at the cliff base, S_c is the compressive strength of cliff material, ρ is the density of water, g is the gravitational acceleration, and A and B are nondimensional constants. The constants A and B may be indicative of the effects of (1) beach sediment acting as an abrasive and (2) discontinuities in cliff material, respectively, if these effects are successfully quantified. Equations (5.4) and (5.5) yield

$$\frac{dx}{dt} = \varkappa \left[\Gamma + \ln\left(\frac{\rho g H}{S_c}\right)\right] \quad (5.6)$$

where Γ is a nondimensional constant: $\Gamma = \ln(A/B)$, and $\varkappa$ is a constant with units of $[LT^{-1}]$. Equation (5.6) relates the rate of wave-induced basal erosion to the two major controlling factors: (1) wave height at the cliff base and (2) compressive strength of the cliff-forming material (Sunamura, 1977).

EROSION RATES AND MEASURING TECHNIQUES

Cliff recession in the real world is essentially an episodic and localized process strongly associated with storm waves (e.g. Sunamura, 1973; Cambers 1976; Griggs and Johnson, 1979; Bokuniewicz and Tanski, 1980, Kuhn and Shepard, 1980; Dick and Zeman, 1983). Figure 5.7 is a schematic illustration of the temporal change in cliff erosion distance at a given site along a stretch of a coast. The erosion rate, dx/dt, is an instantaneous rate which is expressed in terms of the slope of the tangent line to the erosion distance–time curve, $\tan \theta'$. The shape of the curve depends on the temporal variation in wave erosive forces which is directly related to the occurrence mode of storm waves. In reality, however, such a curve has never been drawn in the field due to extreme difficulties in continuous monitoring of cliff-base erosion during a storm event.

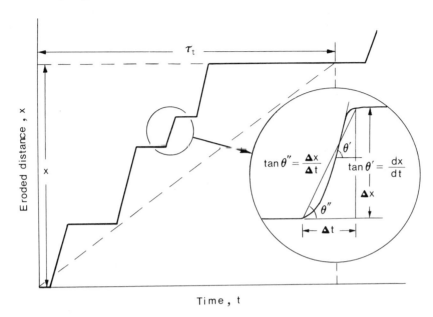

Figure 5.7 Schematic representation of temporal change of cliff recession at a given site on a coast

The erosion rate, $\Delta x/\Delta t$, is the average value of the recession distance, Δx, over a short period of time, Δt, during which erosion actually occurred (Figure 5.7). Only a limited number of measurements have been conducted of the recession distance within a very short-term interval, i.e. a tide duration (Robinson, 1977b) or a few days (Williams, 1956; Davies et al., 1972; McGreal, 1979b; Norrman, 1980). The actual erosion rate $\Delta x/\Delta t$ is quite different from the long-term average erosion rate, x/τ_t, where τ_t is the length of time interval under consideration, say, $10-10^2$ years. The former rate is much greater than the latter. For example, at Shareham on the north shore of Long Island, New York, the hurricane of 14 September 1944 cut back a bluff of glacial deposits a horizontal distance of over 12 m in a single day (Davies et al. 1972), producing $\Delta x/\Delta t = 12$ m/day, which is equivalent to a mean annual rate of 4380 m/year. This provides a striking contrast to an 80-year average of 0.5 m/year for this coast (Bokuniewicz and Tanski 1980), i.e. $x/\tau_t = 0.5$ m/year.

Erosion rates documented or recorded in many countries in the world are tabulated in Appendix 2. Almost all erosion-rate data are the values averaged over (1) the period of 1 to 100 years and (2) a certain longshore distance of the coast. It is clear that these time- and space-average rates are extremely variable. Orders can be summarized on a lithological basis as: 10^{-3} m/year for granitic rocks; 10^{-3} to 10^{-2} m/year for limestone; 10^{-2} m/year for flysch and shale; 10^{-1} to 10^{0} m/year for chalk and Tertiary sedimentary rocks; 10^{0} to 10^{1} m/year for Quaternary deposits; and 10^{1} m/year for unconsolidated volcanic ejecta. These results show that lithology and cohesiveness of cliff material greatly controls the erosion rate. It should be recalled, however, that wave action originally induces cliff recession (Figure 5.1).

Various techniques have been employed for measuring cliff recession distances. The most popular technique is to compare the position of the cliff crest using different editions of maps or plans (e.g. So, 1967; Quigley and Di Nardo, 1980; May and Heeps, 1985; Jones and Williams, 1991). Other techniques similar to this are to use (1) old and recent terrestrial photographs (e.g. Byrne, 1963; Zenkovich et al., 1965) and (2) old and new aerial photographs (e.g. Kirk, 1975; Carter, 1976; Gatto, 1978). A combination of old maps (or plans) and recent field measurements or surveying (e.g. Valentin, 1954; Robinson, 1980a) and a comparison of early surveys and recent air photographs (e.g. Gelinas and Quigley, 1973) have also been commonly employed. Changes in the cliff-top position (e.g. Pringle, 1985), the cliff base (e.g. Carter and Guy, 1988), or both (Bird and Rosengren, 1986) have been measured in actual field situations. A rather unusual technique is to use dated inscriptions or graffiti carved on cliffs (Emery, 1941; Emery and Kuhn, 1980). Direct measurements of the exposure of steel pegs or nails driven into cliffs have been conducted (e.g. Hodgkin, 1964; Kaye, 1967; Lee et al., 1976; Bird et al., 1979). A microerosion meter (MEM), an instrument applying a dial gauge installed on a triangular base plate with three legs, was originally developed by High and Hanna (1970) and later improved by Trudgill et al. (1981) to measure directly the change on bare rock surfaces. The technique was modified for coastal studies (Robinson, 1976), and has been applied to cliff recession studies (e.g. Trudgill, 1976; Robinson, 1977b). Large-scale air-photogrammetric maps have been used for the accurate measurement of cliff profile changes (e.g. Horikawa and Sunamura 1967; Sunamura, 1973; Norrman, 1980). This technique is especially powerful for coasts where human access is difficult because of high cliffs with no beaches at their base. When cliff faces are very steep, some expedients may be employed in the air-photographing technique to obtain accurate data of the cliff base (Sunamura and Horikawa, 1972).

TEMPORAL VARIATIONS IN EROSION RATES

Equation (5.6) suggests that there exists a critical height of waves at a cliff base that is necessary for initial basal erosion. This critical wave height, $H_{\text{crit.}}$, can be derived by setting $dx/dt = 0$; this results in

$$H_{\text{crit.}} = \frac{S_c}{\rho g} e^{-\Gamma} \tag{5.7}$$

Along a stretch of a coast with little significant spatial variations of quantities (1) the assailing force of waves, and (2) the resisting force of cliffs, $H_{crit.}$ as obtained from Eq. (5.7) gives one representative value for the overall coast.

The foregoing conditions are found to be satisfied on the Byobugaura and Okuma cliffs, which are located 200 km apart on the Japanese Pacific coast in a microtidal environment. Byobugaura cliff is 10–60 m high, on a nearly straight coastline 9 km long. Pliocene mudstone is exposed in the lower half of the cliff, and there is no beach. Okuma cliff is 30–35 m high with a longshore length of 1 km. Interbedded Pliocene mudstone and sandstone strata form the lower part of the cliff, which is fronted by a narrow, sandy beach. Large-scale mass movements do not occur on either cliff. Using the data of (1) cliff strength, (2) wave climate, and (3) erosion rates of two time intervals, and taking account of water-level rise at the cliff base (i.e. wave set-up), Sunamura (1982b) determined the constants Γ and $\varkappa$ in Eq. (5.6) and $H_{crit.}$ from Eq. (5.7) for the two areas. Analysis of these results enabled him to plot the relationship

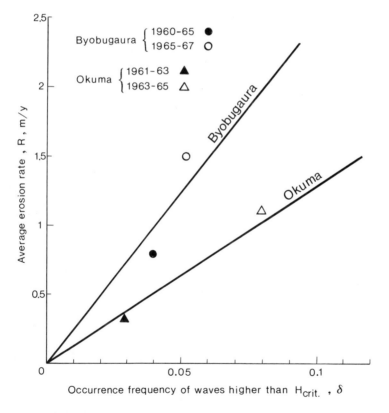

Figure 5.8 Relationship between average erosion rate and frequency of erosion-causing waves for Byobugaura and Okuma cliffs in Japan. From Sunamura (1982b). With permission from *J. Geol.*, copyright © 1982 the University of Chicago

between occurrence frequency of waves and the erosion rate. The plot clearly showed that the larger but rarer waves are solely responsible for the cliff erosion at both sites, as compared with the smaller but more frequent waves.

It has been reported that considerable temporal variations in average erosion rates exist for the short term of 1–10 years (e.g. Sunamura, 1973; Cambers, 1976; McGreal, 1979b; Richards and Lorriman, 1987; Carter and Guy, 1988). One would anticipate that the more frequently large waves attack a coast during a certain period of time, the more severely the cliff recedes to produce a higher erosion rate. Because only the waves which exceed a threshold value [Eq. (5.7)] can cause erosion, the short-term average erosion rate, R, should be related to the frequency of waves higher than $H_{crit.}$, δ. Figure 5.8, constructed using the data obtained from the Byobugaura and Okuma coasts, illustrates that R is linearly related to δ for each location (Sunamura, 1982b). Short-term annual erosion rates are thus strongly influenced by the frequency of the erosion-causing waves at the cliff base.

The assailing force of waves acting on the cliff base is obviously dependent on wave energy level in the offshore region. However, (1) water level (tides and storm surges), (2) the beach fronting the cliff or nearshore bottom topography, and (3) the size and amount of beach material residing at the cliff base, greatly control wave assailing force. Fluctuations of these controlling factors, therefore, produce temporal variations in cliff erosion rates.

Selecting four bluffs composed of glacial deposits and one of shale from the Ohio coast of Lake Erie, Carter and Guy (1988) made detailed observations and measurements of cliff recession process during five years. Accurate surveys were conducted every two weeks and after storms to evaluate the bluff base erosion. Figure 5.9a exemplifies discrete erosion events that occurred at Helen Drive bluff, one of the five sites, and illustrates the cumulative erosion distance for this site, which is shown by the dotted line. The temporal variation in beach width is also plotted; the gaps in the beach-width curve represent intervals of debris and/or ice covers at the base of the bluff. It is seen that no cliff erosion took place during these intervals. Such a protective effect of ice has been observed at the other sites along this coast, and also reported from the eastern coast of Lake Michigan (Davis, 1976; Birkemeier, 1981).

Figure 5.9b shows lake level fluctuations and storm surges. A close look of this diagram, together with Figure 5.9a, reveals a good correlation between bluff erosion and water level: the occurrence of erosion events was closely related to lake levels and/or the magnitude of storm surges. Dependency of average erosion rates on lake levels had already been examined along some of the coasts of the Great Lakes (e.g. Kolberg, 1974; Birkemeier, 1980; Quigley and Zeman, 1980).

A marked water level rise may occur on tidal coasts when storm surges superimpose on high tides. Such an event happened along the east coast of Britain on the night of 31 January to 1 February 1953 (e.g. Summers,

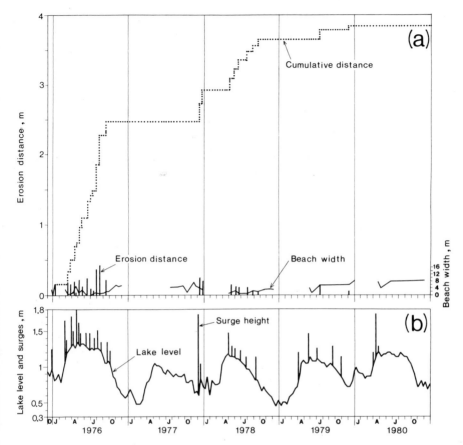

Figure 5.9 (a) Bluff erosion events and beach width variation at the Helen Drive site, and (b) lake level change and storm surge height on the Ohio shore of Lake Erie. The length of vertical lines in (a) and (b) indicates the magnitude of erosion events and storm surges, respectively. After Carter and Guy (1988), *Marine Geol.*, by permission of Elsevier Science Publishers. The dotted line in (a) representing cumulative erosion distance, is drawn based on Carter and Guy's data

1978, and references cited therein); this is known as the 1953 North Sea storm surge. Drastic cliff erosion associated with this event was well documented at Covehithe, Suffolk (Williams, 1960, pp. 82–7), where the storm surge with a maximum height of 2.3 m occurred at high tide yielding a total sea level rise of 3.5 m above MSL. A glacial-sand cliff 9 m high was cut back 9 m due to the action of severe storm waves during about two hours at the peak of the surge. This amount of erosion corresponds to three-quarters of the total distance of the cliff receded during the preceding 17 months from 28 August 1951 to 31 January 1953.

The influence of changes in nearshore bottom morphology on the long-term temporal variation in cliff erosion rate has been illustrated by Robinson's (1980a) work on the coastal area between Dunwich and Thorpeness in Suffolk, England. Glacial sand cliffs at Dunwich have suffered severe erosion over a long period. Using well-documented historical data and a more recent survey, Robinson evaluated mean annual erosion rates in different periods: these were 1.6 m/year from 1589 to 1753, 0.85 m/year from 1753 to 1824, 1.5 m/year from 1824 to 1884, 1.15 m/year from 1884 to 1925, and 0.15 m/year from 1925 to 1977, the last showing an extremely low value. This drop in the erosion rate during the recent fifty years can be attributed to the reduction in wave energy arriving at the coast, due in turn to a significant divergence in wave refraction, produced by a changing bottom configuration associated with the steady northward growth of Sizewell Bank which took place during the past century.

A beach in front of a cliff is an ideal natural defence against the attack of waves, if the beach is too wide for waves to reach the cliff base. The beach profile varies depending on waves, tides, and sediment balance, the last being much influenced by (1) construction of engineering structures such as groynes, jetties, and breakwaters, and (2) the dumping or dredging of beach materials. The beach profile level is of prime importance in controlling assailing force of waves reaching the cliff base, and hence the magnitude of cliff erosion, as reported from a glacial deposit cliff on Outer Cape Cod, Massachusetts (Giese and Aubrey, 1987).

Using a peg-line technique, McGreal (1979b) conducted weekly measurements over a two-year period of erosion distance at the base of glacial drift cliffs exposed to low wave energy near Kilkeel in Northern Ireland. He found a close relationship between temporal variations in beach elevation and the basal erosion rate, and explained that beach lowering, produced under certain wave and tidal conditions, provides a favourable setting to promote cliff base erosion.

Even if the beach elevation is high enough to prevent wave action on the cliff base under normal weather conditions, it is often observed that sediments forming the beach are temporarily moved offshore during severe storm-wave attack. Considerable reduction of beach profile level occurs, allowing waves to reach the cliff base and cut back the cliff. Concurrently shore platforms which are usually hidden beneath the beach are exposed, and suffer erosion. Appreciable lowering of the platform surface takes place if they are composed of the soft rocks (Chapter 6).

Everts (1991) examined the relationship between the the rate of erosion on cliffs composed of Tertiary sediments in southern California and the width of fronting beaches, and found that the beach width strongly influences cliff recession. The recession rate decreased as the beach became wider: it was greatly reduced when the beach width exceeded 20 m, and erosion ceased when it attained 60 m.

An ord, which is a characteristic feature on the Holderness coast, East Yorkshire, England, is a linear depression on the beach fronting the glacial till cliff. The depression extends obliquely at a low angle to the coastline from the base of the cliff, and the shore platform cut in glacial drift is exposed in the depression. These ords move to and fro along the shore. Ord movement produces a temporal variation in beach levels in front of the cliff at a given site along this coast. Pringle (1985) examined the influence of the ord migration on cliff erosion on the south Holderness coast within a distance of 7 km between Withernsea and Easington. The ord that she monitored moved southward at an average speed of 0.5 km/year during six years from 1977 to 1983. Lowering of beach level at the cliff base attained as much as 3.9 m, allowing most high-water neap tides to penetrate to the cliff. This showed a striking contrast to the inter-ord beach section where only high-water spring tides reached the cliff. Figure 5.10 shows the relationship between cliff-top erosion and ord position. Occurrence of marked erosion at or near the ord position can be recognized, but an exact correspondence was not found. Reading this diagram downwards along a vertical line through a specific site gives the temporal variation in erosion rates of the site. The erosion-rate variation at Old Hive, for example, can be well explained by the movement of the ord.

An example of cliff erosion triggered by an artificial factor can be taken from Hallsands in South Devon, England. The fishing village of Hallsands was located on a cliff-base platform cut into micaschist with numerous quartz veins. A wide beach of flint shingle developed in front of the seaward cliff of the platform. During the period of five years from 1897 to 1902, approximately half a million cubic yards of shingle were removed from the foreshore for the construction of a dockyard at Plymouth. As a result, the entire beach was lowered by as much as 3.6 m (Robinson, 1961). This shingle removal, plus a combination of 13 m storm waves, gales, and spring tides, were responsible for the disaster which occurred in 1911 (Hails, 1977, pp. 316–62). Subsequent cliff erosion led to the ruin of the fishing village (Robinson, 1961; Hails, 1975). Although accurate erosion data are unavailable at this site, it is reported that the adjacent micaschist cliff receded by as much as 6 m during the half century following 1907 (Robinson, 1961).

The temporal variation in cliff erosion rate associated with beach-level change was demonstrated in Sunamura's (1976) wave-tank experiment, a brief explanation of which has already been given. The experiment was conducted by applying waves with constant energy in deep water to a model cliff made of cemented sand. The waves started to erode the cliff, and as erosion continued, sand supplied from the cliff formed a beach (Figure 5.5). Figure 5.11 shows the relationships between time and the erosion rate (denoted as $\Delta V/\Delta t$, where V is the eroded volume) or the beach elevation measured at the cliff base. The erosion rate abruptly increased and attained a maximum as the beach elevation increased, but the erosion rate dropped as the beach height

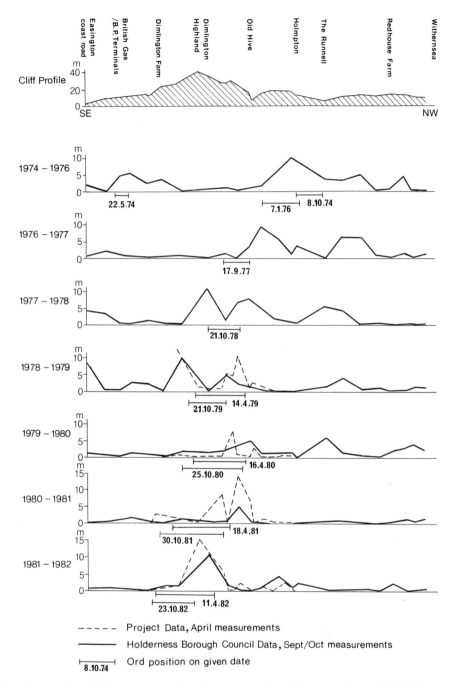

Figure 5.10 Cliff-top erosion in relation to ord position on the south Holderness coast, England. From Pringle (1985). *Earth Surface Processes Landforms*, copyright © 1985, reprinted by permission of John Wiley & Sons Ltd

Processes of Cliff Erosion

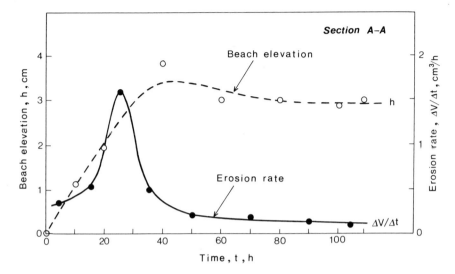

Figure 5.11 Temporal variations in erosion rate, $\Delta V/\Delta t$, and beach elevation, h, expressed by the thickness of sand deposited at the cliff base. From Sunamura (1976). With permission from *J. Geol.*, copyright © 1976 the University of Chicago

further increased. It was found that the increase in erosion rate was caused by the intensified erosive force of waves due to the abrasive action of sand, whereas the decrease was brought about by moderated force due to the accumulation of protective sand. Analyses of the experiment, applying feedback control theory, enabled Sunamura to plot Figure 5.12 in which $\mathbf{F}(t)$ denotes the erosive force including the effect of sand and $\mathbf{F}_0$ indicates the initial erosive force which has no sand effect. For $0 \leq t < 40$ hours, $\mathbf{F}(t)/\mathbf{F}_0 \geq 1$, i.e. $\mathbf{F}(t) \geq \mathbf{F}_0$, which indictes that the erosive force is intensified due to sand grains acting as an abrasive; $\mathbf{F}(t)/\mathbf{F}_0 = 4.5$ at its maximum, that is, the intensified erosive force attains 4.5 times the initial force having no sand effect. On the other hand, for $t \geq 40$ hours, $\mathbf{F}(t)/\mathbf{F}_0 < 1$, i.e. $\mathbf{F}(t) < \mathbf{F}_0$, which implies that sand acted as an obstacle to cliff erosion.

The wave assailing force F_W at the cliff/beach junction greatly depends on the size and amount of beach sediment entrapped in waves. These factors vary with time; this leads to temporal variations in F_W, resulting in erosion rate fluctuation. No quantitative studies on this topic have been performed in the field due to the difficulty in evaluating these factors, but a laboratory study has attempted to explore this problem (Sunamura, 1982a).

If we take a long time span of, say, more than 20 years, then the average coastline erosion rate would mask the influence of the short-term variation in wave assailing force, and show a characteristic value for the location, i.e. a

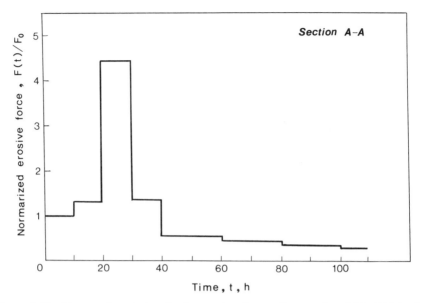

Figure 5.12 Temporal variation in the erosive force of waves in which $\mathbf{F}(t)$ is the erosive force including the effect of sand and $\mathbf{F}_0$ in the initial force with no sand effect. From Sunamura (1976). With permission from *J. Geol.*, copyright © 1976 the University of Chicago

time-average value of the relative intensity of wave action to the cliff resistance. This site-specific value can obey a form similar to Eq. (5.4):

$$\bar{R} = K \ln\left(\frac{\bar{F}_W}{\bar{F}_R}\right) \tag{5.8}$$

where $\bar{R}$ is the long-term average erosion rate, $\bar{F}_W$ is the long-term average assailing force, and K is a constant with units of $[LT^{-1}]$. In this equation the cliff resisting force F_R is considered to be time-independent. On a coast where the cliff material is prone to weather, however, F_R should be treated as a time-dependent variable. The site-specific erosion rate is again affected by temporal changes in water level, beach and nearshore bottom morphologies, if they occur with a long period of fluctuations, as in the case of Dunwich cliff erosion (Robinson, 1980a) as already described.

SPATIAL VARIATIONS IN EROSION RATES

As discussed before, the short-term erosion rate fluctuates greatly in relation to time. In order to eliminate this fluctuation, let us consider the long-term cliff recession first. Suppose that several cliffs with rocks of differing resistance are exposed to a similar wave climate. Because the wave assailing force $\bar{F}_W$ is

Processes of Cliff Erosion

space-constant and the rock-resisting force F_R is variable in this case, Eq. (5.8) is rewritten using Eq. (5.5b) as

$$\bar{R} = K(\Gamma_1 - \ln S_c) \qquad (5.9)$$

where $\Gamma_1 = \ln(\bar{F}_W/B) = $ constant. This indicates that a linear relationship should be found between the long-term erosion rate and compressive strength in a semilogarithmic expression. This is clearly shown in Figure 5.13. The data plotted here were collected from the Pacific coast of eastern Honshu, Japan, where the spatial variation in $\bar{F}_W$ appears to be small. Erosion rates over a fairly long term (18 to 45 years) were employed. Figure 5.13 shows that the erosion rate decreases with an increase of compressive strength, i.e. the so-called 'harder' cliffs exhibit lower erosion rates.

Another example of the cliff strength vs. erosion rate relationship can be taken from the Pacific coast of Japan. Along a coastal strip (17 km long) of the Atsumi Peninsula, Yamanouchi (1977) used old and recent maps to assess

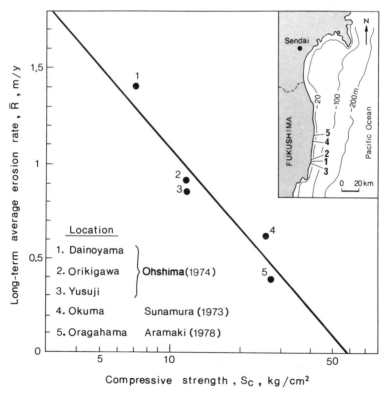

Figure 5.13 Relationship between long-term erosion rate and rock strength on Tertiary cliffs on the Fukushima coast, eastern Honshu, Japan

the long-term (70 or 80 years) erosion rate of the cliff composed of weakly consolidated Pleistocene deposits. A beach fronts the cliff. Low marine stacks of hard Palaeozoic rocks occur on the shore in some places. In an attempt to evaluate the cliff strength, Yamanouchi measured the degree of hardness of the cliff-forming material using a pocket-size penetrometer (Chapter 4), the readings of which are indicative of the ease of penetration: this penetration value, denoted as P_v, increases as the rock strength decreases. Input wave characteristics are considered to be uniform along this coast. Yamanouchi found that (1) the erosion rate increased with increasing P_v values; and (2) the erosion rate in the region of marine stack development was lower than that in the other region, suggesting that the marine stacks hindered wave erosion. Because P_v is assumed to be inversely related to the cliff resisting force F_R, i.e. $F_R = B'/P_v$ (B' = const.), and the wave assailing force $\bar{F}_W$ is uniform alongshore, Eq. (5.8) yields

$$\bar{R} = K(\Gamma_2 + \ln P_v) \qquad (5.10)$$

where $\Gamma_2 = \ln(\bar{F}_W/B')$ = constant. Using Yamanouchi's data, this relation is plotted on semilogarithmic graph paper (Fig. 5.14). The best-fit line to the data points indicate that $\Gamma_2 = 2.48$ for the region with marine stacks and $\Gamma_2 = 3.01$ for the region without them. These results enable us to evaluate the relative intensity of $\bar{F}_W$ between the two regions using the relationship $\bar{F}_W = B'e^{\Gamma_2}$ because B' is assumed constant. The assailing force of waves acting on the cliff behind the marine stacks is found to be reduced to 60% of that of waves reaching the cliff directly.

Assuming that cliffs with similar lithology are subjected to the different wave climate, the erosion rate should be higher where wave conditions are more severe, if other controlling factors are constant. This is seen on the Holderness coast of England. Eroding cliffs cut in glacial drift deposits extend for 55 km from Flamborough Head in the north to Spurn Head in the south. The entire coast receded at an approximate annual rate of 1.2 m/year during the century between 1852 and 1952 (Valentin, 1954). Valentine's 100-year data clearly indicate that the recession rate increased in a southerly direction as a whole, although it dropped locally due to the effect of shore protection structures such as sea walls and groynes. A general trend for southward-increasing erosion was also recognized in Robinson's (1980b) data for the period between 1952 and 1980. This trend is consistent with increasing wave exposure towards the south. According to the wave-height prediction made by Fleming (1986), waves are smallest in the northern part of this coast which is sheltered from the more severe northern storms by Flamborough Head, and waves become larger but uniform in the central part down to Easington in the south; this wave-height increase is due probably to an increase in nearshore bottom slope. Assuming that the wave assailing force $\bar{F}_W$ is directly related to the nearshore wave height, H', i.e. $\bar{F}_W = A'H'$ (A' = const.), the Holderness situation

Processes of Cliff Erosion

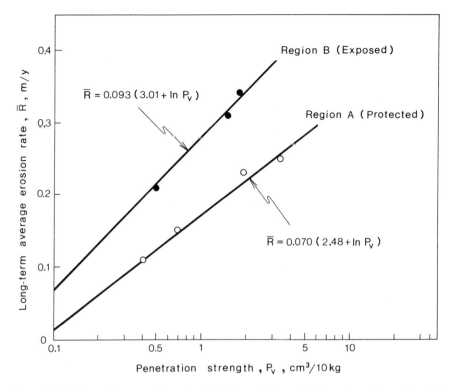

Figure 5.14 Long-term erosion rate plotted against penetration strength on the cliff along the southern coast of the Atsumi peninsula, Japan. (Data from Yamanouchi, 1977) Region A: Section with low marine stacks in front of cliff; Region B: Section with no marine stacks

allows us to rewrite Eq. (5.8) as

$$\bar{R} = K(\Gamma_3 + \ln H') \qquad (5.11)$$

where $\Gamma_3 = \ln(A'/F_R) =$ constant. The long-term erosion rate is plotted against the wave height H' in Figure 5.15. Valentin's (1954, Figure 3) erosion data were utilized for $\bar{R}$, which is a space-average value around the location where the wave data were available. Nearshore significant wave heights with a return period of one year (Fleming, 1986, Table 2) were employed for H'. Figure 5.15 indicates that Eq. (5.11) holds, although the data were limited.

A similar situation to the Holderness coast is found in the central part of the north shore of Lake Erie, where cliffs cut in glacial drift have developed with no marked alongshore variations in their geotechnical properties, but different exposures to waves. There are no wide beaches to protect the cliff from wave erosion. Comparing old survey data with recent maps plotted from

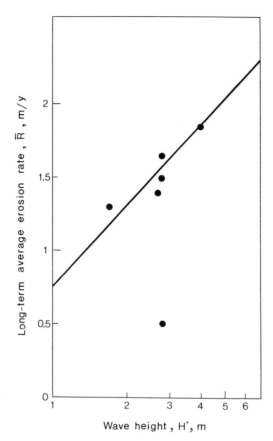

Figure 5.15 Cliff erosion rate related to wave height on the Holderness coast, England. The outlying data point denotes the value obtained near Sand le Mere, where the erosion rate was much reduced due to the effect of shore protection structures (Valentin, 1954, Figure 3)

air photographs, Gelinas and Quigley (1973) obtained the long-term (150-year) erosion rate of the cliff in a section between Rondeau and Long Point, a 120 km alongshore distance, and plotted the erosion rate against long-term wave power. (For wave power, see Chapter 2.) Through straight-line regression analysis of the erosion rate $\bar{R}$ versus the wave power $\bar{P}$, they obtained $\bar{R} = 0.28 + 0.96\bar{P}$. In an attempt to construct an erosion model of a cliff cut in soft glacial drift, Kamphuis (1987) theoretically derived a simple $\bar{R}$ vs. $\bar{P}$ relationship, which can be expressed by $\bar{R} \propto \bar{P}^{1.4}$, and examined the validity of this expression using his own data acquired from the same area as Gelinas and Quigley's (1973). Again using Eq. (5.8), and assuming $\bar{F}_W = A''\bar{P}$ ($A'' = $ const.), we have

$$\bar{R} = K(\Gamma_4 + \ln \bar{P}) \tag{5.12}$$

where $\Gamma_4 = \ln(A''/F_R)$ = constant. Fitting this relation to Gelinas and Quigley's long-term data gives the straight line in Figure 5.16 with determined values for K and Γ_4. Gelinas and Quigley's relationship is also plotted in this figure, together with Kamphuis' result of re-analysis of their data. Quigley and Zeman (1980) suggested that minimum wave power exists to initiate bluff erosion on the Lake Erie north shore. The existence of such a threshold is shown by the straight line in Figure 5.16, but not by the two curves. The scatter of the data points is due mainly to local differences in geological conditions (Gelinas and Quigley, 1973).

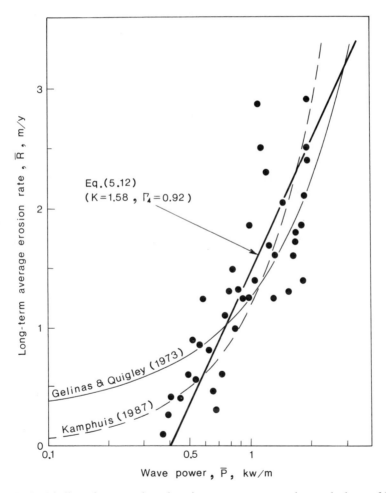

Figure 5.16 Bluff erosion rate plotted against wave power on the north shore of Lake Erie. Data from Gelinas and Quigley (1973)

On the south shore of Lake Erie, wide beaches develop in front of the cliff in some areas, but none in others. The width of beaches greatly controls the assailing force of waves acting on the cliff base. Alongshore changes in beach width give rise to spatial variations in erosion rate, if other factors are uniform. Along the Lake Erie Ohio shore 286 km long, Carter et al. (1986) examined the effect of beach width on the long-term (1876 to 1973) cliff recession using maps and air photographs. Cliffs are composed of glacial till, sand, shale, or glaciolacustrine clay, with the first occupying almost half the total shore length. The analyses indicated that the annual recession rate for clay cliffs was 1.3–2 m/year and that for till was 40 cm/year at locations where the beach has a width of 0 to 6 m; both cliffs reduced their erosion rates with increasing beach width; and clay cliffs had annual erosion rates of only a few to ten centimetres and till cliffs had 10 to 15 cm/year on the shore with wide beaches (>24.3 m). A less consistent erosion-rate reduction with beach width was found for sand and shale cliffs.

If the respective alongshore variations of the wave assailing force, F_W, and the cliff resisting force, F_R, are uniform on a coastal strip, then Eq. (5.4) gives the same erosion rate at any location along the coast, i.e. there is no spatial variation in erosion rate. Hence the coefficient of variation, σ_s/R, should be zero, in which R is the space-average value of erosion rates and σ_s is the standard deviation of R. This implies parallel recession of the cliff line. Figure 5.17 is a plot of σ_s/R against τ_t, the length of interval, using the data from Byobugaura, Japan, where neither F_W nor F_R shows notable alongshore variations (Sunamura, 1973). This graph illustrates that σ_s/R decreases with increasing τ_t, suggesting that the degree of erosion-rate variation becomes smaller as the time scale lengthens. On a coast with little alongshore variation

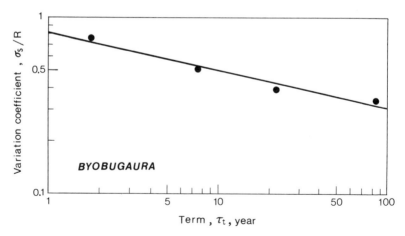

Figure 5.17 Variation coefficient of erosion rate, σ_s/R, plotted against the length of interval τ_t, for Byobugaura cliff, Japan. From Sunamura (1973)

Processes of Cliff Erosion

of input-wave characteristics and cliff strength, parallel recession of the coastline will occur over a long period.

A diagram by De Boer (1977) shows that parallel recession has taken place since Roman times along the central part of the Holderness coast (about 25 km) where the two main factors are assumed uniform. Parallel cliff retreat has also been found on the east coast (cliff material: base-surge deposits, term: 1815–1968, alongshore length: 3 km) of Niijima Island, Japan (Sunamura, 1987, Figure 5); the Kilkeel coast (glacial deposit, 1834–1962, 3 km) in Northern Ireland (McGreal, 1979a, Figure 2); the Dunwich coast (glacial sand, 1589–1977, 1.5 km) in Suffolk, England (Robinson, 1980a, Figure 5); and the Beltinge coast (London Clay, 1872–1959, 1.2 km) near Reculver, Kent, England (So, 1967, Figure 3); with the last showing a less marked parallel recession compared with the other three.

As shown in Figure 5.17, spatial variations in the erosion rate increase with decreasing time span. Considerable alongshore fluctuations are found in the short-term (3–5 year) erosion pattern of the cliff at Easington on the Holderness coast (Richards and Lorriman, 1987, Figure 10.7). Pringle's (1985) plot also shows large spatial variations for the Holderness cliff between Withernsea and Easington, see Figure 5.10. Glacial drift cliffs of the Norfolk coast, England, have demonstrable alongshore variations in the erosion rate obtained on a year by year basis (Cambers, 1976).

This field evidence suggests that the value of F_W/F_R considerably varies alongshore in a short time period. Actually, the F_W-value fluctuates greatly along the coast, with F_R being approximately uniform due to similarity of cliff materials. One of the causes for F_W-fluctuations is alongshore changes in the beach profile level at the cliff foot.

Beach material is transported offshore during storm events. Rip currents (Chapter 2) accelerate offshore transport, so that the shoreline in the lee of rip currents recedes more rapidly than the area midway between the adjacent rips. An embayment is therefore formed shoreward of each rip, with a cusp in between (Figure 2.15). Such an embayment may cause easy landward penetration of storm waves, resulting in erosion of a cliff located behind it. Localised and episodic cliff erosion associated with rip-current embayments has been reported from south of Lincoln City on the northern Oregon coast where Pleistocene terrace sands form sea cliffs (Komar and McDougal, 1988; Komar and Shih, 1991).

Along the cliffed coast composed of glacial deposits, (1) locations with a lower beach elevation have a higher erosion rate as suggested from McGreal's (1979b) work on the Kilkeel coast, Pringle's (1985) investigation on the Holderness coast, and Carter and Guy's (1988) study on the Lake Erie Ohio shore; and (2) locations with a smaller volume of beach material have a higher erosion rate as suggested from Jones and Williams' (1991) work on the west Wales coast. In other words, cliff erosion decreases with increasing size of beaches. This suggests that the beach works as a wave-energy dissipator, and that no significant mechanical action induced by the beach sediment (Figure 5.2) is likely to be involved in this erosion. Hydraulic action alone is enough to produce erosion for the case of very soft cliff materials. This can be easily understood by supposing cliffs made of poorly consolidated sand, for example, dune bluffs, as an extreme case.

The relationship between erosion rate and beach elevation varies with the strength of cliff material, on which mechanical action plays a dominant role. Based on measurements using a micro-erosion meter, Robinson (1977b) reported that at Whitby on the northeast Yorkshire coast of England, erosion rates of Upper Lias shale cliffs fronted by beaches were 15–18.5 times higher than those with no beaches. Sunamura (1982b) examined two eroding cliffs on the Pacific coast of Japan, Byobugaura and Okuma, both composed of Pliocene sedimentary rocks, the former having a compressive strength of 14.9 kg/cm^2 and the latter 26.4 kg/cm^2. Through evaluation of the value of A in Eq. (5.5a), he found that the wave assailing force acting on the Okuma

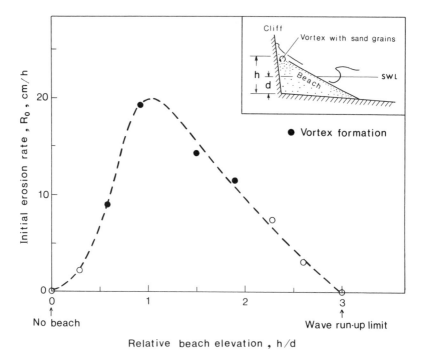

Figure 5.18 Experimental results showing the effects of beach sand accumulation on cliff erosion rate. In this figure, R_0 is the erosion rate immediately after wave action, h is the height of the cliff–beach junction, and d is the water level at the base of the cliff with no beach. After Sunamura (1982a). *Earth Surface Processes Landforms*, copyright © 1982, reprinted by permission of John Wiley & Sons Ltd

cliff which is fronted by a beach is 1.8 times greater than that on the Byobugaura cliff with no beach.

Figure 5.18 is a plot of the erosion rate against relative beach elevation, obtained from a wave-tank experiment (Sunamura, 1982a). The experiment was performed under constant wave conditions, by systematically changing the amount of beach sand at the foot of steep model cliffs of the same slope and strength. The duration of wave action was two hours for each run. Cliff erosion occurred when the beach material at the cliff–beach junction was moved by waves. Figure 5.18 shows that (1) no erosion took place when there is no beach, indicating inability of hydraulic action, and (2) the greatest erosion occurred when $h/d \approx 1$, viz., when the beach top abutting the cliff was located at about still-water level (SWL). At the cliff–beach junction, the bores running up the model beach produced a vortex rotating around a horizontal axis. Sand grains entrapped in this vortex scraped down cliff material working as an abrasive tool. The formation of such a vortex was found to be closely associated with severe cliff erosion (Figure 5.18). An analysis of the experimental results showed that the assailing force of the sediment-laden water masses is proportional to the square of the bore speed immediately in front of the cliff (Sunamura, 1982a).

The field and laboratory findings are used in Figure 5.19, which is a conceptual diagram illustrating the dependence of the erosion rate on the beach elevation when other variables are assumed constant. Considering the effect of beach elevation at the cliff base, Dick and Zeman (1983) have attempted to construct a model for assessing erosion rate of silty or clayey bluffs. Unfortunately field data which permit to quantify this effect have not been available.

Another cause of large alongshore variation in F_W is the presence or absence of debris material supplied from the cliff as a result of slope failure. Cliff undercutting is halted until the debris is removed by wave and current action. More time is needed for the removal of larger debris masses. This leads to the concept that higher cliffs have smaller erosion rates than lower cliffs, if a coastal strip with similar lithology but varying height is subjected to the same wave conditions alongshore (e.g. Shepard and Grant, 1947; Kawasaki, 1954; Sparks, 1986, p. 236). This cliff-height–erosion-rate relationship may hold in a short interval of time as suggested from (1) Williams' (1956) record of the fact that during the 1953 storm surge a glacial sand cliff 12 m high was eroded 12 m overnight, whereas a cliff 3 m high was cut back 27 m, at Covehithe, Suffolk, England; and (2) Suwardi and Rosengren's (1983) study conducted at Sertung, Krakatau Island, that showed cliffs made of pyroclastic deposits had receded at a maximum rate of 3 m/month where the cliff height was 5.7 m, and 1.4 m/month where it was 9.7 m, during the monsoon season between 1981 and 1982. A definite relation between cliff height and erosion rate becomes less pronounced as the time span lengthens, as shown by Bird and Rosengren's (1984, Table 1) data obtained from pyroclastic cliffs (3–14 m

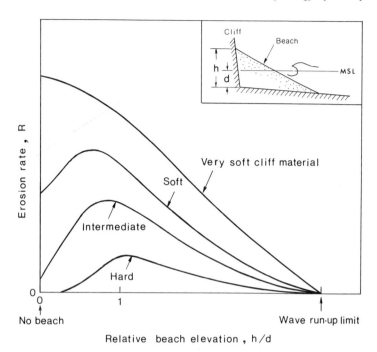

Figure 5.19 Conceptual sketch illustrating the effect of beach elevation on cliff erosion rate. Different curves are plotted, depending on the strength of cliff material

high) on the south coast of Anak Krakatau between 5 November 1981 and 28 June 1983. This strongly suggests that the fallen debris can be rapidly removed from the cliff base. According to Griggs and Johnson (1979, Figure 12), a considerable amount of material, including large blocks (approximately 4 m in diameter) of sandstone and siltstone, falling from a Tertiary cliff (25 m high) at Santa Cruz, California, resided only a few years at the foot of the cliff.

No correlation exists, over a long period of time, between cliff height and erosion rate for a cliff with similar lithology and exposed to the same wave condition, because such a cliff has a tendency toward parallel retreat with increasing time, as already illustrated by the case of Byobugaura cliff (Figure 5.17) which ranges from 10 to 60 m in height. A plot of cliff height against erosion rate during 100 years for the Holderness cliff (4–43 m high) indicates a considerable data scatter with no definite trend (Doornkamp and King, 1971, Figure 11.8). No positive correlation between bluff height and recession rate has also been reported from the Lake Michigan shore, cut in weakly consolidated Quaternary deposits (Buckler and Winters, 1983).

CHANGES IN CLIFF PROFILES

Basal erosion by waves is of great importance for coastal slope instability, eventually leading to intermittent mass movement (Figure 5.1). The occurrence of mass movement is affected by subaerial factors such as rain wash or precipitation (e.g. Kaye, 1967; May, 1971; McGreal, 1979a; Kuhn and Shepard, 1980; Bird and Rosengren, 1986), ground water level (e.g. McGreal and Craig, 1977; McGreal 1979a; Bryan and Price, 1980; Edil and Vallejo, 1980), groundwater seepage (e.g. Buckler and Winters, 1983; Buttle and von Bulow, 1986; Eyles et al., 1986; Hutchinson, 1986; Leatherman, 1986, Zeman, 1989), wind (Cambers, 1976), and frost action (Kaye, 1967; May, 1971; Hutchinson, 1986).

Mass movement occurring on a rocky coast is of four primary types: falls, topples, slides, and flows (Figure 5.20). These types depend mainly on lithological factors of the coast-forming material which are geological structures, stratigraphic features, and geotechnical or rock-strength properties. Intermediate types exist between two or more of these types.

Falls (Figure 5.20a) denote movement of a mass that travels most of the distance through the air as a freely falling body. They are subdivided into rock falls, debris falls, and earth falls, according to the type of cliff material before movement. Cliff retreat associated with rock falls has been reported from the coast from Ogmore-by-Sea and Barry in the Vale of Glamorgan in Wales (Williams and Davies, 1980, 1987), where precipitous cliffs made of densely jointed Lias limestone and mudstone have receded at an average annual rate of 6.8 cm/year (Williams and Davies, 1987). Jones and Williams (1991) described cliff erosion associated with rock falls which occurred on a cliff cut in the Aberystywth Grits (mostly greywacke) near Craig Ddu on the west Wales coast.

Topples are different from falls in that little free-fall movement takes place because rotation of a block around a fixed hinge dominates during the motion (Figure 5.20b). Topples are common on bluffs composed of vertical-jointed hard rocks as occurred on Lower Lias cliffs in South Wales (Davies et al., 1991), but the London Clay cliffs at Warden Point on the Isle of Sheppey, England, often suffer this type of failure after wave-cut notches have been formed at the cliff base (Hutchinson, 1986).

In *slides*, shearing displacement occurs on a distinct slip surface, and a sliding mass exhibits block movement (Figure 5.20c). There are two major movements: planar slides and rotational slides. The former has an almost linear sliding surface, whereas the latter is of circular plane. Descriptions of rotational slides are often prefixed by 'shallow-seated' or 'deep-seated' depending on the depth of sliding surface. Slides are commonly observed on most eroding coastal cliffs. This type of failure will be described later in detail.

Flows move with increasing velocity towards the upper part of a moving body: no block movement is present due to differential shearing within the body. No clear boundary can be drawn at the base of the moving mass. Flows

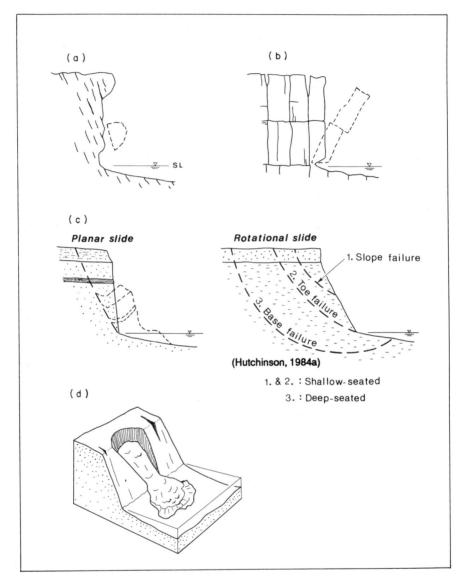

Figure 5.20 Mass movement: four primary types: (a) fall; (b) topple; (c) slide; (d) flow

occurring on some coasts cut in soft clayey materials are called mudflows, and are frequently initiated by mudslides. The sliding mass may disintegrate during its movement: the flow component increases, resulting in the generation of mudflows in the latter phase of the event. However, the distinction between mudslides and mudflows is not easy in the field. Many studies show that these two failure modes contribute significantly to recession cliffs cut in soft materials such as clay and till (e.g. Prior et al., 1968, Hutchinson, 1970, 1973, 1983, 1984a; Brunsden, 1974; Brunsden and Jones, 1980; Bryan and Price, 1980; Grainger and Kalaugher, 1987).

Mass movement always acts to render the cliff face profile stable: a more gently sloping profile appears. Simultaneously debris masses are supplied to the base of the cliff (e.g. May and Heeps, 1985; Everts, 1991). Waves remove the debris, and again undercut the cliff base, so that the overall cliff profile becomes steep and unstable. Mass movement will ensue. As seen in this recurrence, a cyclic change between steep and gentle profiles occurs during cliff recession processes.

To illustrate this, simple models for sliding failure will be presented which are based on the Culmann theory (e.g. Taylor, 1948, pp. 453–5; Carson, 1971, pp. 100–6). This theory makes the assumption that the failure occurs along a linear plane passing through the base of the slope. Figure 5.21 illustrates the sequential profile change of a cliff with a constant height, C_H. A stable profile with an angle of α (Stage 1) is cut back due to wave-induced basal erosion (Stage 2). With increasing erosion distance, x, the overall slope becomes steeper and shear stress within the slope increases: cliff instability augments to finally cause failure. Denoting the effective slope angle by i and the angle of potential failure plane by α_f (Stage 2), we have (e.g. Carson, 1971, pp. 116–18)

$$\alpha_f = \tfrac{1}{2}(i + \phi) \qquad (5.13)$$

where ϕ is the angle of internal friction of cliff material. The weight of the wedge $ACB'A'$, acting on the potential failure plane $A'B'$, should be equivalent to that of the triangle $A'DB'$. This yields

$$i = \tan^{-1}\left(\frac{C_H^2}{x^2 \tan \alpha - 2C_H x + C_H^2/\tan \alpha}\right) \qquad (5.14)$$

This equation indicates that i increases from α to $\pi/2$ with increasing x from zero to $C_H/\tan \alpha$.

When the shear stress arising on the plane $A'B'$ at Stage 2 (driving force) exceeds the shear strength (resisting force), the sliding occurs. At the critical condition,

$$\underset{\text{(driving force)}}{W \sin \alpha_f} = \underset{\text{(resisting force)}}{S_s L' + W \cos \alpha_f \tan \phi}$$

where S_s is the cohesive strength or shear strength of slope material, L' is the length of $A'B'$, which is given by $L' = C_H/\sin \alpha_f$, W is the weight of the wedge $A'ACB'$, or the triangle $A'DB'$, which can be expressed by $W = \gamma C_H^2 (\tan^{-1} \alpha_f - \tan^{-1} i)/2$, in which γ is the unit weight of slope material.

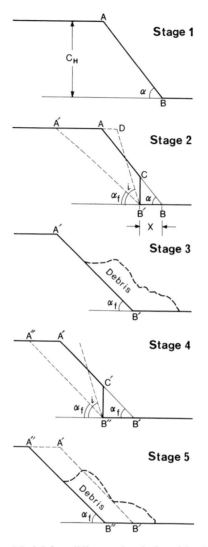

Figure 5.21 Model for cliff recession induced by basal erosion

Transformation of the above balancing equation yields

$$\frac{\gamma C_H}{S_s} = \frac{4 \sin i \cos \phi}{1 - \cos(i - \phi)} \tag{5.15}$$

When i satisfies this relation, failure takes place (Stage 3); and the angle of failure plane α_f is given by Eq. (5.13). Once the debris mass residing at the foot of the cliff is removed, the cliff base is again exposed to wave action and starts

Processes of Cliff Erosion

to recede (Stage 4). When the slope angle i meets the requirement of Eq. (5.15), failure resumes at a slope angle of α_f (Stage 5).

Figure 5.21 shows that a cliff profile with the same slope angle appears repeatedly as the cliff recedes (e.g. profiles $A'B'$ and $A''B''$), suggesting that parallel retreat of cliff profile takes place in a long time span. An example to indicate such parallel retreat can be taken from the area of Taito-misaki on the Pacific coast of Japan, where a Pleiocene mudstone cliff (Figure 5.22) of 2.5 km alongshore length is found (Sunamura, 1973). A possible period for occurrence of parallel profile is estimated ten years for this coast (Sunamura, 1983). This type of slope failure, i.e. planar slide, has also been reported from till bluffs along Kilkeel Bay area in Northern Ireland (McGreal, 1979a,c), and near Aberarth in west Wales (Jones and Williams, 1991).

If the right-hand side of Eq. (5.15) is greater than the left side when $i = \pi/2$, then no slope failure occurs even if the cliff face slope becomes vertical. Under this situation, cliff instability is induced by the notch formed by waves at the base of the cliff (Figure 5.23). With growth in the size of the notch x, the slope stress increases, leading eventually to collapse. Two models shown in Figure 5.23 assume that the potential failure plane with an angle of α_f passes through the deepest part of the notch.

In Model I, the failure plane extends to the cliff top (Stage 2). Considering the balance of two forces, sliding force and resisting force both acting on the plane $A'B'$ one obtains

$$\frac{\gamma x}{S_s} = 2 - \frac{1}{2}\left(\frac{1-\sin\phi}{\cos\phi}\right)\frac{\gamma C_H}{S_s} \qquad (5.16)$$

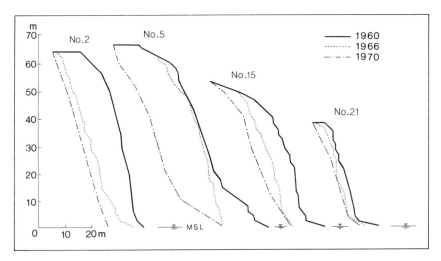

Figure 5.22 Some examples showing profile change of Taito-misak cliff, Japan. The cliff profiles are based on air-photogrammetric maps. From Sunamura (1973)

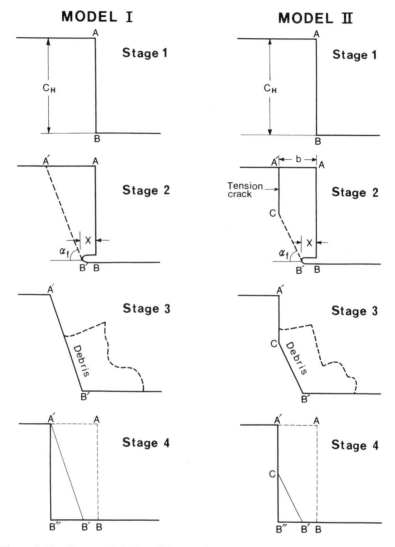

Figure 5.23 Two models for cliff recession associated with notch development

When x reaches the value which fulfils this relation, failure occurs with an angle of α_f (Stage 3). Substituting $i = \pi/2$ for this case into Eq. (5.13), we have

$$\alpha_f = \frac{\pi}{4} + \frac{\phi}{2} \qquad (5.17)$$

Incessant wave action clears away the debris mass temporarily covering the base of the cliff, and as waves resume their attack on the cliff base, a vertical

Processes of Cliff Erosion 113

profile appears (Stage 4). The model assumes no notch development between Stages 3 and 4. This assumption does not always hold in the field.

In Model II, a tension crack developing at a horizontal distance of b from the cliff-top edge extends down to the failure plane (Stage 2). Equating (1) the shear force on the plane CB' and (2) the shear resistance along this plane, and applying Eq. (5.17), the following relation can be written:

$$x = b + \frac{2}{\gamma}\left[S_s - \sqrt{\left(S_s^2 + \frac{\gamma^2 b C_H(\cos(\phi/2) - \sin(\phi/2))}{2(\cos(\phi/2) + \sin(\phi/2))}\right)}\,\right] \quad (5.18)$$

A notch grows, and increases in depth. When it reaches x given by Eq. (5.18), the slope collapses (Stage 3). After debris removal by waves, the cliff base is again subjected to erosion, resulting in the appearance of a vertical cliff (Stage 4). Incidentally, a cliff is likely to suffer toppling failure at Stage 2 when a notch grows, and a tension crack extends further below, or both.

On a geotechnical basis Hutchinson (1972) described how this type of failure occurred on a chalk cliff at Joss Bay on the Isle of Thanet in England (Figure 5.24). The cliff is 15 m high and the chalk is almost uniform with nearly horizontal bedding and vertical joints. Shear failure was preceded by the extension of a tension crack to almost half the height of the cliff. It is evident that the formation of a wave-cut notch about 0.5 m deep facilitated cliff

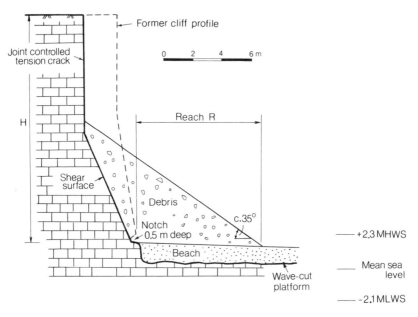

Figure 5.24   Collapse of the chalk cliff in early 1966 at Joss Bay, Kent, England. From Hutchinson (1984a), by permission of Japan Landslide Society

failure (Hutchinson, 1984a). Hoek and Bray (1981, pp. 194–5) discussed a possible failure which will occur in the future at this site, and estimated that the depth of a notch attains 0.9 m on the verge of a subsequent failure. Notch-induced cliff collapse has also been found on some till cliffs on the Holderness coast (Hutchinson, 1986).

A temporal cliff-profile change shown in Figure 5.25 is an example taken from a cliff cut in Miocene sedimentary rocks on the Pacific coast of Japan. Vertically jointed massive mudstone layers occupy most of the cliff and weakly consolidated sandstone layers are exposed in the cliff base, the latter being suitable for notch formation (Aramaki, 1978). The notch has induced cliff failure. Figure 5.25 indicates that the cliff recession process occurring here is similar to Model II (Figure 5.23); this is conjectured from the presence of blocks leaning against the cliff face. Figure 5.25 also illustrates parallel retreat of the cliff profile.

Another characteristic mode of failure frequently observed on severely eroding cliffs is rotational slides, a general treatment of which is provided, for example, by Hoek and Bray (1981, pp. 226–56), Bromhead (1986, pp. 135–77), and Nash (1987). Figure 5.26 is a schematic illustration of cyclic occurrence of deep-seated rotational slides found on coasts composed of London Clay. This diagram was constructed by Hutchinson (1973) from his

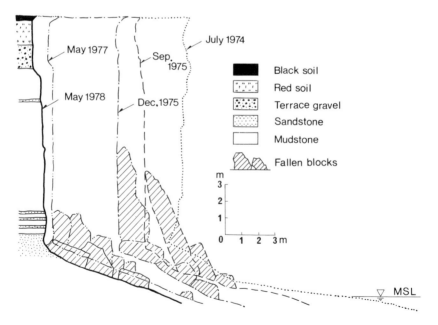

Figure 5.25 Cliff profile changes at Kohriyama on the Fukushima coast, Japan. From Aramaki (1978), by permission

Processes of Cliff Erosion

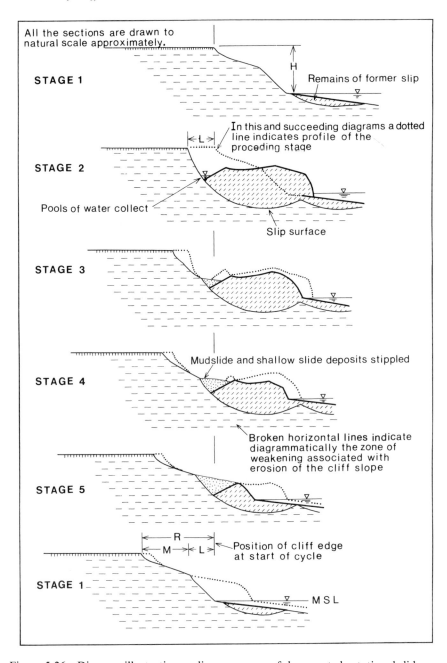

Figure 5.26 Diagram illustrating cyclic occurrence of deep-seated rotational slides on a London Clay cliff. From Hutchinson (1973), by permission of Istituto Geologia Applicata e Geotecnica, Universita di Bari

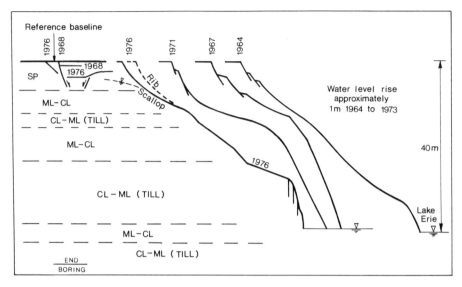

Figure 5.27 Temporal changes in bluff profiles near Port Bruce, on the north shore of Lake Erie. From Quigley et al. (1977), by permission of National Research Council Canada. Soil description is based on the Unified Soil Classification System (SP = poorly graded sand, ML = silt of low plasticity, and CL = clay of low plasticity)

study of the north coast of the Isle of Sheppey, England, where intense wave erosion has taken place to cut back London Clay cliffs at average rates of 0.9 to 2.1 m/year. This figure shows that a similar cliff profile resumes after each sliding mass has been completely removed. The time necessary to complete this cycle at Warden Point on the Isle of Sheppey is estimated to have been 30–40 years during the past century (Hutchinson, 1986).

Shallow-seated rotational slides are common on the till cliff along the Great Lakes shore (e.g. Quigley and Gelinas, 1976; Quigley et al., 1977; Edil and Haas, 1980; Edil and Vallejo, 1980). Some till bluffs in the Kilkeel Bay area of Northern Ireland also exhibit this failure mode (McGreal, 1979a). Cyclic cliff instability induced by basal erosion is widely recognized, and parallel recession of cliff profiles is exemplified by Figure 5.27, which shows temporal profile changes of a receding till cliff at a rate of 2.2 m/year near Port Bruce on the north shore of Lake Erie (Quigley et al., 1977).

As long as the intensity of marine erosion does not vary with time, a cyclic change in cliff geometry and the resultant parallel-profile occurrence will continue. The recurrence of parallel cliffs depends on (1) the frequency of waves causing basal erosion, (2) the amount of debris supplied from the retreating cliff, and (3) the debris-removing ability of waves and currents.

Chapter Six
Underwater bedrock erosion

INTRODUCTION

Wave action on rocky coasts gives rise not only to horizontal erosion at the base of subaerial cliffs but also to vertical erosion of the sea floor profile extending from the cliff base. Elucidation of both processes is essential for adequate understanding of evolution of rocky coasts. In contrast to the large body of knowledge on the process of coastal cliff erosion (Chapter 5), available information on the process of downward erosion of underwater bedrocks is limited. The main reasons for this are (1) the much smaller magnitude of morphological changes occurring on the nearshore sea floor, and also (2) the difficulties encountered in underwater investigations.

On tidal coasts, the gently sloping or nearly horizontal part of the intertidal zone is much easier of access than the subtidal area which is always under water. Therefore, intertidal studies have been conducted more intensively than subtidal studies. This chapter will provide a summary of studies on downward bedrock erosion in intertidal and subtidal environments.

FACTORS AFFECTING BEDROCK LOWERING

Multiple factors are involved in erosion of intertidal and subtidal bedrock as illustrated in Figure 6.1, which shows a relationship similar to that of cliff erosion shown in Figure 5.2. The occurrence of bedrock erosion is ultimately determined by the relative intensity of two forces: the wave-induced assailing force, F_W, the lithology-related resisting force, F_R. If $F_W > F_R$, erosion occurs, resulting in the lowering of the bedrock surface; otherwise, no erosion takes place.

Assailing force

Wave-induced oscillatory movement of water particles characterizes fluid motion on the sea floor in the shallow region extending from the cliff base to

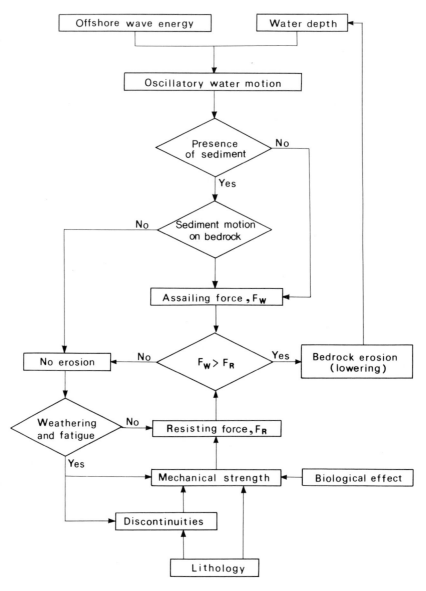

Figure 6.1 Interaction of factors affecting downward erosion of bedrocks at a given point on the sea floor

Underwater Bedrock Erosion

a certain depth beyond the wave breaking point, although much turbulence is also introduced into surf and swash zones. In addition to the oscillatory motion, the direct pounding of a water mass on the sea floor occurs beneath plunging breakers, but this type of action is very localized. The primary assailing force responsible for the lowering of sea floor bedrock originates from the to-and-fro water motion, the intensity of which depends on (1) input-wave energy level and (2) water depth.

If sea floor bedrock is not covered with sediment, then the oscillatory water motion is directly related to the assailing force F_W, which may produce repeated shear stress in the upper thin portion of the bedrock. The wave-induced shear stress acting on the bedrock surface, τ_B, can be expressed in terms of

$$\tau_B = \tfrac{1}{2}\rho C_f u_B^2 \qquad (6.1)$$

where u_B is the maximum near-floor horizontal orbital velocity, C_f is a wave friction factor (Jonsson, 1966; Kamphuis, 1975), and ρ is the density of water. Applying linear wave theory, we have

$$u_B = \frac{\pi H}{T \sinh(2\pi h/L)} \qquad (6.2)$$

where H is the wave height, T is the wave period, h is the water depth, and L is the wavelength at a water depth of h. This is obtained through substitution of $z = -h$, $k = 2\pi/h$, $\sigma_* = 2\pi/T$, $a = H/2$, and $\sin(kx - \sigma_* t) = 1$ into Eq. (2.20a).

On most rocky coasts the bedrock is sparsely covered with sediment. If waves can move the layer of sediment particles on the bedrock surface, then the assailing force is greater because of the combined effect of the repeated shear stress and the abrasive action of sediment moving across the bedrock (e.g. Wood, 1968; Kirk, 1977; Robinson, 1977c). When the oscillatory water motion fails to mobilize the lowermost sediment particles, then no assailing force can extend to the bedrock, $F_W = 0$, and no erosion will occur.

The depth of sediment cover is therefore a crucial controlling factor for bedrock erosion (Bradley and Griggs, 1976; Robinson, 1977c). Evaluation of the depth of layer of sediment mobilized by waves, i.e. the thickness of horizontal moving layer of sediment, is required. Several predictive relationships have been proposed for the thickness of moving layer in the surf zone of sandy beaches (King, 1951; Komar and Inman, 1970; Gaughan, 1978; Greenwood and Hale, 1980; Inman et al., 1980; Kraus, 1985; Sunamura and Kraus, 1985). All these are described in breaking wave height, but Sunamura and Kraus (1985) differ in that a sediment grain-size factor is incorporated in their relationship, which is expressed by

$$\frac{z_s}{D} = 81.4\,(\Psi_B - \Psi_C) \qquad (6.3)$$

where z_s is the average depth of moving layer, D is the grain size of the sediment, Ψ_B is the Shields parameter at the wave breaking point, and Ψ_C is the critical Shields number for oscillatory flow. The Shields parameter Ψ (e.g. Raudkivi, 1967, pp. 20–3) is given by $\Psi = \tau_B/(\rho_s - \rho)gD$, where ρ_s is the density of the sediment and g is the acceleration due to gravity. Equation (6.3) has been calibrated using field data obtained from tracer experiments.

Equations which enable one to estimate the depth of the moving layer in the swash zone and in the offshore region beyond wave breaking point are not yet available. Even if the depth estimation becomes possible in the entire nearshore region, there is little knowledge of: (1) the moving speed of the fluid containing sediment grains u_B^*, (2) the density of the sediment-laden fluid, ρ^*, and (3) a coefficient of friction between this fluid and the bedrock surface, C_f^*. The present situation does not allow us to quantify shear stress induced by the sediment-laden oscillatory flow, τ_B^*, because this may be obtained through the relation:

$$\tau_B^* = \tfrac{1}{2}\rho^* C_f^* u_B^{*2} \tag{6.4}$$

Quantification of τ_B^* will enable us to assess the assailing force F_W using

$$F_W = A^* \tau_B^* \tag{6.5}$$

where A^* is a coefficient representing the abrasive effect which moving sediment grains can exert on the bedrock. It is needless to say that thorough evaluation of F_W requires a quantified value of A^*. Actually, the thickness of sediment varies in time and space due to: (1) changes in wave and tidal conditions (e.g. Inman and Filloux, 1960; Hattori, 1982), (2) the migration of large-scale bedforms such as bars (e.g. Hayes, 1972; Sunamura and Takeda, 1984; Coakley et al., 1986), and (3) changes in nearshore current intensity and pattern. Sediment grain size, which influences bedrock abrasion (Robinson, 1977c), is also a time- and/or space-dependent factor. We will be a long way towards achieving our goal when we can quantify F_W.

Resisting force

Lithology principally determines the resisting force of sea floor bedrock, F_R, through its mechanical strength (Figure 6.1). The primary action of the assailing force on the bedrock is shearing due to wave-induced to-and-fro water motion, as described before, and this action is characterized by the tractive abrasion when a favourable condition for effective sediment motion on the bedrock is established. Therefore, the most appropriate strength parameter for expressing F_R would be shear strength. Discussion by Coakley et al. (1986) indicates that, for consolidated cohesive sediments such as glacial tills, F_R can be described by the shear strength if geotechnical properties such as clay content, plasticity index, and water content are constant. When a measure of shear strength is unavailable for some reason, the compressive strength can replace it (Chapter 4).

Discontinuities in rocks, such as cracks, joints, faults, and stratifications, always act to reduce their mechanical strength (Chapter 4); the same is true of shear strength. However, there have been few studies of the effects of discontinuities on shear strength, even in rock and soil mechanics, and no attempt has been made to evaluate the discontinuities effect on the strength of underwater bedrocks subject to fluid-induced shear force.

Biological factors can also influence the bedrock strength. Removal of lithic substrate from the bedrock surface by marine organisms may bring about direct bedrock lowering, as in the case reported from the northwest coast of County Clare, Ireland, where the surface of intertidal platforms made of Carboniferous limestone has been eroded by the action of a species of boring echinoid (Trudgill *et al.*, 1987). Aside from such bioerosion, activities of grazing and boring organisms indirectly lead to erosion of intertidal or subtidal bedrock through reduction in their mechanical strength due to furrowing and/or honeycombing. Based on his compressive strength tests, Healy (1968b) has reported a significant strength reduction (20–35%) of the riddled part of sandstone and siltstone which form intertidal shore platforms near Auckland, New Zealand. Another indirect influence is that organism-caused ruggedness of the bedrock surface may increase the value of the friction factor of fluid; this in turn augments shear force (Eq. (6.1)), and facilitates erosion.

Grazing molluscs include certain species of gastropods and chitons (e.g. McLean, 1967; Healy, 1968b; Craig *et al.*, 1969), and other boring organisms are algae (Swinchatt, 1969; Schneider, 1976), sponges (Neumann, 1966), barnacles (Ahr and Stanton, 1973; Trudgill, 1976), bivalves (Trudgill and Crabtree, 1987), and echinoids (Healy, 1968b; Trudgill *et al.*, 1987). Molluscan grazing is marked on intertidally exposed rocks. Most boring activities are limited to the intertidal zone and a very shallow water region, but some species of boring algae (Swinchatt, 1969) and bivalves (Jehu, 1918) extend their activities to a considerable depth offshore, say, a few tens of metres. Marine biological grazing and boring processes are pronounced on carbonate-rock coasts, especially in tropical environments. The processes are also found on some temperate coasts, even when the rocks are not carbonates.

If the nearshore bedrock is composed of materials susceptible to weathering, then a subaerial part adjacent to the cliff base or the portion that is exposed at low tide on tidal coasts is subjected to alternate wet/dry weathering, slaking (e.g. Suzuki *et al.*, 1970, 1972; Takahashi, 1975, 1976; Robinson, 1977c), which may initiate cracks or joints on the surface of a certain rock type to finally yield considerable reduction in overall bedrock resistance. The frost action (e.g. Trenhaile and Mercan, 1984; Robinson and Jerwood, 1987) may also facilitate bedrock deterioration, resulting in rapid erosion as seen on the intertidal zone of London Clay platforms at Clacton,

England (Harris and Ralph, 1980). In many subarctic, and occasionally more temperate regions, frost weathering and ice action are two main processes for bedrock erosion (e.g. Dionne and Brodeur, 1988). Chemical solution (e.g. Emery, 1946; Kaye, 1959; Mii, 1962; Emery and Kuhn, 1980) may play a dominant role in accelerating bedrock lowering on some areas of calcareous-rock coasts.

On some types of lithology, fatigue due to repeated stresses generated on the sea floor by oscillatory water motion may decrease the mechanical strength of the bedrock. According to Davidson-Arnott (1986a,b), who investigated downward erosion of a nearshore till profile in Lake Ontario, wave-induced fatigue failure was probably occurring in a thin surface layer of subaqueous till, reducing its mechanical strength. Deterioration by fatigue was comparable to that caused by wetting and drying, or frost action working on the subaerial part of the till profile in this area.

Biological activities, weathering, and fatigue, which are all time-dependent factors which cause deterioration in the mechanical properties of nearshore bedrock, bring about a gradual decrease in their resisting force. When the resisting force becomes smaller than the wave assailing force, this leads to bedrock lowering (Figure 6.1).

Basic relationships for bedrock lowering

In their study of erosion of a submarine profile cut in Pliocene sedimentary rocks, Horikawa and Sunamura (1970) intuitively assumed

$$\frac{dz}{dt} \propto \tau_B \tag{6.6}$$

where dz/dt is the rate of bedrock erosion (lowering). Laboratory erodibility tests of subaqueous till samples taken from the Lake Ontario area indicated that this relationship holds approximately, although a considerable scatter of data was seen (Coakley et al., 1986, Figure 4). Zeman (1986, Figure 12) also obtained a similar relationship, from his tests on Lake Erie tills.

As a more generalized expression, inclusive of the assailing and the resisting forces, the following equation can be written from an analogy of Eq. (5.4):

$$\frac{dz}{dt} = \zeta \ln\left(\frac{F_W}{F_R}\right) \tag{6.7}$$

where ζ is a constant with units of $[LT^{-1}]$. If $F_W > F_R$, then erosion occurs and the erosion rate is given by this equation; whereas if $F_W \leq F_R$, then no erosion takes place $dz/dt = 0$.

RATES OF LOWERING

Substrate removal or erosion due to the activities of marine organisms have been intensively studied in the nearshore zone (Spencer, 1988). The removal rate has been assessed using biological methods (North, 1954; McLean, 1967; Evans, 1968; Trudgill and Crabtree, 1987; Trudgill et al., 1987). Chemical analyses were adopted for the estimation of rates of bedrock lowering induced by solution on limestone coasts (Emery, 1946; Revelle and Emery, 1957). Erosion rates obtained by these biological or chemical means do not generally indicate the total rate of actual bedrock lowering, because the lowering is an integrated form of wave-induced mechanical, biological, and chemical actions, one of which is usually dominant, depending on environmental conditions. To quantify such a total rate, the only way is to take direct measurements of vertical lowering of bedrock surface over a certain period of time.

Direct measurements have been attempted from the early 1960s (Table 6.1), notably by using a micro-erosion meter (MEM) technique. This technique, originally developed by High and Hanna (1970) and later modified for coastal studies by Robinson (1976), has enabled precise measurements of erosion of intertidal bedrock to be made over a short period. As Trenhaile (1987, p. 235) pointed out, however, this technique has an inherent limitation that it cannot measure the bedrock lowering at a place where plucking or quarrying of large rock fragments or jointed blocks frequently occurs.

A modification of the MEM technique was made by Askin and Davidson-Arnott (1981), with the intention of applying it to the sea floor environment, and this modified technique led to Davidson-Arnott's (1986b) measurements off the Canadian shore of Lake Ontario. In West Germany, Wefer et al. (1976) had already attempted to measure underwater bedrock erosion off Bokniseck on the Baltic Sea coast, using a specially designed instrument with a microcaliper, which is similar to the MEM.

Another method for measurement of sea floor bedrock change is to use a high-resolution acoustic transducer which is fixed at both ends of the horizontal bar of a submerged T-shaped frame positioned on the sea floor to measure the vertical distance (Coakley et al., 1986). This method seems to be less accurate than the MEM, but it has advantages in that continuous monitoring can be made even at a location where plucking occurs.

Zenkovich (1967, p. 168) related the lowering rate of subaqueous bedrocks, dz/dt, to the rate of horizontal erosion of a subaerial cliff, dx/dt, by

$$\frac{dz}{dt} = \frac{dx}{dt} \tan \beta \qquad (6.8)$$

where $\tan \beta$ is the gradient of bedrock profile. Although this relationship includes the assumption that the overall nearshore profile shows parallel retreat, it gives a reasonable approximation to average lowering rates over a

Table 6.1 Measured or calculated vertical erosion rates

Researcher(s)	Location	Rate (mm/y)	Method	Lithology	Environment	Dominant process	Remarks
Agar (1960)	North Yorkshire, UK	0.254	—	Lias shale	Intertidal	Wave action	Lowering of landing step (106–122 years)
Hodgkin (1964)	Norfolk Is., Australia	0.66–0.76	—	Limestone	Intertidal	—	1904–61
So (1965)	Isle of Thanet, England	25.4	Survey	Chalk	Intertidal	Wave action	
Zenkovich (1967, p. 168)	Black Sea coasts, the former USSR	Order of 0.1	Calculation	Neogene calcareous rocks	Cliff base	Wave action (?)	
Horikawa and Sunamura (1970)	Byobugaura, Chiba, Japan	Order of 10	Calculation	Clays	Cliff base	Wave action (?)	Long term (2000 years) Cliff retreat rate: 0.7 m/y
		16.5 (water depth: 0 m; −1.8 (7 m)	Calculation	Pliocene mudstone	Subtidal	Wave action	
	Okuma, Fukushima, Japan	19.0 (water depth: 0 m; −3.5 (10 m)	Calculation	Pliocene mudstone/sandstone	Subtidal	Wave action	Long term (2000 years) Cliff retreat rate: 0.63 m/y
Trenhaile (1974)	Vale of Glamorgan, Wales	0.64	Calculation	Lias limestone	Intertidal	Wave action	Long term Cliff retreat rate: 1.27 cm/y
Trudgill (1976)	Aldabra Atoll, Indian Ocean	2–4	MEM	Reef limestone	Intertidal	Wave action	1969–71 Most exposed site (Site 5)
Wefer et al. (1976)	Bokniseck, Kiel Bay, Germany	21 (water depth: 1.7 m) −1.6 (10 m)	Micro-caliper	Boulder clay	Subtidal	Wave action	1972–73
Kirk (1977)	Kaikoura, NZ	0.4–1	MEM	Tertiary limestone	Intertidal	Wave action	Bimonthly (1973–75)
		0.6–2.5	MEM	Tertiary mudstone	Intertidal	Wave action	

Reference	Location	Rate (mm/a)	Method	Lithology	Environment	Process	Period
Robinson (1977c)	Northeast Yorkshire, England	0.25–14.7 (Ramp) 0.13–2.25 (Plane)	MEM MEM	Lias shale Lias shale	Intertidal Intertidal	Abrasion Wet/dry weathering	Bimonthly (1970–72)
Emery and Kuhn (1980)	La Jolla, California	0.03–0.6	Photographs	Cretaceous sandstone	Intertidal	Solution	35 years
		10	Photographs	Cretaceous sandstone	Intertidal	Abrasion	
Harris and Ralph (1980)	Clacton, Essex, England	300 mm/a few weeks	—	London clay	Intertidal	Frost action	Winter of 1962/63 Kinley's observation
Trudgill et al. (1981)	Country Clare, Ireland	0.145–0.383	MEM	Limestone	Intertidal	—	Table 1.4 (mean value)
Gill and Lang (1983)	Otway, Victoria, Australia	0.3–0.9	MEM	Greywacke and siltstone	Intertidal	Wave action	Bimonthly (1978–80)
Viles and Trudgill (1984)	Aldabra Atoll, Indian Ocean	1.27 (Ramp edge)–2.20 (Ramp foot)	MEM	Reef limestone	Intertidal	Wave action	1971–82 (Table 1-c)
Spencer (1985b)	Grand Cayman Is., West Indies	0.29–0.61 (Reef-protected) 1.8–3.7 (Exposed) 0.10–0.31 (Surf platform)	MEM MEM MEM	Reef limestone Reef limestone Reef limestone	Intertidal Intertidal Intertidal	Biological action Wave action Wave action	1978–79 (Table 2)
Coakley et al. (1986)	Southwestern shore, Lake Ontario	<20	Acoustic waves	Till	Subaqueous	Wave action	1984–85
Davidson-Arnott (1986b)	Eastern shore, Lake Ontario	35 (water depth: 2.3 m) –11 (6.4 m)	MEM	Till	Subaqueous	Wave action	1980–84
Hutchinson (1986)	Warden Point, Isle of Sheppey, England	70–214	—	London Clay	Intertidal	Wave action	1983–86

long period as Hutchinson (1986) stated. Estimates are only possible on a coast where a long-term cliff recession rate (i.e. a time-independent value—see Chapter 5) and the adjacent bedrock profile are known. Some values of lowering rate shown in Table 6.1 (Zenkovich, 1967, p. 168; Horikawa and Sunamura, 1970; Trenhaile, 1974) were obtained in this way. Horikawa and Sunamura estimated underwater values according to Eq. (6.8), by multiplying the long-term cliff erosion rate by the sea floor gradient at an arbitrary depth.

The work of Trenhaile (1974), using Eq. (6.8), indicated that the ratio of vertical lowering at the cliff base to horizontal recession of the cliff was about 2% for chalk platforms in England, and 5% for Lias platforms in Wales. Kirk (1977), who also applied this equation, reported that the ratio was 3% for Tertiary mudstone and limestone platforms on the Kaikoura Peninsula coast in New Zealand. Horikawa and Sunamura (1970) showed that on the Byobugaura coast (cut in Pliocene mudstone) it had a value of approximately 2% and on the Okuma coast (Pliocene mudstone/sandstone) 3% (Table 6.1). It is seen from Eq. (6.8) that these values represent the sea floor gradient of the bedrock profile from the base of the cliff.

An attempt to check the applicability of Eq. (6.8) was made by Kirk (1977), based on data obtained through MEM measurements over two years. Such short-term data seem inappropriate for the applicability test.

Table 6.1 shows that data on erosion rates in the intertidal zone are more abundant than in the underwater environment. These intertidal data indicate that there is a general tendency for the rate of lowering to increase as the hardness of bedrock diminishes: for example, 10^{-1} to 10^0 mm/year for limestones and 10^1 to 10^2 mm/year for unconsolidated materials such as London Clay. This suggests a lithological control. However, there is considerable spatial variation in erosion rates, even on the same rock type at the same location, as Robinson (1977c) reported from the northeast Yorkshire coast of England, based on his measurements using the MEM. Temporal variations have also been found by Kirk (1977), Robinson (1977c), and Viles and Trudgill (1984), these studies all using MEM measurements. A comparison between the short-term (2 years) and long-term (11 years) erosion rate data obtained from Aldabra Atoll, Indian Ocean, led Viles and Trudgill to warn that the use of short-term data should be made with caution.

PROCESSES OF LOWERING

One of the well-known modes of downward erosion of bedrock is plucking or quarrying, which refers to the removal by wave action of rock fragments bounded by geological discontinuities such as joints, faults, and bedding planes (e.g. Gill, 1971, Figures 1 and 2; Trenhaile 1972, Figure 4; Short, 1982b, Figure 1). It is frequently observed after major storm events that angular blocks are thrown up on a shore, with some being honeycombed or having attached kelp. This is evidence of plucking. This type of erosional mode becomes pronounced when the bedrock has more abundant, pre-existing discontinuities, and when highly turbulent water acts on them, as in the swash–surf zone.

A similar process has been commonly found on rocks susceptible to slaking although the size of fragments to be removed is much smaller. Suzuki *et al.* (1970, 1972) were probably the first to treat this topic on the basis of the measured strength characteristics of coast-forming materials. Their investigation sought to elucidate the formative process of corrugated relief on intertidal platforms on the Miura Peninsula coast of Japan, which are cut into a rhythmic alternation of steeply dipping Miocene mudstone and tuff layers. Field observations of the characteristics of joints on the platform surface, and laboratory tests of the various mechanical strengths of mudstone and tuff, including wet/dry weathering (slaking) experiments, led to the conclusion that mudstone subjected to slaking results in multiple irregular-shaped joints with average intervals of 1–5 cm; and waves can easily erode many minor mudstone flakes separated by such weathering joints. As a result, the surface of mudstone is lowered much more rapidly than that of tuff.

Applying similar methodologies to intertidal platforms around Aoshima Island, Japan, Takahashi (1975, 1976) investigated the lowering of platform composed of an alternation of gently sloping Pliocene sandstone and mudstone layers, the former constituting ridges and the latter furrows. He emphasized the role of wet/dry slaking in the lowering of the mudstone surface, on which an abundance of small flakes (about 1 cm across) form, and are easily removed by waves. In these studies in Japan, no data on actual lowering rates has so far been provided.

Using MEM measurements, Robinson (1977c) examined flake removal on Lias shale platforms on the Yorkshire coast of England. Flakes produced by alternate wetting and drying on well-drained areas of the platform were in the form of polygons about 2 cm in diameter, and these flakes were readily removed by waves. Erosion was most rapid during summer, the season of moderate wave action; it averaged 1 mm/year, ranging from 0 to 9 mm/year. Cracking and flaking have also been reported from intertidal platforms cut in London Clay at Warden Point, Isle of Sheppey, England (Hutchinson, 1986) and from mudstone platforms on the Kaikoura coast, New Zealand (Kirk, 1977).

A similar mode of lowering is found on shores exposed to frost action, which leads to disintegration of bedrock surfaces, followed by rapid removal of material by waves. Harris and Ralph (1980) reported that the frost-disintegrated surface of a London Clay platform near Clacton, Essex, England was lowered 30 cm in a few weeks during the harsh winter of 1962–63.

Another mode of bedrock lowering is abrasion by the grinding and wearing action of sand and gravel moved by, or entrapped in, waves. A polished and smoothed bedrock surface is produced by such abrasion, as commonly seen on hard rocks (Zenkovich, 1967, Figures 42 and 44; Davies, 1972, Figure 56). Abrasion has attracted much attention, but quantitative studies based on instrumentally measured erosion rates have been few. Two studies were conducted in the same year in New Zealand (Kirk, 1977) and in England (Robinson, 1977c), both applying the MEM technique. According to Kirk's study, carried out on seven intertidal platforms cut in Tertiary mudstone and limestone around the Kaikoura Peninsula, abrasion was the dominant process operating on the two mudstone platforms showing highest erosion rates, 2.4 to 2.5 mm/year. Robinson found that abrasional lowering of Lias shale platforms in northeastern Yorkshire depended on the depth of sediment working as an abrasive tool: the erosion rate was 3.94×10^{-2} mm/tide for the zone where the sediment depth was 0 to 5 cm, 3.26×10^{-2} mm/tide for 5 to 13.5 cm, and 1.13×10^{-2} mm/tide for greater than 13.5 cm. He also suggested that the grain size of sediment controlled erosion rates, and stated that wave energy was the most important variable if the sediment was thicker than 13.5 cm. Robinson's study also showed that abrasion-dominated erosion may give considerable spatial variation in bedrock lowering rates.

Zenkovich (1967, pp. 147–51) has reviewed Russian studies of sea floor abrasion on the Black Sea and other coasts. On the Caucasian coast of the Black Sea, steeply dipping flysch beds (alternations of marl and shale) form the nearshore bottom area. Near the shoreline, especially where pebbles and boulders are present, the underwater bedrock surface is smooth as a result of abrasion by these sediments. With increasing water depth, however, the more resistant marl beds gradually project to form ridges above the shale beds; the ridges have a relative height of up to 3 m at depths of the order of 20 m. In a slightly embayed area between Tsemes Bay and Gelendzhik Bay, steeply inclined alternating flysch layers are exposed in the nearshore region, on which clastic materials are patchily deposited. In this region, the bedrock surface has been smoothed by wave action armed with these sediments, and the bottom elevation is noticeably lower than in adjacent areas of the same lithology but devoid of bottom sediments, where the profile is of corrugated surface features with harder layers projecting (Zenkovich, 1967, Figure 53).

The importance of abrasion in the lowering of sea floor bedrock has also been pointed out in a recent study of till erosion by Davidson-Arnott (1986a), conducted near Grimsby on the southwestern shore of Lake Ontario. This included precise measurements of bedrock elevation using the MEM and observations by scuba diving. Another important process was found to be erosion of individual grains by shear associated with wave-induced water motion.

An investigation by Coakley *et al.* (1986) at Stoney Creek, several kilometres west of Davidson-Arnott's study site, included acoustic monitoring of elevation change in sea floor till during a period of 13 months. They found that wave-induced shear stress was responsible for the lowering of the till surface at a water depth of 6.9 m. Based on wave data of this term, they predicted erosion rates assuming Eq. (6.6): the predicted value (10 cm/year) was an order of magnitude greater than the actually measured rate, which was less than 2 cm/year.

In order to elucidate erosional processes on till profiles, a wave-flume experiment was attempted by Nairn (1986) using a clay/silt mixture for modelling coasts of till. The nearshore slope (with a uniform gradient of $\frac{1}{10}$ or $\frac{1}{20}$), with or without an overlaying veneer of sand, was exposed to random waves. The influence of sand deposits, i.e. whether they accelerate erosion working as an abrasive tool or decelerate it acting as a protective layer, were discussed. In the tests with sand no marked erosion occurred due to its armouring effect, whereas in the tests without sand overall lowering occurred in the surf zone.

Croad (1981) investigated at a microscopic level the mechanics of erosion of cohesive soils in laboratory tests in which various clay samples were exposed to Couette flow (e.g. Schlichting, 1968, p. 6). Croad found that erosion of stiff clays is characterized by a process of formation of fine cracks initiated by pressure fluctuations associated with turbulent shear flow, followed by the development of cracks and the plucking of clay fragments which occurs at the ejection phase of the turbulent cycle. Based on Croad's findings, Nairn et al. (1986) constructed an erosion model applicable to the downcutting of cohesive shores, assuming that the intensity of turbulence near the bed can be represented by the rate of wave-energy dissipation in the surf zone.

A very localized lowering of intertidal platforms cut in glacial till including erratics has been reported from the Holderness coast of England (Hutchinson, 1986). It was found that the protrusion of erratics from the platform surface induces scouring of till around them. One of the causes for this scouring is the presence of marked discontinuities in erosion resistance as pointed out by Hutchinson, and another cause is the occurrence of three-dimensional vortices (e.g. Lugt, 1983, pp. 81–7) induced by protruding erratics.

Whatever mode of erosion takes place, bedrock lowering brings about increase in water depth, which in turn leads to a reduction in the intensity of the assailing force acting on the sea floor. This decelerates downcutting for erosion rates decrease with increasing water depth. Ultimately, erosion ceases when the assailing force becomes equal to or less than the resisting force of the bedrock. Thus, the erosional process forms a negative feedback system (Figure 6.1). This suggests the existence of a maximum water depth beyond which no bedrock erosion can occur, leading to the concept of a 'wave base' which will be discussed later. Prior to discussion of this topic, it seems appropriate to examine the relationship between erosion rate and water depth.

Figure 6.2 shows the relationship between erosion rate of till and water depth, plotted by use of data from Grimsby on the southwestern shore of Lake Ontario. The open symbol denotes data obtained through the MEM

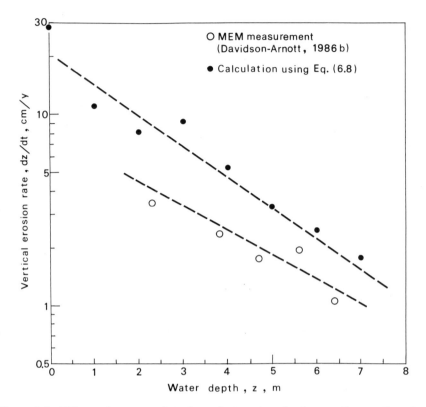

Figure 6.2 Till erosion rate plotted against water depth, on the southwestern Canadian coast of Lake Ontario

measurement by Davidson-Arnott (1986b) between 1980 and 1984. The solid symbol indicates the results of a calculation using Eq. (6.8). In this calculation, the bottom gradient (tan β) was measured after smoothing out the nearshore bedrock profile of Davidson-Arnott (1986b, Figure 1). As a long-term cliff recession rate for this site (dx/dt), a value of 1.4 m/year was used, which was an average rate over 40 years from 1934 to 1973, estimated from aerial photographs by Coakley and Boyd (1979).

The calculated values are considerably greater than the measured values. One of possible reasons for this discrepancy is that measurements were made in a much shorter period than the 40 years during which the bluff recession rate was measured. Short-term bedrock lowering may be strongly affected by temporal variability in controlling factors such as wave climate and the thickness of surficial sediments. Although the data points in Figure 6.2 are scattered, both measurements and calculations show a general trend of erosion rate decreasing linearly with increasing water depth in a semilogarithmic

Underwater Bedrock Erosion

expression: this indicates that the erosion rate decreases exponentially with water depth.

Using data from near Port Burwell on the north central shore of Lake Erie (Philpott, 1986), the erosion rate of till is plotted against water depth (Figure 6.3). Erosion rates denoted by the open symbol were directly obtained by measuring vertical changes between old (1896) and recent (1979) nearshore bottom profiles provided by Philpott (1986, Figure 7). The measurement was made after averaging and smoothing these profiles. The solid symbol indicates data estimated, through Eq. (6.8), using the 1979 profile and applying a cliff recession rate of 2.8 m/year during the period of 83 years from 1896 to 1979. The two data sets thus obtained show a general trend as indicated by the dashed line in Figure 6.3. Again, there exists an exponential relation between erosion rate and water depth.

The linear relation as shown in Figures 6.2 and 6.3 can be given by an exponential decay function:

$$\frac{dz}{dt} = A_s e^{-B_s z} \tag{6.9}$$

where dz/dt is the downward erosion rate, z is the water depth, A_s is the erosion rate at the water edge, $z = 0$, and B_s is a constant. Such an exponential

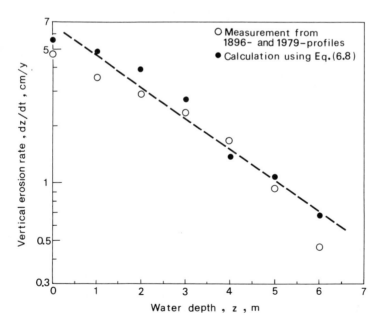

Figure 6.3 Till erosion rate plotted against water depth, on the north central shore of Lake Erie

relation has already been reported from two sites on the Pacific coast of Japan, Byobugaura and Okuma, both cut into weakly consolidated Tertiary sediments (Horikawa and Sunamura, 1970). These findings from Canadian and Japanese shores strongly suggest that, on a coast made of cohesive soils or weak rocks, nearshore bedrock erosion occurs at a rate that decreases exponentially with water depth.

WAVE-BASE PROBLEMS

Gulliver (1899, pp. 176–7) used the term 'wave base' to denote the ultimate water depth of submarine bedrock erosion. Later, Fenneman (1902) applied this term to describe the maximum depth to which waves can move bottom sediments. Since then, some workers have used the term to denote the greatest depth of bedrock erosion (e.g. Twenhofel, 1950, p. 231; Rich, 1951; Zenkovich, 1967, p. 18; Garner, 1974, p. 709), and others the critical depth for sediment movement (e.g. Gary, et al., 1974, p. 788; King, 1974, p. 75; Holmes and Holmes, 1978, p. 503); whereas Birkeland and Larson (1978, p. 442) and Snead (1982, p. 233) gave both definitions. It is clear that bedrock erosion and sediment movement are different phenomena, although the latter may influence the former, depending on the thickness of the sediment layer, the grain size of the sediment, and the intensity of wave action, as discussed before.

The term 'wave base' has another definition which refers to the water depth equal to half the wavelength, i.e. the water depth at which waves begin to 'feel the bottom' (e.g. Longwell et al., 1969, p. 662; Friedman and Sanders, 1978, p. 469; Allen, 1982, p. 36; Skinner and Porter, 1987, p. 737; Carter, 1988, p. 50). More physically, this is the water depth at which the behaviour of waves starts to alter from deep-water to intermediate-water characteristics (Chapter 2). This depth has no direct relation to either bedrock erosion or sediment motion. Incidentally, 'wave base' has been used in stratigraphic discussions to refer to the upper limits of medium- or fine-grained sediments that lack evidence of wave action (Dietz and Fairbridge, 1968).

One of most important elements for the study of rocky coast evolution is the limiting depth beyond which no waves can erode the bedrocks, as Bird (1976, pp. 76–8) pointed out. This book will use the term 'wave base' to denote *the ultimate water depth for bedrock erosion*, paying respect to Gulliver (1899) who first introduced this term.

An actual value for the wave base was not given by Gulliver (1899), but Johnson (1919, pp. 80–3) gave it as 180 m; since then various estimates have been presented by many workers (Table 6.2). This table shows that the values range from the order of 10^0 to 10^2 m. A value of approximately 10 m was proposed in the 1950s by some investigators (Dietz and Menard, 1951; Dietz, 1952; Fairbridge, 1952; Bradley, 1958) for the limiting depth of effective abrasion of bedrocks, and Dietz (1963) termed this depth the 'surf base'. The surf

Table 6.2   Previous estimations of wave base

Source	Depth (m)	Area
Johnson (1919, pp. 80–3)	180	—
Barrell (1920)	90	—
Rode (1930)	45–90*	Santa Cruz, California
Miller (1939)	30	—
Von Engeln (1942, p. 525)	180	—
Tayama (1950)	60	Shikoku, Japan
Dietz and Menard (1951)	9†	—
Dietz (1952)	9†	—
Fairbridge (1952)	Few fathoms	—
Yoshikawa (1953)	40–50	Around Japanese islands
Shumway et al. (1954)	15–18	Southern California
Yoshikawa and Saito (1954)	10–15	Chikura, Japan
Kaizuka (1955)	20	Southern Kanto, Japan
Newell and Imbrie (1955)	7	Great Bahama Bank
Bradley (1958)	9†	Santa Cruz, California
Russell (1958)	11	—
Kuenen (1960, p. 228)	150	—
Mii (1962)	8	Tanabe Bay, Japan
Dietz (1963)	9–18†	—
King (1963)	9–12	—
Toyoshima (1967)	6–7	San-In coast, Japan
Zenkovich (1967, p. 150)	5–12	Lake Baikal
Zenkovich (1967, p. 168)	50–60 / 10–20	Open coasts / Black Sea coasts
Toyoshima (1968)	10	Shionomisaki, Japan
Longwell et al. (1969, p. 333)	10	—
Hoshino (1975, p. 29)	10	—
Sunamura (1978a)	40*	Byobugaura, Japan
Sunamura (1987)	40*	Niijima Is., Pacific Ocean

*Calculated value
†Vigorous abrasion

base is considered to correspond to the water depth where waves break under stormy conditions. In the surf zone, much more turbulence is produced by wave breaking and hence more vigorous abrasion, as compared with the zone seaward of the wave breaking point. Thus, the surf base is an important physical parameter. Values of approximately 10 m, listed in Table 6.2, may be taken for the surf base for open coasts. It is obvious that the value for surf base is smaller than the wave base.

The most reasonable value for the wave base is a problem to which no definite solution has been offered. Because submarine bedrock erosion is primarily caused by wave action, this problem should be considered in terms of wave parameters. A pioneering work by Rode (1930) attempted to estimate the wave base for Santa Cruz area in California using linear wave theory with local wave conditions taken into account.

A theoretical approach was taken by Sunamura (1978a) to calculate the wave base in a generalised form. It was assumed that abrasive action of the bottom sediment when moved on the bedrock surface strongly affects bedrock lowering. According to Rode (1930), the minimum grain size of the sediment which can act as an abrasive was considered 1 mm in diameter, whereas Twenhofel (1945) thought that it was 0.5 mm. Selecting the latter value, and applying the relationship presented by Sato *et al.* (1962) to predict critical water depth for sediment movement (in traction) by waves, Sunamura (1978a) obtained the following equation, which provides the wave base, h_a:

$$\left(\sinh \frac{2\pi h_a}{L}\right)\left(\frac{H_0}{H}\right) = 9.33 \, L_0^{1/3} \left(\frac{H_0}{L_0}\right) \tag{6.10}$$

where H_0 and L_0 are the deep-water wave height and wavelength, respectively, and units are metres. This equation indicates that h_a can be calculated from the deep-water wave parameters by applying Eqs (2.18) and (2.33). The calculation is somewhat laborious. Figure 6.4 shows a nomograph for Eq. (6.10).

Data of the largest waves occurring in an area under consideration should be used for the wave-base estimation. The largest waves are important because they have potentially to affect the deepest area (e.g. Zenkovich, 1967, p. 153; Bradley and Griggs, 1976), although the duration of action of such waves is extremely short as compared with that of moderate waves having little influence on the wave base.

Sunamura (1990a) has also tackled wave-base problems taking another study approach. Assuming that $F_W \propto \tau_B$ (shear stress) and $F_R \propto S_s$ (shear strength), and using Eqs (6.1) and (6.2), he derived from Eq. (6.7) the following relationship:

$$z = \frac{\Gamma_s}{\lambda}(1 - e^{-\zeta \lambda t}) \tag{6.11}$$

where z is the depth of bedrock erosion, t is the time, Γ_s and ζ are constants, and

$$\lambda = \frac{10.7}{L_0} \tag{6.12}$$

Equation (6.11) indicates that the erosion depth increases rapidly with time at the initial stage, but it later tends to approach a critical value, i.e. wave base. Therefore, the wave base h_a is given by

$$h_a = \lim_{t \to \infty} z = \frac{\Gamma_s}{\lambda} \tag{6.13}$$

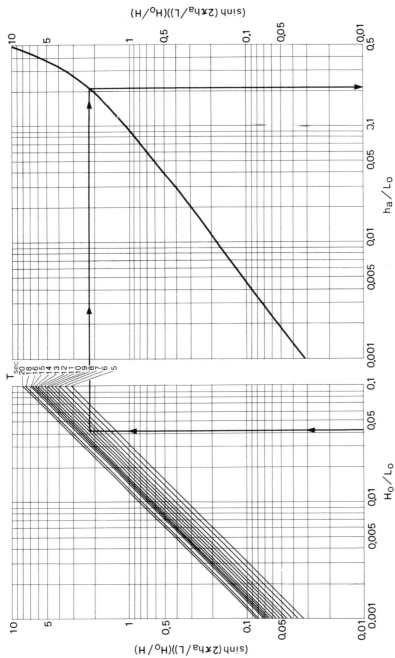

Figure 6.4 Nomograph for Eq. (6.10). Estimation of the limiting abrasion depth, h_a, is possible by knowing deep-water wave properties, H_0, and T (or L_0); the units used in the estimation are metres and seconds. From Sunamura (1978a). *Geol. Soc. Am. Bull.*, **89**, 504–10, by permission of the Geological Society of America

The time required for z to reach 95% of h_a was defined as the equilibrium time, t_e:

$$t_e = \frac{3}{\zeta\lambda} \qquad (6.14)$$

Norrman (1980), who has monitored sea floor erosion on truncated tephra-cone islands in the Surtsey area of Iceland, discussed the temporal changes in erosion depth. Assuming $H_0 = 9$ m and $T = 14$ s as representative values for the largest significant waves occurring in this area on the basis of data presented by Viggósson and Tryggvason (1985), Sunamura (1990a) obtained $\lambda = 0.038$ m^{-1} from Eq. (6.12). A best fit of Eq. (6.11) with Norrman's data (Figure 6.5) resulted in $\Gamma_s = 1.9$ and $\zeta = 7.4$ m/year. Calculation of Eqs (6.13) and (6.14) led to $h_a = 50$ m and $t_e = 10.7$ years. This wave-base value was found to be close to the value estimated through Eq. (6.10) using the representative wave data, i.e. $h_a = 54$ m.

Differentiation of Eq. (6.11) gives

$$\frac{dz}{dt} = \zeta(\Gamma_s - \lambda z) \qquad (6.15)$$

On coasts made of erodible materials such as clays, tills, unconsolidated volcanic ejecta, and weak sedimentary rocks, the rates of lowering of sea floor bedrock can be described by Eq. (6.15) for the deeper region up to the wave base, and by Eq. (6.9) for the shallower region to the water edge. It is reasonable to consider that these two equations should meet at the surf base, h_a^*,

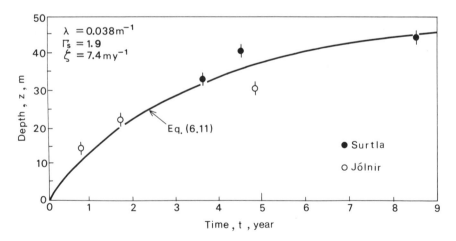

Figure 6.5 A best fit of Eq. (6.11) with Norrman's (1980) data obtained at Surtla and Jólnir in the Surtsey area of Iceland. After Sunamura (1990a)

Underwater Bedrock Erosion

which may correspond to the breaking depth of the largest waves in the area (Figure 6.6). Substituting $z = h_a^*$ into both equations leads to

$$A_s e^{-B_s h_a^*} = \zeta(\Gamma_s - \lambda h_a^*) \tag{6.16}$$

Naturally, $dz/dt = 0$ at $z = h_a$: Eq. (6.15) reduces to

$$\Gamma_s = \lambda h_a \tag{6.17}$$

which is the same as Eq. (6.13).

This idea, represented by Figure 6.6, can be applied to the Byobugaura area of Japan, where the submarine bedrock is Pliocene mudstone. Using Eq. (6.10) the wave base in this area was estimated at 40 m: $h_a = 40$ m (Sunamura, 1978a). The largest significant waves in this area can be assumed to be $H_0 = 8$ m and $T = 12$ s (Sunamura, 1978a). Using this wave data, the breaking depth of the waves can be evaluated at 12 m from Figure 2.12. This is regarded as a value for the surf base: $h_a^* = 12$ m.

Data plotted in Figure 6.7 were obtained by Horikawa and Sunamura (1970) through calculations using Eq. (6.8). Based on these data, they expressed the relationship between erosion rate and water depth as:

$$\frac{dz}{dt} = 1.9e^{-0.35z} \tag{6.18}$$

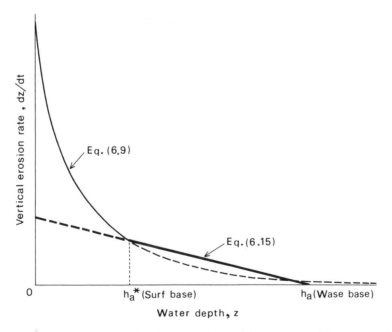

Figure 6.6 Schematic graph showing the relationship among surf base, wave base, Eqs (6.9) and (6.15)

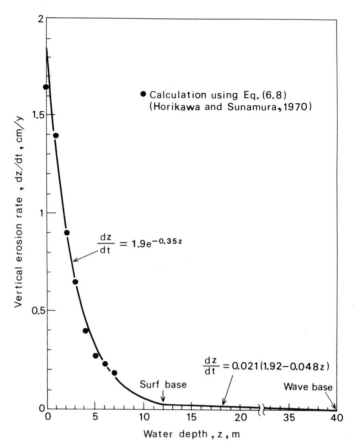

Figure 6.7 Underwater erosion rate of Pliocene mudstone in the Byobugaura area, Japan

where the unit of dz/dt is in cm/year and that of z in metres. In Figure 6.7, the curve representing Eq. (6.18) is plotted extending its end to the surf base. Substituting $A_s = 1.9$ cm/year, $B_s = -0.35$ m^{-1}, $h_a^* = 12$ m, $h_a = 40$ m, and $\lambda = 0.048$ m^{-1} with the last being determined from Eq. (6.12) using the wave data, into Eqs (6.16) and (6.17), one can obtain $\Gamma_s = 1.92$ and $\zeta = 0.021$ cm/year. Equation (6.15) with these determined values is depicted in Figure 6.7.

The equilibrium time for this area is estimated from Eq. (6.14) as $t_e = 3 \times 10^5$ years. This is very large compared with the value for the Surtsey area (11 years), which clearly reflects the great difference in resistivity of the bedrock, because the wave climate in both areas is similar.

Chapter Seven
Shore platforms

MAJOR CONTEMPORARY MORPHOLOGIES

Except for regions that are tectonically unstable or subjected to a strong influence of glacial isostasy, the Holocene transgression brought the sea to its present general level about 6000 years ago. Since then an almost stable sea level condition has been established although minor relative sea level fluctuations have occurred. Marine processes have been working on rocky coasts at present sea level to create characteristic shore-zone morphologies. They can be categorized into three major kinds (Figure 7.1): two types of platforms and a plunging cliff. There are many variations reflecting (1) lithological factors, i.e. rock types and geological structures (discontinuities) such as bedding planes, joints, and faults, (2) weathering properties of rocks, (3) tides, (4) degree of exposure to wave assault, and (5) the inheritance of minor relative changes in land and sea levels.

Platforms developing around the present sea level are largely classified into: (1) gently sloping platforms without a significant topographic break, extending from the base of a cliff to the nearshore sea floor below low tide level (Figure 7.1a); and (2) nearly horizontal platforms with a marked drop at their seaward edge (Figure 7.1b). There are also steeply descending cliffs that pass far below sea level without any shore platform (Figure 7.1c), called plunging cliffs by Davis (1928, p. 152). This classification is based on that of Bird (1976, Figure 59) who subdivided the subhorizontal-platform type (Figure 7.1b) into two according to the platform elevation: high-tide and low-tide platforms.

For sloping platforms (Figure 7.1a), the literature has employed various terms: abrasion platforms (Johnson, 1919, p. 162; Wefer et al., 1976); beach platforms (Wood, 1968); benches (Zenkovich, 1967, p. 245); coastal platforms (So, 1965); shore platforms (Bird and Dent, 1966; Wright, 1967; Davidson-Arnott, 1986b; Gill, 1972a; Bird, 1976, p. 74; Robinson, 1977a; Trenhaile, 1987 p. 192); submarine platforms (Healy and Wefer, 1980); wave-cut benches (Barrell, 1920; Longwell et al., 1969, p. 334; Lee et al., 1976); wave-cut platforms (Bradley, 1958; King, 1963; Sorensen, 1968; Bradley and Griggs,

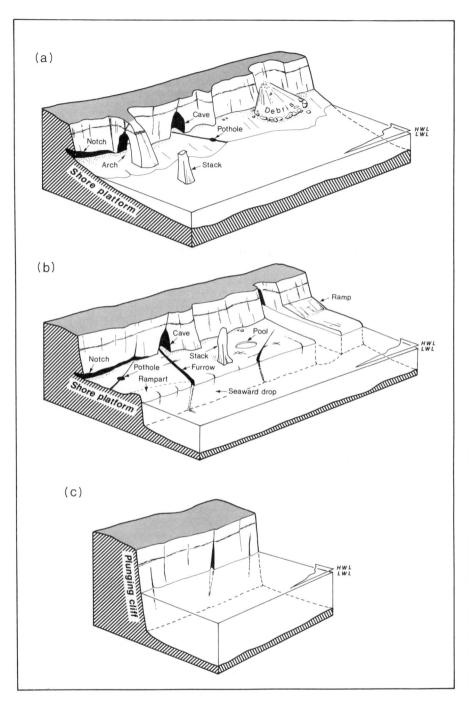

Figure 7.1 Three major morphologies on rocky coasts and characteristic erosional features: (a) sloping shore platform; (b) horizontal shore platform; (c) plunging cliff

1976; Sparks, 1986, p. 243); wave-cut terraces (Von Engeln, 1942, p. 522, Leet and Judson, 1958; p. 311); and wave ramps (Hills, 1972).

There are also various terms applied to describe horizontal or subhorizontal platforms (Figure 7.1b): abrasion or denuded benches (Zenkovich, 1967, p. 136); coastal platforms (Guilcher, 1958, p. 67); low rock terraces or platforms (Emery, 1960, p. 21–22); marine benches (Wentworth, 1938; Cotton, 1963); rock platforms (Bartrum, 1916); shore benches (Stearns, 1935; Wentworth and Hoffmeister, 1939); shore platforms (Bartrum, 1924; Jutson, 1939; Hills, 1949; Edwards, 1951; Cotton, 1960, p. 413, 1963; Bird and Dent, 1966; Healy, 1968a; Wright, 1967; McLean and Davidson, 1968; Sanders, 1968a; Gill, 1972a; Trenhaile, 1972; Takahashi, 1974; Bird, 1976, p. 74; Sunamura, 1978b; Tsujimoto, 1987); storm-wave platforms (Edwards, 1941); storm terraces (Emery, 1960, pp. 21–2); wave-cut benches (Short, 1982c); and wave-cut platforms (Johnson, 1931; Edwards, 1958; Fairbridge, 1968; Short, 1982c).

Some of the nomenclature includes a genetic connotation such as 'wave-cut', 'abrasion', or 'storm-wave'. Because the platform genesis and formative process have not yet been fully elucidated, purely descriptive terms are preferable as suggested by Bird (1976, p. 60), Pethick (1984, p. 203), and Trenhaile (1987, p. 192). The term 'shore platforms' is appropriate, and widely used in the modern literature. It is then necessary to separate the two types of shore platform, sloping and horizontal: 'Type-A' and 'Type-B' will be prefixed to 'shore platforms' to designate the former and the latter, respectively, according to Sunamura (1983).

Three morphological types commonly found on the present-day rocky coast are: Type-A platforms, Type-B platforms, and plunging cliffs. Their longitudinal profiles are schematically illustrated in Figure 7.2.

Recession of coastal cliffs is essential for Type-A or Type-B platform development. Cliff erosion can be determined by the magnitude of the assailing force of waves, F_W, relative to the resisting force of rocks, F_R (Figure 5.2). The latter force is much influenced by weathering and fatigue effects, and also by the biological factors described in Chapter 5. The tide is also important in that (1) water level determines the elevation of wave action and controls the type of waves arriving at the cliff, and (2) the tidal range may influence weathering and biological activities, although the tide itself has no force to erode the cliff.

Full understanding of these two forces, F_W and F_R, and their quantification are of vital importance for furthering platform research on a dynamic basis. Compressive strength, which is a representative strength parameter for F_R (Chapter 5), was first employed in the platform study of Edwards (1941), who discussed the degree of shore platform development in Victoria and Tasmania using measured values of compressive strength, and his study indicated that Type-B platforms develop only on coasts within a certain range of rock

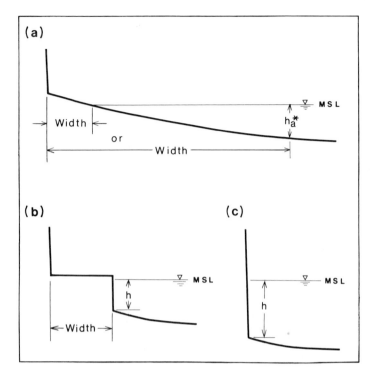

Figure 7.2 Schematic cross-sections of three major morphologies on rocky coasts: (a) Type-A shore platform; (b) Type-B shore platform; (c) plunging cliff

strength. Surprisingly, this work had been overlooked for more than forty years: no platform research with an attempt to quantify the strength of coastal rocks had been made until Tsujimoto (1985) used a compressive-strength parameter in his study of shore platforms on the Pacific Chiba coast of Japan. This work also showed that a certain favourable strength condition exists for Type-B platform development.

The evaluation of wave assailing force F_W in previous platform studies had been much more retarded. This retardation was due to general lack of wave data available for rocky coast areas (Trenhaile, 1987, p. 211) and also to lack of suitable parameters for representing F_W. Tsujimoto (1987) considered that F_W can be represented by wave pressure. He discussed the platform initiation with quantified two forces, F_W and F_R, using data collected from twenty-five Japanese coastal sites, mostly located in a microtidal environment, with various rock strengths, and exposed to different wave climate.

Plunging cliffs, as described before, descend into a considerable depth of water far below low-tide level without any development of shore platforms (Figures 7.1c and 7.2c). Figure 7.3 shows some profiles of typical plunging

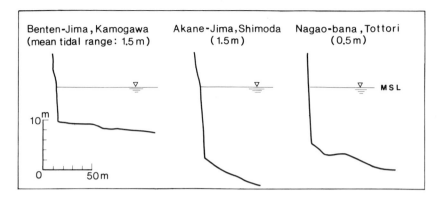

Figure 7.3 Selected profiles of plunging cliffs in Japan. From Tsujimoto (1987), by permission

cliffs on the Japanese coast. Studies of plunging cliffs are not numerous as compared with those of shore platforms due probably to (1) less attractive landforms and (2) much harder access for investigations.

Cotton (1949, 1951, 1952a, 1967, 1968) studied plunging cliffs. Along the coast of the Banks Peninsula near Christchurch, New Zealand, sheer plunging cliffs were described first by Davis (1928, pp. 151–4). These cliffs, composed of hard basaltic rocks, are bordered by a considerable water depth, attaining 20 to 30 fathoms (about 36 to 54 m) on the the southeastern sector of the peninsula, and little accumulation of rock fragments has taken place at the cliff base (Cotton, 1951). Waves approaching the cliffs are reflected due to the large water depth at the cliff base and the precipitous cliff face (Cotton, 1949, 1951, 1952a). Plunging cliffs on the Banks Peninsula coast have resulted from drowning of steep cliffs originally formed by wave action at the time of low sea level (Cotton, 1952a, 1968).

Most plunging cliffs found on hard-rock coasts of various places in the world are of a similar origin, i.e. drowning of pre-existing, wave-formed cliffs due to subsidence of the land and/or sea level rise associated with the Holocene marine transgression. Examples are seen at the following places: Tahiti (lithology: basaltic rocks) in the Pacific Ocean (Davis, 1928 pp. 254–5); Kauai (basaltic rocks) in Hawaii (Hinds, 1930, p. 43); the west coast of Moresby Island (intrusive rocks), British Columbia, Canada (Clague and Bornhold, 1980); St Helena Island (basaltic rocks) in the Altantic Ocean (Daly, 1927); Pointe du Raz (granite) and outer coast of Belle Ile (Precambrian schist), both in Bretagne, France (Guilcher, 1958, p. 62); Cape St Vincent (limestone) at the south western tip of Portugal (Sparks, 1986, pp. 222–3); Wilsons Promontory (granite), Victoria, Australia (Bird, 1976, p. 93); Campbell Island (basaltic rocks) (Cotton, 1951) and Auckland Islands (basaltic rocks) (Cotton, 1967), both located south of New Zealand; Benten-Jima and

Shinyashiki (both made of basalt), Kamogawa, Japan and Akane-Jima Island (andesite), Shimoda, Japan (Tsujimoto, 1987); Nagao-bana (andesite), Tottori, Japan (Toyoshima, 1967); and Oniga-jyo (rhyolite) on the Kii Peninsula coast, Japan (Toyoshima, 1973).

Other kinds of plunging cliffs are associated with (1) faulting, as seen on the west coast of Port Nicholson, Wellington, New Zealand (Cotton, 1952b) and the southwestern coast of Angel de la Guarda Island, Gulf of California (Guilcher, 1958, p. 164; Cotton, 1967); and (2) recent volcanic activity, as on the island of Hawaii (Bird, 1976, p. 93). Plunging cliffs also border on ria or fiord coasts.

Quantitative data on the frontal depth of plunging cliffs are limited: it is 8 to 9 fathoms (about 14 to 16 m) at the headlands in the northern area of the Banks Peninsula (Cotton, 1949, Figure 1); 40 fathoms (76 m) at Auckland Islands (Cotton, 1968); 12 fathoms (22 m) on the west coast of Port Nicholson (Bird, 1976, p. 93); 8 m at Benten-jima, 6 m at Shinyashiki, and 14 m at Akane-jima (Tsujimoto, 1987, Table 5); 10 to 11 m near the tip of Nagao-bana Point (Toyoshima, 1967, Figures 16-4 and 16-5); and 12 to 15 m at Oniga-jyo (Toyoshima, 1973, Figure 3). The available data on the cliff angle are more limited; the approximate angle is 55° at Port Nicholson (Bird, 1976, p. 93), 40° to 50° in the subaerial part and more than 50° in the subaqueous portion at Akane-jima and 50° at Nagao-bana (Tsujimoto, 1987), and more than 45° on the southern shore of the Crimea and on the Murmansk coast of the Barents Sea (Zenkovich, 1967, p. 137).

It has been considered that no significant erosion has occurred on the cliffs now plunging (e.g. Davis, 1928, pp. 151–4; Baulig, 1930; Cotton, 1949, Toyoshima, 1973; Tsujimoto, 1987). Based on his study of Banks Peninsula cliffs, Cotton (1951) ascribed this immunity from wave erosion to (1) generation of reflected waves, (2) lack of accumulation in shallow water of debris that could have been used in abrasion, and (3) high resistance of cliff-forming rocks. Aside from these three favourable conditions for the survival of plunging cliffs, Bird (1976, p. 93) stated that the length of time elapsed since the sea attained its present level is too short.

As to the survival of plunging cliffs, an attempt will be made to give a physical explanation based on wave force at the cliff base. As discussed in Chapter 5, the assailing force of waves consists of two kinds of action: hydraulic and mechanical (Figure 5.2). Let us first consider the vertical distribution of the hydraulic action of waves on plunging cliffs; this can be approximated by the distribution of dynamic pressure along a vertical wall. This pressure pattern indicates that it attains the maximum at or slightly above sea level and decreases abruptly with increasing water depth below sea level or elevation above it irrespective of the types of waves occurring in front of the wall (Figures 2.18, 2.19, and 2.21), although the maximum value differs considerably among the wave types.

Mechanical action of waves seems likely to assume a vertical distribution, if sand and/or gravel residing on a sea floor in front of plunging cliffs are available for abrasive tools. When breaking waves occur just in front of a plunging cliff, they may pick up bottom sediments and cast them against the cliff face at or slightly above sea level, exerting mechanical action on this very limited portion of the cliff. Otherwise, waves cannot exert abrasive action on the cliff even if they succeed in suspending sediment, because they induce only an up-and-down water motion along the face of the cliff without giving any driving force to direct the suspended sediment towards the cliff. This is just like sandpaper without sand. Such waves cannot incise notches at the base of resistant cliffs. If notches are found at the foot of the present-day plunging cliff, then they must be relics of notches created when the relative sea level was lowered.

In spite of these characteristic vertical distributions of the wave assailing force, the overall profile of plunging cliffs is almost straight with no major discontinuities in the vicinity of sea level. This fact clearly indicates that little erosion has taken place on plunging cliffs in the most recent time during which no significant relative sea level change has occurred. It can be easily deduced from the vertical distribution of the wave force that no parallel recession in overall cliff profile has taken place; this strongly contrasts with the parallel recession occurring on sea cliffs with a base located around sea level (Chapter 5). However, some minor modification along joints or faults on plunging cliffs may have occurred to form notches or caves.

Relative subsidence of pre-existing escarpments is a *necessary* condition for the occurrence of most plunging cliffs. However, plunging cliffs would not occur if rapid erosion has taken place during submergence or after the attainment of a stationary sea level. Therefore, another condition should be considered for preservation of plunging cliffs: the strength of the cliff must be high enough to resist the most vigorous wave action which the cliff would have experienced during a relative sea level rise or under a stable sea-level condition. As the water depth in front of the cliff increases with rising sea level, the wave type varies generally from broken to standing waves through the stage of breaking waves, which can exert the largest assailing force on the cliff. If the resisting force of rocks, F_R, is always greater than the assailing force of breaking waves, $(F_W)_b$, then the plunging cliff will occur independently of the relative rate of drowning. If the range of frontal water depth favourable for creating the erosion-causing condition, i.e. $F_R < (F_W)_b$, is wide enough, then the rate of drowning is a controlling factor for the existence of plunging cliffs: if the drowning rate is much lower than the cliff erosion rate, then plunging cliffs will not develop.

Dynamic conditions for the occurrence of plunging cliffs will be quantitatively explained under the assumption that plunging cliffs develop if the resisting force of cliff material, F_R, is greater than the wave assailing force, F_W, under contemporary wave and tidal conditions. Otherwise, shore platforms develop. The boundary between the two distinct types of landforms should be given by

$$F_W = F_R \qquad (7.1)$$

Tsujimoto (1987) assumed that F_W and F_R can be expressed, respectively, in terms of a linear function of wave pressure, p, and of compressive strength including the effect of discontinuities in rocks, S_c^* (Chapter 4), i.e. $F_W = a_1 p$ and $F_R = B^* S_c^*$, where a_1 and B^* are dimensionless constants. Using Eq. (7.1), he obtained

$$p = c_1 S_c^* \qquad (7.2)$$

where $c_1 (= a_1/B^*)$ is an unknown dimensionless constant. This equation indicates that the landform-type demarcation should be made by a straight line with an inclination of 45° on logarithmic graph paper. In order to examine this, Tsujimoto selected from the Japanese coast 25 sites (Figure 7.4) where

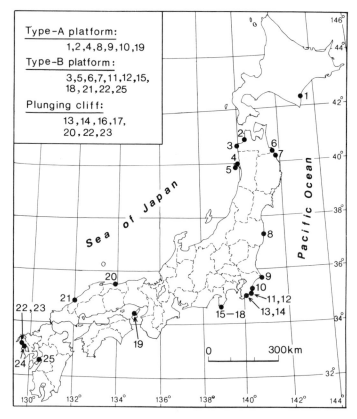

Figure 7.4 Geographical location of coastal sites selected in the study of Tsujimoto (1987)

Shore Platforms

no marked tectonic movement has occurred after sea level had become stable. The sites selected show a wide variety of rock types ranging from poorly consolidated Quaternary deposits (compressive strength: 13 tonnes/m^2) to hard basalt (20 000 tonnes/m^2) and they are exposed to different wave climate, some being located on sheltered coasts and others on open coasts. The sites are in the microtidal environment (mean tidal range: 0.4–1.5 m) except three which are on mesotidal coasts (2.6–4.0 m). Using data of shore-zone morphology and wave climate, and taking account of the type of waves which can produce the maximum wave pressure at each site, Tsujimoto calculated the maximum value of p. In his calculation, Eqs (2.51), (2.55), and (2.58) were employed for three different wave types: standing, breaking, and broken waves, respectively. Quantification for S_c^* was based on the measurements of (1) compressive strength for intact rock specimens and (2) sonic velocities through these specimens in the laboratory and through rock masses in the field [Eq. (4.27)]. Figure 7.5 is a logarithmic expression for p and S_c^*; the open symbol denotes the site where shore platforms develop and the solid symbol indicates the site of plunging cliffs. The two landform types can be well delimitated by the solid line which is given by Eq. (7.2) with $c_1 = 0.081$. On the left-hand side of this line, $F_W > F_R$, which results in erosion, so that shore platforms develop. On the right-hand side, on the other hand, $F_W < F_R$: no erosion takes place, resulting in plunging cliffs.

Another approach will be taken here to demarcate shore platforms and plunging cliffs in an easier way but with less physical exactness than

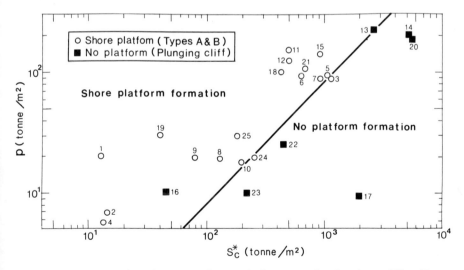

Figure 7.5 Demarcation between shore platforms and plunging cliffs. From Tsujimoto (1987), by permission. The numbers attached to the symbols denote the locations shown in Figure 7.4

Tsujimoto's (1987) approach. It would be reasonable to assume that a representative value of F_W for a coast can be expressed by an F_W-value obtained based on the largest waves occurring in the area, because larger waves play a more significant role in the development of rocky coasts. From analogy with Eq. (5.5a), we have $F_W = a_2 \rho g H_l$, where H_l is the height of the largest waves, ρ is the density of water, g is the acceleration due to gravity, and a_2 is a constant. As to the resisting force of rocks, F_R, Eq. (5.5b) can be applied: $F_R = BS_c$, where S_c is the compressive strength of rocks and B is a constant. Substitution of these two equations into Eq. (7.1) leads to

$$\rho g H_l = c_2 S_c \qquad (7.3)$$

where $c_2 (= a_2/B)$ is a dimensionless constant. Using available data, $\rho g H_l$ is plotted against S_c on logarithmic graph paper (Figure 7.6). The former quantity increases with increasing degree of exposure to wave attack and the latter increases with increasing hardness of coastal rocks. Figure 7.6 shows that the boundary between shore platforms and plunging cliffs can be reasonably described by the solid line, $\rho g H_l = 0.0017 S_c$, although considerable overlapping of data points is seen. It is found that occurrence or non-occurrence of plunging cliffs can be determined by a relative intensity of rock resisting force, F_R, to wave assailing force, F_W, and also that the rocks factor alone can determine the existence of plunging cliffs if the coast is composed of resistant rocks with a compressive strength of more than 3000 tonnes/m^2 (= 300 kg/cm^2 = 4270 lb/in^2). Gill (1972a) noticed a similar relationship, although it is qualitative. Figure 7.6 indicates that even on cliffed coasts made of less resistant rocks, if located in closed or sheltered environments, plunging cliffs can be preserved as long as the condition of 'no erosion', $F_R > F_W$, is fulfilled.

The length of time is an important factor to affect the existence of plunging cliffs, as suggested by Bird (1976, p. 93). The longer a stable sea level lasts, the more a subaerial portion of the plunging cliff will become susceptible to weathering and will gradually reduce the resisting force. However, the plunging cliff will persist as long as $F_R > F_W$. A physical setting favourable for this condition is the occurrence of standing waves for most plunging-cliff coasts because of (1) steep cliff slopes and (2) larger water depth at the cliff base than the breaking depth of waves approaching the shore (e.g. Zenkovich, 1967, p. 137). Standing waves have much smaller assailing force as compared with the other two types of waves, breaking and broken waves (Sunamura, 1975). Even if breaking waves, which possess the largest value of F_W, occur and act on the cliff, plunging cliffs develop when $F_R > F_W$ due to high resistance of rocks. According to Tsujimoto (1987), cliffs at Benten-jima, Shinyashiki, and Nagao-bana, all located on the open coast in Japan, are exposed to breaking waves during storms in which deep-water waves attain 5 to 6 m in height. In spite of the action of breaking waves, these locations exhibit a typical shape of plunging cliffs (Figure 7.3), due to highly resistant

Shore Platforms

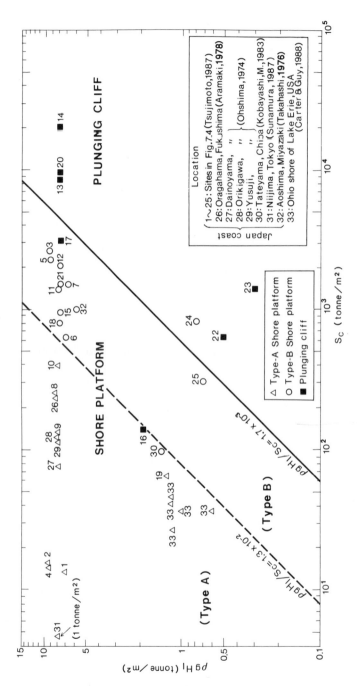

Figure 7.6 Another demarcation between shore platforms and plunging cliffs

cliff-forming rocks which have a compressive strength of 9000–20 000 tonnes/m^2.

As discussed in Chapter 6, the rate of downward erosion of submarine bedrock decreases exponentially as the water depth increases. It is reasonable to assume that little lowering occurs on the sea floor in front of the plunging cliff. Bedrock erosion must have taken place during submergence, the erosion rate having probably been higher at its earlier stages when the water depth was smaller, if the wave-induced assailing force acting on the surface of submarine bedrock exceeded their resisting force. Unfortunately, no quantitative studies of the lowering process on plunging-cliff coasts have been made. It would be worth attempting this kind of research backed by data of measured erosion rates, but there will be difficulties in collecting such data, because the bedrock lowering is so slow.

Investigations on (1) morphologies of plunging cliffs and the fronting sea floor and (2) dating of their initial landforms are inadequate. This is because most geomorphologists' attention has been directed to changing coasts. Plunging cliffs form a unique morphological unit that is fixed in time and space as far as the recent several thousand years are concerned during which sea level remains almost stable. This unit should be taken into account when we study the evolution of rocky coasts from a broader viewpoint that has not been taken in previous geomorphology. For furthering research on the origin and the developmental process of plunging cliffs, cliff morphologies including sea floor, preferably their three-dimensional characteristics, must be scrutinized first. Data of wave climate and rock strength are also indispensable.

DEMARCATION OF PLATFORM TYPES

A marked difference exists between Type-A and Type-B platforms: the former type lacks a seaward drop, whereas the latter type has one (Figure 7.2). Such a geomorphological difference can be attributed to the spatial (cross-shore) difference in the relative magnitude of wave assailing force, F_W, to rock resisting force, F_R. The tide directly controls the magnitude and the elevation of action of F_W and indirectly influences the value of F_R through weathering and biological activities. According to Trenhaile (1987, p. 192) Type-A platforms are most common in macrotidal environments and Type-B in meso- or microtidal regions. Gill (1972a) suggested that Type-A platforms well develop in softer rocks; and Bird and Dent (1966) found that Type-B platforms give place to Type-A platforms at a place where locally derived shingle is available as an abrasive tool for waves. Edwards (1951) and Hills (1972) recognized that Type-B platforms are well developed on headlands, with Type-A in intervening embayments, composed of less resistant material. It has been noted that geological structure, such as joints and bedding planes, strongly reflects platform geometry on some coasts (e.g., Gill, 1972c; Bird, 1976, p. 79; Trenhaile, 1971, 1987, p. 215; Sparks, 1986, Pl. 28).

Let us review Figure 7.6. Platform data plotted in this figure are all obtained from less structural-controlled coasts located mostly in microtidal environments of Japan and some taken from nontidal regions along the Lake Erie shore of the United States. The two types of platforms can possibly be delimited by a dashed line, although there is considerable lack of data of Type-B platforms in the lower area of this diagram. It is clearly shown that Type-B platforms develop within a certain range of rock strength on open or exposed coasts (see the upper part of this diagram).

A more process-oriented approach has been taken by Tsujimoto (1987) to demarcate the two platform types. Based on his own experimental results and Sunamura's (1975) showing that a seaward drop is immune from wave erosion during platform cutting, Tsujimoto assumed that the seaward topographic break characterizing Type-B platforms in the full-scale environment is a relic of the initial profile of steep drowned coasts. If the downward bedrock-erosion concomitant with platform growth is so active as to lower the platform surface eventually leading to disappearance of a seaward drop, then Type-A platforms result. Otherwise, Type-B platforms develop. These assumptions can be rephrased as that Type-A platforms appear when the assailing force of waves to degrade the platform surface, $(F_W)_s$, is greater than the rock resisting force against this, $(F_R)_s$, i.e. $(F_W)_s > (F_R)_s$; whereas the development of Type-B platforms requires $(F_W)_s < (F_R)_s$. Therefore, a limiting condition for the two types of landforms should be given by

$$(F_W)_s = (F_R)_s \qquad (7.4)$$

Tsujimoto (1987) assumed that

$$(F_W)_s = a_3 \tau \qquad (7.5)$$

and

$$(F_R)_s = B_s^* S_s^* \qquad (7.6)$$

where τ is the wave-induced shear force acting on the platform surface, S_s^* is the shear strength including the influence of discontinuities in a rock mass, and a_3 and B_s^* are dimensionless constants. Use of Eqs (7.4), (7.5), and (7.6) yields

$$\tau = c_3 S_s^* \qquad (7.7)$$

where $c_3 (= B_s^*/a_3)$ is a constant to be determined empirically. If these two quantities, τ and S_s^*, are properly selected, then the platform-type demarcation must be made by this equation. A few assumptions were made in his work to obtain $\tau = C_f \rho g H$, where C_f is a friction factor. Applying $C_f = 0.15$, a value obtained from the study of Kohno et al. (1978) of wave-height attenuation on a fringing coral reef, and replacing H with H_b on the basis of an idea that water motion associated with breaking waves exerts maximum shear force, Tsujimoto (1987) finally obtained

$$\tau = 0.15 \rho g H_b \qquad (7.8)$$

where H_b is the height of breaking waves just in front of the cliff. Evaluation of the shear strength S_s^* was made using the following relationship:

$$S_s^* = (V_{pf}/V_{pl})S_s \qquad (7.9)$$

where V_{pf}/V_{pl} is the discontinuity index (Chapter 4), and S_s is the shear strength of intact rock specimens. This equation is analogous to Eq. (4.27). He calculated the values of τ and S_s^* through Eqs (7.8) and (7.9), respectively, using data collected from eighteen sites where shore platforms develop (Figure 7.4), and plotted the relationship between τ and S_s^* (Figure 7.7). This figure shows that two types of platforms are clearly demarcated by the solid line, which is Eq. (7.7) with $c_3 = 0.005$. On the left of this line, $(F_W)_s$ is in excess of $(F_R)_s$ so that Type-A platforms are formed. On the right, lowering of the platform surface does not progress due to $(F_W)_s < (F_R)_s$, resulting in the formation of Type-B platforms.

Other important factors influencing platform geometry are tidal range and geological structure; quantification is easy for the former factor, but not for the latter. Even the effect of tidal range on platform morphologies has not been fully examined on a dynamical basis, much less the effect of geological structure.

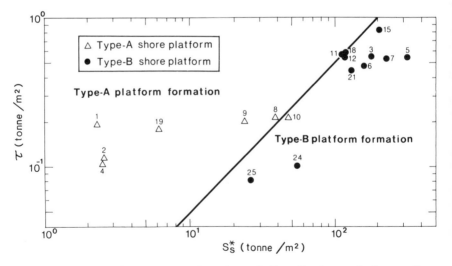

Figure 7.7 Demarcation between Type-A and Type-B shore platforms. From Tsujimoto (1987), by permission. The numbers attached to the symbols denote the locations shown in Figure 7.4

TYPE-A PLATFORMS

Many models for the development of this type of platforms have been diagrammatically illustrated in existing literature (e.g. Davis, 1909; Johnson, 1919, p. 213; Barrell, 1920; Challinor, 1949; Keunen, 1960, p. 305; Twenhofel, 1950, p. 232; Garrels, 1951, p. 103; Bradley, 1958; Machatschek, 1959, p. 165; Zenkovich, 1967, p. 153; Louis, 1968, p. 321; Sparks, 1986, p. 244). These models are slightly different from each other, but similar in that seaward sloping platforms grow with the recession of coastal cliffs.

The first mathematical description on Type-A platform evolution under the stable sea-level condition was made by Flemming (1965). Taking account of wave energy expended in erosion of the cliff and its adjacent sea floor, and assuming that (1) a cliff is made of uniform rocks with no subaerial weathering and (2) no beach develops, he fabricated a model; this indicates that the rate of cliff recession decreases with time, but does not reach zero when time tends to infinity. This result is not in accord with the concept of equilibrium.

Under the same assumptions as Flemming's, a physical model was constructed using Eq. (5.6) by Sunamura (1978c). On this type of platforms, all waves break before approaching the cliff and a surf zone is formed, due to (1) very shallow water at the cliff foot, and (2) the seaward sloping bottom. Using a relation for the surf-zone wave-height attenuation, which is similar to Eq. (2.43), he transformed Eq. (5.6) to obtain the long-term cliff recession distance, x, which is described as

$$x = (K_*/\beta_*)(1 - e^{-\beta_* \varkappa t}) \qquad (7.10)$$

where $K_* = \Gamma - \beta_* W_0 + \ln(\rho g H_b/S_c)$, $\beta_* = \alpha_* \sqrt{h_b}/(\sqrt{g}) h_a L_b^2$, t is the time, $\varkappa$ is a constant, H_b is the breaker height, h_b is the breaking depth, L_b is the wavelength at the breaking point, α_* is a wave-height attenuation coefficient, h_a is the wave base (Chapter 6), and W_0 is the initial platform width which is defined as the horizontal distance from the position providing h_a to the cliff base at $t = 0$ (Figure 7.8). In this model, (1) a platform with uniform gradient is formed, and (2) both H_b and h_b are assumed to be independent of time and the gradient of sea floor. The rate of cliff recession, i.e. platform development, dx/dt, can be obtained from Eq. (7.10):

$$dx/dt = K_* \varkappa e^{-\beta_* \varkappa t} \qquad (7.11)$$

This equation indicates that $dx/dt \to 0$ as $t \to \infty$. While waves at the cliff foot have sufficient force to cause cliff erosion, a platform continues to grow with a rapid rate at the early stage. In time, however, the growth rate decreases as the platform becomes wider due to the interaction between waves and platform width. Finally, platform development halts when the wave assailing force becomes equal to the resisting force of the cliff.

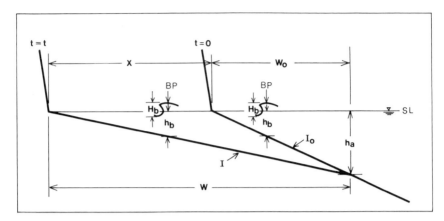

Figure 7.8 Definition sketch for the model of Type-A platform development

The platform width, W, and the platform gradient, I, both defined in Figure 7.8, are given by $W = x + W_0$ and $I = h_a/W$; and they have the limit W_e and I_e, respectively, as time approaches infinity:

$$W_e = \lim_{t \to \infty} W = \frac{(\sqrt{g})h_a L_b^2 [\Gamma + \ln(\rho g H_b/S_c)]}{\alpha_* \sqrt{h_b}} \tag{7.12}$$

and

$$I_e = \lim_{t \to \infty} I = \frac{\alpha_* \sqrt{h_b}}{(\sqrt{g}) L_b^2 [\Gamma + \ln(\rho g H_b/S_c)]} \tag{7.13}$$

These two equations indicate that the ultimate platform, which is not influenced by the initial topographic conditions, becomes wider and flatter when the coastal rock is weaker, if the other factors remain constant.

No effect of weathering was included in this modelling. The temporal reduction of the rock resisting force (here represented by S_c) due to weathering should be quantified for the construction of more realistic models. In addition to this, the amount of (1) debris supplied from the cliff by subaerial processes and (2) sediment transported laterally by longshore currents should be taken into account. These materials can accelerate or hinder erosion of the sea floor (Chapter 6), so that they much influence the platform geometry.

Actual platform profiles cannot be represented by a straight line as depicted in Figure 7.8. Figure 7.9 shows five profiles on actively eroding coasts: three from Japan and two from the United States, all facing the Pacific Ocean. Figure 7.10 illustrates some examples taken from receding shores in the closed environment of the Great Lakes and the western Baltic Sea. Long-term average cliff recession rate, denoted by $\bar{R}$, is attached to each profile. It is seen that all these profiles have a distinct concavity in the inshore zone.

Shore Platforms

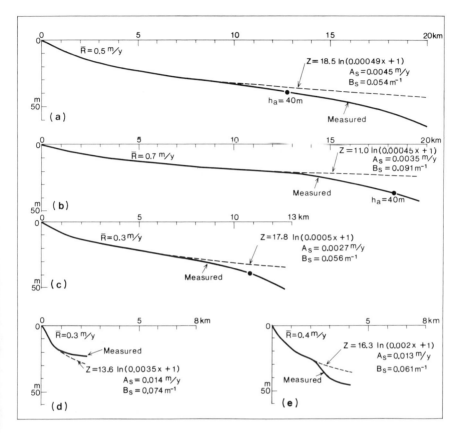

Figure 7.9 Selected profiles of Type-A platforms on open coasts: (a) Omika, Fukushima, (b) Byobugaura, Chiba and (c) Akabane, Aichi, from Japan (Sunamura, 1972); (d) Montara, California and (e) Santa Cruz, California, from the United States (Sorensen, 1968). Profile deeper than 30 m in Santa Cruz area is based on hydrographic chart (No. 18685) published by NOAA, USDC. The dotted line shows calculated profiles using Eq. (7.14)

Dean (1977) presented a simple relationship to describe the concave profile of sandy beaches: $z = A_0 x^{2/3}$, where z is the vertical axis taken positively downwards from the origin positioned at the water edge, x is the horizontal axis directed offshore, A_0 is a coefficient depending on grain size of the sediment. Studies by Nairn et al. (1986) and Kamphuis (1987) have attempted to apply Dean's relationship to represent the inshore sea-floor profile cut in glacial till on the Lake Erie shores. Application of such a relationship as expressed in terms of $z \propto x^n$ ($0 < n < 1$) to actual platform profiles seems

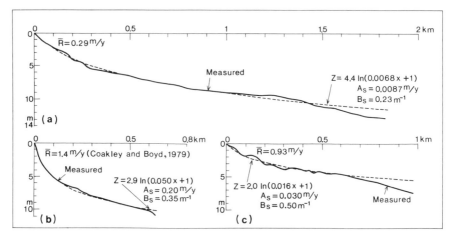

Figure 7.10 Selected profiles of Type-A platforms in closed waters: (a) from Bokniseck, Kieler Bucht, Western Baltic (Healy and Wefer, 1980); (b) from Grimsby, Lake Ontario (Coakley et al., 1988); (c) from east of Port Bruce, Lake Erie (Kamphuis, 1987). The dotted line shows calculated profile using Eq. (7.14)

inappropriate, because the gradient of sea floor at the water edge takes on an unrealistic value, i.e. $dz/dx = \infty$ at $x = 0$. A precise profile representation is required at the water edge which, subjected not only to vertical but also to horizontal erosion, moves landwards resulting in the platform development.

On rapidly retreating coasts, the rate of downward erosion of sea floor can be expressed by Eq. (6.9). Integration of this equation with $z = 0$ at $t = 0$, and transformation of the integrated equation assuming a linear relationship between cliff recession distance, x, and time, t, i.e. $x = \bar{R}t$, yields the following equation:

$$z = \frac{1}{B_s} \ln\left(\frac{A_s B_s}{\bar{R}} x + 1\right) \quad (7.14)$$

where A_s is the lowering rate of bedrock at the water edge, and B_s is a constant. This equation indicates that (1) the sea floor profile is concave upwards, and (2) the gradient at the water edge is given by a definite value:

$$\left(\frac{dz}{dx}\right)_{x=0} = \frac{A_s}{\bar{R}} \quad (7.15)$$

This equation shows that the bottom slope at the shoreline can be expressed by the ratio of the rate of vertical topographic change to that of horizontal one (Chapter 6).

Figures 7.9 and 7.10 show that the actual profile and the calculated profile using Eq. (7.14) with appropriate values for A_s and B_s are in good agreement for the inshore zone. As is clearly seen in Figure 7.9, the difference between the two profiles becomes more notable with increasing water depth. For three Japanese sites, the wave base, h_a, was estimated at 40 m (Sunamura, 1972). It is natural for the calculated profile not to be extended up to the wave base, because Eq. (7.14) is derived from Eq. (6.9) which has been presented to describe the lowering rate of bedrock shallower than the surf base, h_a^*. According to the traditional view that the surf base corresponds to breaking depth of storm waves (Chapter 6), wave climate for these three sites (Sunamura, 1972) allows us to estimate $h_a^* \approx 10$ m. A similar value (30 ft) has been considered as h_a^* for both Montara and Santa Cruz sites (Sorensen, 1968). The actual nearshore concave profiles in Figure 7.9 can be divided into two, steeper inshore and gentler offshore segments, although this segmentation is not always strict. The average water depth of the gentler segment is much larger than the value of $h_a^* (\approx 10$ m). This difference, which has been already pointed out by Sorensen (1968), suggests that lowering of sea floor is extending far below the surf base. The surf base is thus not the ultimate water depth for bedrock erosion (Chapter 6). Figure 7.11 is a schematic diagram showing platform development, illustrating the possibility that a wider and flatter erosion surface forms with time at a place considerably below the surf base.

Even the family of concave profiles (Figures 7.9 and 7.10) shows the presence of various platform shapes. This diversity reflects (1) wave climate, (2) lithology, especially rock structures, (3) availability of clastic sediment on the sea floor, and (4) the history of Holocene (relative) sea level change. Investigations on these points should be furthered to elucidate platform concavity on a more dynamical basis.

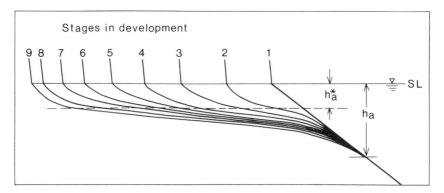

Figure 7.11 Schematic diagram showing the development of Type-A platform under stationary sea level conditions. After Sunamura (1972)

A strong contrast to the concave profiles is the existence of convex platforms as found on a coast cut in flysch beds near Cape Doob in the Black Sea (Zenkovich, 1967, Figure 53A). The convex profile is characterized by the minimum gradient occurring at the water edge and increasing gradient with increasing offshore distance. Zenkovich (1967, p. 153) presented a model in which a convex platform grows, keeping up with the recession of a sea cliff, and convexity reduces with time; Zenkovich stated that the shape of platforms is convex when produced purely by abrasion.

Some studies (e.g. King, 1963; Flemming, 1965; Zenkovich, 1967, p. 163; Bradley and Griggs, 1976) have discussed the minimum value of the platform gradient without giving a clear definition for it. Because the actual platform profile cannot be represented by a straight line, quantitative discussions without definition seem less fruitful. The same is mentioned of discussions on the width of platforms without defining the seaward limit. Platform width can be defined either as (1) the horizontal distance from mean (or low) water shoreline to the cliff base, or as (2) the horizontal distance from the position providing h_a^* (surf base) to the foot of a cliff (Figure 7.2a). The definition should be chosen according to the problem to be tackled.

A close investigation, on a short time-scale, of the platform near the shoreline may indicate that morphological changes are not as simple as depicted in Figure 7.11. No accurate and consecutive measurements of platform changes were made until Robinson (1977a,c) applied the micro-erosion meter technique (Chapter 6) to the northeastern Yorkshire coast, where the mean tidal range is about 4 m and shore platforms are cut into Lower Jurrassic rocks (Lias shale). He found three modes of development of intertidal platforms in the area with little lithological influence on platform morphologies. These developmental modes are schematically illustrated in Figure 7.12.

Class-1 profile (Stage 1 in Figure 7.12a) is characterized by a junction of a vertical cliff and a slightly seaward-inclined platform, described as the 'plane' in this figure, is located at around 1 m above Ordnance Datum (O.D.); the lowering rate of the platform is on the order of 0.01 to 0.2 cm/year except for the cliff base where it attains 0.1 to 1 cm/year. As the cliff recedes, the platform grows with a slight change in its gradient or in the altitude of the cliff base. The profile change may be produced by the size or amount of debris supplied either from the adjacent coast by longshore currents or from the cliff by mass movement activities. If a sandy or pebbly beach develops with sediment movable by waves, Class-2 profile (Stages 4–6 in Figure 7.12a) results in, having a concave ramp with a minimum slope of 2.5° and showing the altitude of a cliff/ramp junction increasing to higher than 4 m above O.D. Erosion rates of the ramp are of the order of 0.1 to 3 cm/year. When the influx of sediment is reduced, the size of the beach on the ramp of Class-2 platform (Stage 1 in Figure 7.12b) is also diminished. The ramp is lowered with a minor recession of the cliff, which leads to the development of Class-1

Shore Platforms

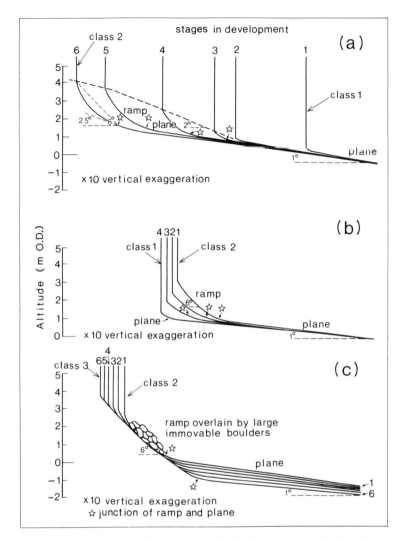

Figure 7.12 Development of three classes of platform profile: (a) Development of class 2 platform from class 1; (b) Development of class 1 platform from class 2; (c) Development of class 3 platform from class 2. From Robinson (1977a), *Marine Geol.*, by permission of Elsevier Science Publishers

platform (Stage 4 in Figure 7.12b). If much debris composed of large immovable boulders accumulates on the ramp of Class-2 platform (Stage 1 in Figure 7.12c), erosion rates on the ramp are substantially reduced. Cutting back of the cliff and lowering of the platform produce Class-3 platform which has a longer ramp.

Robinson (1977a) suggested that changes in the size or amount of the debris may lead to changes in the platform profile. The influence of the amount of beach sediment on spatial variation in platform elevation is noted in the work of Wood (1968) who examined the platform morphology on a rapidly eroding Chalk coast near Minnis Bay located on the Isle of Thanet, England, where mean tidal range is about 4 m. Figure 7.13 shows that the elevation of platforms located in the bays is greater than that off headlands. A similar relationship is found in the adjacent area, near Grenham Bay, where the difference in altitude attains as much as 3.6 m between bay-head and headland, only 25 m apart (Wright, 1970), but a reverse relationship is present at Epple Bay, just east of Grenham Bay, and in some other locations on the Isle of Thanet (So, 1965). It should be noted that, even under stationary sea level conditions, platform elevation is not constant, but varies greatly along the coast.

According to Wood (1968), platforms in the bays near Minnis Bay are covered by beach sand, whereas platforms around the headlands consist of bare bedrock. As he suggested, waves exert a force intensified by the abrasive

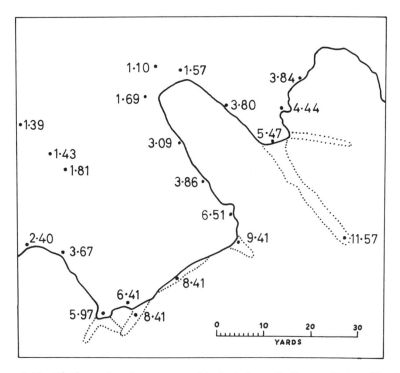

Figure 7.13 Platform elevation, measured in feet above Ordinance Datum (Newlyn), east of Minnis Bay, Kent, England. From Wood (1968), *Z. Geomorph.*, by permission of Gebrüder Borntraeger. The dotted lines show caves below the cliffs, and heights of cave floors are given

Shore Platforms

action of sand on a bay-head cliff at a higher level when they hit the cliff after running up on the beach.

For continual recession of cliffs cut into soft glacial drift in the nontidal environment such as the Great Lakes, Philpott (1984) considered that downcutting of the bedrock immediately in front of the cliff must occur prior to the cliff recession: increasing water depth in front of the cliff will reduce the dissipation of wave energy arriving at the cliff base, which facilitates cliff erosion. This erosional mode, which is named the 'vertical-erosion antecedent' model, is schematically drawn in Figure 7.14a, assuming that no cliff retreat occurs during bedrock lowering. In this view, the altitude of a cliff/platform junction fluctuates with time. Focusing on this mode of downcutting on which the abrasive action of clastic sediment is assumed to play an important role, Kamphuis (1987) constructed a model which relates cliff erosion to wave power (i.e. wave energy flux—Chapter 2), and applied it to the north shore of Lake Erie. His result is plotted as the curve of Kamphuis (1987) in Figure 5.16.

Marked lowering of the platform surface at the cliff base was demonstrated in a wave-flume experiment by Nairn (1986) who used a clay–silt mixture to simulate a cohesive coast composed of glacial till (Figure 7.15a). In this experiment, broken random waves acted on the cliff. Nairn attributed this lowering to the action of reflected waves from the cliff. It should be noted that this cliff

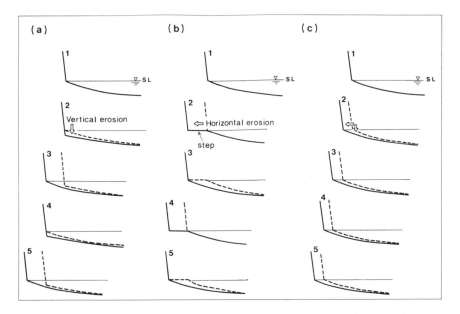

Figure 7.14 Three models for illustrating erosional modes at the cliff base of Type-A platforms: (a) vertical-erosion antecedent model; (b) horizontal-erosion antecedent model; (c) simultaneous-erosion model

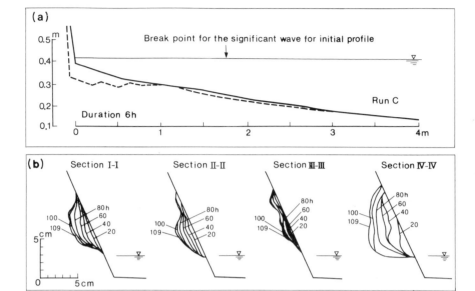

Figure 7.15 Profile changes observed in laboratory experiments: (a) from Nairn (1986), by permission of National Research Council Canada, and (b) from Sunamura (1973)

material is soluble in water, the solubility leading to a gradual reduction in strength of the submerged portion. The strength reduction must be involved in the lowering observed in this test. Reduction in sea-floor strength through fatigue (Davidson-Arnott, 1986a,b) or chemical reaction may occur on the actual coast composed of clayey glacial drift. Quantitative research is needed concerning strength reduction.

Figure 7.15b exemplifies a result of cliff-erosion experiments using a wave basin (Sunamura, 1973). A sand/cement mixture was used as the cliff material. The strength of this material does not vary with time (Chapter 5). The model cliff, 3 m in alongshore length, was exposed to the action of obliquely incident, broken waves. Wave reflection from the cliff occurred, as in Nairn's experiment. Cliff profile changes at four sections, 50 cm apart along the cliff, clearly show that (1) erosion is limited above a certain elevation and (2) no sea floor lowering takes place (Figure 7.15b). These results strongly suggest the presence of a different mode of the platform development from the 'vertical-erosion antecedent' model shown in Figure 7.14a. Storm waves, usually accompanying a rise of water level (due to wave set-up and/or storm surge) which allows waves to reach the cliff without losing much energy, may cut back the cliff leaving a horizontal step at its foot. This mode is called the 'horizontal-erosion antecedent' model, which is plotted in Figure 7.14b. The cutting back of the seaward edge of the step and/or the lowering of the step

Shore Platforms 163

surface follow due to the abrasive action associated with the swash–backwash process, or the hydraulic action represented by plucking if the rock is densely jointed, or both. The cliff may recede during the down-cutting of the step. Recurrence of an ephemeral step morphology characterizes this model. The height of the step depends on location. The Byobugaura coast (mean tidal range: 1.4 m) has a receding cliff (9 km long) composed of Pliocene mudstone layers with nearly horizontal bedding showed, an approximate recession rate of 0.7 m/year (Sunamura, 1973). Along this coast, a typical Type-A shore platform develops with a cliff–platform junction located at about LWL except for some places where narrow steps (a few metres wide) are found at altitudes ranging between MSL and HWL. These steps are eroded away gradually, and new steps form at different locations as the cliff recedes. Some of these step morphologies are much controlled by lithological structures, subhorizontal bedding planes in the coastal rock.

There is another mode, which is illustrated in Figure 7.14c. This shows that the platform develops with no time-lag between the cliff recession and the sea-floor lowering: the 'simultaneous erosion' model. The overall profile only shifts landwards with time, and the elevation of a cliff–platform junction remains unvaried.

Wright (1970) reported from the south coast of England that alongshore variations in the altitude of a cliff–platform junction generally accord with those in tidal range. This altitude approximately coincides with MHWS with a considerable local variation being exemplified by the coast near Grenham Bay as described before. On Russian tidal coasts also the foot of the cliff is normally located at high-water level, and along nontidal sea shores it is usually above still water level, although the occurrence of a cliff–platform junction at various levels is suggested (Zenkovich, 1967, pp. 154–5). Because the height of this junction varies in time and in space depending on many factors shown in Figure 6.1 plus the tidal factor, its exact prediction is not easy in the present state of our knowledge. Quantitative studies based on measurements, as performed by Robinson (1977a,c), are necessary for a deeper understanding of contemporary processes of platform growth. Accumulation of data obtained from such work will enable us to gain an insight into Type-A platform evolution.

TYPE-B PLATFORMS

Shore platforms of this type have a characteristic shape with a marked drop at their seaward edge (Figure 7.2b), and are found on many rocky coasts around the world, especially in the Australasian region: New South Wales, Australia (Jutson, 1939; Bird and Dent, 1966; Gill, 1972a), Tasmania (Edwards, 1941; Sanders, 1968a, 1970), Victoria (Edwards, 1941; Hills, 1949, 1971; Gill, 1972a) Western Australia (Edwards, 1958); Québec, Canada

(Trenhaile, 1978); Japan (Toyoshima, 1967; Takahashi, 1974; Imanaga, 1975; Sunamura, 1978b; Tsujimoto, 1987); New Zealand (Bartrum, 1924, 1926, 1935; Johnson, 1931; Gill, 1950; Cotton, 1963; Healy, 1968a; Wright, 1967; McLean and Davidson, 1968; Kirk, 1977); Portugal (Guilcher, 1957); North Devon, UK (Hills, 1972), Isle of Man (Phillips, 1970); California, USA (Russell, 1970; Bradley and Griggs, 1976), Oregon (Russell, 1970; Johannessen *et al.*, 1982) and Hawaii (Wentworth and Palmer, 1925; Johnson, 1931; Wentworth, 1938). Many studies have been performed since the mid-nineteenth century with an attempt to elucidate formative processes including the genesis or development of Type-B platforms.

Dana (1849), an American geologist visiting New Zealand, first described this type of platform, which is well developed at the tip of spurs and around small islands in a sheltered area of the Bay of Islands, North Island. One of the islands fringing a typical platform of this type has been called 'The Old Hat' because of its shape. The elevation of the platforms in this area is just below HWL. Presenting a sketch of 'The Old Hat' island together with a profile of Type-B platform developing on a coast near Port Jackson, Sydney, Australia, Dana (1875, p. 664) ascribed the formation of these platforms to the action of waves exerted at 'a level of greatest wear' located a little above 'half tide'.

As to the origin of shore platforms along the protected waters in New Zealand, which were termed the 'Old Hat type' by Bartrum (1926), the function of waves was considered only to wash away the weathered debris and the importance of subaerial weathering was emphasized which proceeds down to the level of permanent saturation in the rock, a little below HWL (Bartrum, 1916, 1926, 1935). Thus, the term 'Old Hat' has a genetic connotation which refers to platforms created at the definite saturation level controlled by prior weathering (Trenhaile, 1980). Fairbridge (1952, 1968, p. 862) considered that this weathering-controlled level should be close to LWL, so that Old Hat platforms must have been produced when sea level was higher than the present. Based on the fact that Old Hat platforms are wider on more exposed sides of islands, and yet, on an island at Russell, Bay of Islands, notches develop at the base of the landward cliff of the platform on the side of the Pacific Ocean, Trenhaile (1980) indicated the crucial importance of wave action in platform formation.

The study of platforms by Tricart (1959) on exposed tropical coasts of Brazil indicated that salt-spray weathering is the most effective process in producing Type-B platforms. A similar explanation has been given to account for platforms on the Oregon coast where they are at or above HWL (Johannessen *et al.*, 1982). Incidentally, Ongley (1940) reported from Castlepoint, New Zealand, facing the Pacific, the presence of platforms with an origin ascribed to this type of weathering, although they are located high, between 17 m and 24 m above sea level.

Contrary to this origin, many studies of platforms on exposed sectors of the New Zealand coast (Bartrum, 1924, 1935 ; Cotton, 1963), Australia (Jutson, 1939 ; Edwards, 1941, 1951; Hills, 1949) and Japan (Toyoshima, 1956, Sunamura, 1978b) have ascribed the platform origin to the action of waves. Johnson (1931, 1938) also supported this wave-cut origin, which is essentially the same as the precursory view of Dana (1875, p. 664). The origin of some platforms on open coasts has been attributed to the removal of weathered rocks by wave action (e.g. Edwards, 1958; Mii, 1962; Gill, 1967, Russell, 1971; Bradley and Griggs, 1976). In this case, weathering is a necessary condition for platform formation.

Most of these previous studies have sought to infer the processes responsible for platform genesis based on observations of (1) the subaerial configuration of platforms, (2) the hardness of rocks forming them, (3) the degree of weathering on these rocks, (4) the absence or presence of clastic sediments on the platform surface, and (5) wave and tidal behaviour. This traditional research approach seems to have difficulty concerning the origin of Type-B platforms: At what level, by what kind of waves, and how fast are platforms originated at a place with given topographical, geological, tidal and wave conditions? Holocene sea level fluctuations plus local crustal movements have made the solution of this problem very difficult.

The platform cutting takes place when the assailing force of incoming waves is greater than the resisting force of cliff-forming rocks, irrespective of whether they are heavily weathered or not. Because waves and rocks are crucial factors in platform origination, genetic studies require a deeper understanding of both factors, preferably on a physical basis. Such investigations should be made with these factors appropriately quantified. Quantification of the rocks factor including the effect of weathering is more difficult than that of the waves factor. Even for the latter factor, however, few attempts have been made in the existing platform studies to evaluate the assailing force, and besides, nearshore wave dynamics has not fully been taken into consideration.

The behaviour and intensity of waves directly acting on the cliff are greatly controlled by the nearshore sea floor topography as described in Chapter 2. Bartrum (1926) pointed out the importance of offshore morphology in wave-based platform studies, and almost half a century passed before Toyoshima (1967) conducted a nearshore survey with the aid of scuba divers, on the San-In coast of Japan, in connection with his platform research, although he did not refer to waves. Sanders (1968a) attempted to make echo soundings in front of some platforms on the Tasmanian coast of Australia. Reffell (1978) also applied an echo sounder to obtain nearshore data north of Sydney, Australia. In Japan, Kayanne and Yoshikawa (1986) and Tsujimoto (1987) obtained nearshore topographical data through scuba diving. Some selected profiles from these studies are illustrated in Figure 7.16. Acquisition of such

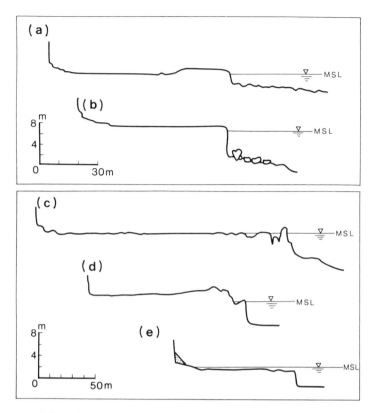

Figure 7.16 Selected profiles of Type-B platforms: (a) North Newport Head, Sydney (Spring tidal range 1.6 m) and (b) Turimetta Head, Sydney (1.6 m) from Australia. After Reffell (1978), by permission; (c) Kominato, Chiba (mean tidal range 1.5 m), (d) Odosezaki, Aomori (0.4 m) and (e) Rikuchu-Noda, Iwate (1.4 m) from Japan. After Tsujimoto (1987), by permission

topographical information is necessary for further research on the origin of platforms from the viewpoint of wave dynamics.

Seaward drops and platform initiation

Figure 7.16 demonstrates the typical morphology of Type-B platforms, which is characterized by the presence of a precipitous slope at the seaward edge of the platform, although all the platforms of this type do not possess such a feature because of geological structure, as seen on the Gaspé and the western Newfoundland coasts of Canada (Trenhaile, 1987, p. 215). This marked drop is frequently called the 'low-tide cliff' (e.g. Edwards, 1941; Hills, 1971; Pethick, 1984, pp. 201–2; Trenhaile, 1987, pp. 215–16), 'low-tidal sea cliff' (Healy, 1968a), or 'low-water cliff' (Kirk, 1977). This terminology seems inappropriate because of its connotation that this type of platforms always develop on tidal coasts, and yet their height should be between HWL and LWL.

Discussions on whether the seaward drop recedes under present sea level conditions, have not been made on the basis of wave dynamics. Many researchers have considered that the recession of seaward scarps may occur, its speed being dependent upon lithological, wave, and tidal factors (e.g. Bartrum, 1926, 1935; Jutson, 1939; Edwards, 1941, 1951; Hills, 1949; Cotton, 1960, p. 416, 1963; Gill, 1972a; Bird 1976, p. 85; Trenhaile and Layzell, 1981; Trenhaile 1987, p. 211). However, explanations on the recession mode are few. Gill (1972a) stated that the seaward drop of aeolianite platforms at Warrnambool, Victoria, Australia, is undercut mainly by solution, and the overhanging part is easily broken by the uplift force of waves, in blocks fractured along the structural weakness. The presence of such fractured blocks in front of the platform is evidence of the recession of seaward cliffs.

It is commonly found that the face of seaward drop remains covered with rich marine flora and/or fauna even immediately after the attack of severe storm waves; this indicates that no erosion takes place there. Results of a recent laboratory experiment by Sunamura (1991) also showed that the submerged part of a homogeneous cliff made of unbedded, unjointed, and insoluble material (i.e. a mixture of Portland cement and fine sand—see Chapter 5) is free of erosion under a stationary water level condition. In this experiment the cliff was subjected to the action of breaking waves, because they are considered to be most responsible for the formation of Type-B platforms in the field (Sunamura, 1978b), and breaking waves occurring just in front of cliffs are found to be most effective in platform cutting in the laboratory (Sunamura, 1975). Figure 7.17 is the result which shows the development of laboratory wave-cut notches. Such notches indicate that platform formation would occur in actual field situations, because the overhanging part of the notches would collapse due to subaerial processes and the effect of discontinuities in rocks. As clearly illustrated in this figure, no erosion occurs below a certain level along the face of the cliff: below this level it remains unchanged. A similar result has been previously noted from another experiment (Sunamura, 1975).

As described in Chapter 2, the dynamic pressure of breaking waves acting on the vertical wall decreases exponentially with increasing water depth as expressed by Eq. (2.56). Assuming that the assailing force of breaking waves is linearly related to the intensity of this dynamic pressure, the wave assailing force also has an exponential downward decrease such as described by this equation:

$$F_W(z) = F_W e^{a_f(z/h)} \tag{7.16}$$

where $F_W(z)$ is the assailing force at a water depth of z (negative value) measured from still water level (SWL), F_W is the assailing force at SWL, h is the

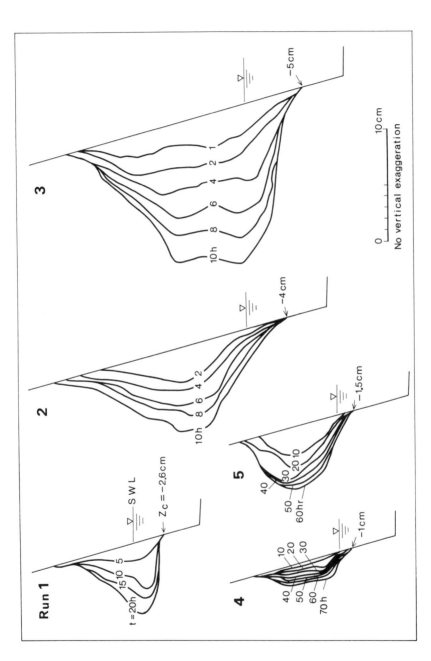

Figure 7.17 Profile changes of laboratory cliffs caused by breaking waves. From Sunamura (1991). With permission from *J. Geol.*, copyright © 1991 the University of Chicago. Experimental conditions are: $S_c = 110$ g/cm^2, $H_b = 4$ cm, and $h = 6.0$ cm for Run 1; $S_c = 110$ g/cm^2, $H_b = 6$ cm, and $h = 7.0$ cm for Run 2; $S_c = 110$ g/cm^2, $H_b = 10$ cm, and $h = 9.0$ cm for Run 3; $S_c = 540$ g/cm^2, $H_b = 5$ cm, and $h = 5.0$ cm for Run 4; and $S_c = 540$ g/cm^2, $H_b = 7$ cm, and $h = 6.5$ cm for Run 5

Shore Platforms

water depth at the base of the cliff, and a_f is a reduction coefficient of F_W. On the other hand, the strength of cliff is space-independent; thus the resisting force of the cliff, $F_R(z)$, is written as

$$F_R(z) = F_R = \text{constant} \tag{7.17}$$

Figure 7.18 is a schematic diagram showing the vertical distribution of these two forces. No erosion occurs in the zone where $F_W(z) < F_R(z)$. This is the reason why the lower part of the cliff is immune from erosion. Such immunity occurs even when broken waves act on the cliff (Sunamura, 1975) due to abrupt reduction in their assailing force similar to the case of breaking waves.

It is reasonable, therefore, to consider that the seaward drop of Type-B platforms in the field does not recede under present sea level, as long as the platforms are cut in insoluble rocks without significant lithological structures. Actually, Cotton (1963) has already suggested that the seaward drop of platforms of Old Hat type and those developed in sheltered regions in New Zealand is a remnant of the initial profile of a steep drowned coast. In his study of shore platforms on a Pacific coast of New Zealand, Gill (1950) also emphasized no apparent retreat of seaward scarps.

In some places, the seaward drop of insoluble-rock platforms is undercut in the form of a subtidal notch at its base. According to observations through laboratory experiments, no notch-forming process takes place even if sand is placed at the base of a seaward cliff as an abrasive tool. This is due to the buffer action of water to protect the cliff face from erosion. Subtidal notches in the real world can be ascribed to ancient undercutting when sea level was

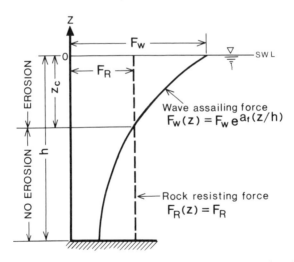

Figure 7.18 Vertical distributions of assailing force of waves and resisting force of rocks. After Sunamura (1991). With permission from *J. Geol.*, copyright © 1991 the University of Chicago

lower than the present. Fairbridge (1952) has considered that a subtidal deep notch at the seaward edge of limestone platforms is also the product of a former lower sea level.

An experimental result by Sanders (1968b) using a mixture of plaster and sand as a cliff material (Figure 7.19) presents a striking contrast to the result of the experiment using a mixture of cement and sand (cf. Figure 7.17). In

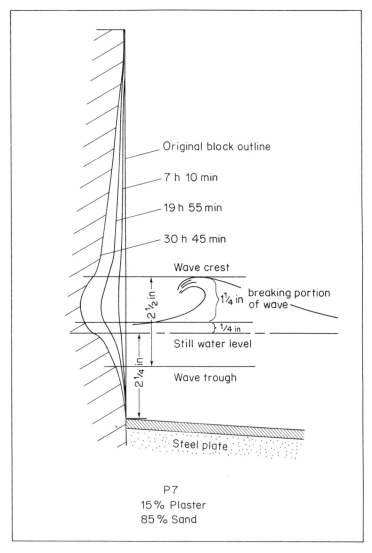

Figure 7.19 Profile change of laboratory cliff made of a plaster–sand mixture. From Sanders (1968b), by permission of Royal Society of Tasmania

his experiment, a model cliff was exposed to the action of 0.65-second breaking waves with a height of 6.4 cm. The cliff material using plaster is soluble in water, so that the cliff strength varies with time and space. The result in Figure 7.19 showed that (1) erosion extended to the base of the cliff and the cliff face below SWL receded with time reducing its slope angle, a hinge point being placed at the base of the cliff; and (2) the eroded area increased with time. These are in strong contrast with the experiment of the cement–sand cliff in which no erosion occurred below a certain level and no significant temporal increase in the eroded area is found. These distinctive differences reflect the chemical and mechanical properties of the cementing material used for making the model cliffs. Cliff experiments using a plaster:sand mixture would be helpful for an understanding of erosional processes occurring on calcareous rocky coasts in the field situation.

Sanders (1968b) argued that, because no horizontal parts were formed in his experiment (Figure 7.19), there was no evidence to suggest that horizontal platforms (i.e. Type-B platforms) are cut by storm waves. This is only because the cementing material that he used was not suitable for modelling insoluble rocky coasts, although Gill (1972b) and Trenhaile (1980) considered that the duration of wave action in his experiment was not enough to produce the horizontal element in the cliff profile. Figure 7.17, the result of insoluble cliff experiments, indicates that Type-B platforms can be created by the action of waves alone.

Concerning the seaward drop of Type-B platforms in the field, one of the important problems that remain unsolved is: Why does the drop maintain its precipitous feature without suffering any significant erosion during the final stages of Holocene marine transgression? This is probably because (1) the rate of cliff erosion during sea level rise was much smaller than the rising rate of sea level which has been estimated at about 1 cm/year at maximum; or (2) the rate of sea level rise was much higher than that estimated. An explicit explanation based on quantified data must await future studies.

It would be reasonable to assume that platform cutting commenced after sea level became almost stationary, about 6000 years ago. If the sea failed to cut platforms, then plunging cliffs have developed according to the mechanism described before.

The morphological study by Reffell (1978) of the sea floor in front of Type-B platforms in the north of Sydney indicates that the water depth at the base of the seaward drop increases from the bay towards the tip of headlands, with increasing exposure to wave attack. She thought that larger water depth off the headland is due to more significant abrasion occurring under contemporary sea level conditions. However, this spatial variation in water depth may be a heritage of a former lower sea level at which the seaward cliffs were active. It is conjectured from the Chalk coast of England (Figure 7.13) that the bedrock elevation off the headland at the study site of Reffell had already been lower than in the bay when sea level was lower. If this is the case, then active abrasion would not have taken place since the sea reached its present level. Sanders (1968a) also reported that there is little evidence of sea-floor abrasion in front of platforms in Tasmania. Further research based on actual observations or measurements is necessary to determine whether the sea floor just in front of platforms is subjected to abrasion.

Platform elevation

As indicated in the statement of E. S. Hills given at the William Smith lecture to the Geological Society of London in 1960 (Cotton, 1963; Gill, 1972a), it is useless to draw conclusions about relative movements of land and sea from spatial distribution of shore platforms without understanding the mechanism of their formation. One of the important morphological elements to be elucidated from this applied point of view as well as from the purely geomorphological viewpoint is the elevation of platforms.

It has been reported that the elevation of platforms differs considerably from place to place along a stretch of the coast (e.g. Bird and Dent, 1966; Phillips, 1970; Gill, 1972a; Takahashi, 1974; Reffell, 1978; Sunamura, 1978b, Figure 4; Gill and Lang, 1983, Table 1). Factors controlling platform height are (1) waves (Johnson 1931; Kirk, 1977; Reffell, 1978), (2) rocks (Sanders, 1968a; Phillips, 1970; Gill, 1972a), (3) lithological structure such as joints, faults, and bedding planes (Bird and Dent, 1966; Kirk, 1977), (4) tides (Johnson, 1931; Sanders, 1968a; Hills, 1972; Trenhaile, 1978), (5) susceptibility of weathering (Bird and Dent, 1966; Hills, 1972), and (6) availability of abrasive tools (Hills, 1972). A positive correlation was established between platform elevation and exposure to wave action on the Kaikoura Peninsula coast of the South Island, New Zealand (Kirk, 1977) and at headlands located north of Sydney, Australia (Reffell, 1978). The work of Gill (1972a) in the southeastern part of the Australian coast indicated that platforms cut in soft rocks such as aeolianite are located around low-tide level, whereas those cut in hard rocks such as Lower Cretacous arkose are in the supratidal zone. Trenhaile (1987, p. 223) also stated that the mean elevation of platforms usually increases with hardness of rocks. On the Izu Peninsula coast of Japan, however, no significant correlation was found between platform altitude and lithology (Sunamura 1978b).

The elevation of platforms observed at present is an integrated result of the effects of these multiple factors. It is difficult to evaluate in an isolated manner the role of each factor played in controlling platform elevation. Late Holocene sea level fluctuations and local tectonic movements complicate this evaluation. The problem of platform elevation has not been discussed with factors appropriately quantified.

Shore Platforms

Sunamura (1991) discuss this issue using data obtained through a simplified laboratory test with two crucial factors, waves and rocks, taken into account. In this experiment a homogeneous model cliff made of a mixture of cement and sand was exposed to the action of breaking waves. The result, already shown in Figure 7.17, indicates that downward erosion extended to some critical level below which no erosion occurred. This level can be regarded as the elevation of platforms.

Platform cutting is thus determined by the relative magnitude of wave assailing force, F_W, to rock resisting force, F_R, both having vertical distribution (Figure 7.18). This figure shows that the critical condition is given by $F_W(z) = F_R(z)$. Denoting platform elevation as z_c, then we can write

$$F_W(z_c) = F_R(z_c) \tag{7.18}$$

From the analogy of Eq. (5.5), F_W and F_R can be expressed by

$$F_W = A\rho g H_b \tag{7.19a}$$

$$F_R = B S_c \tag{7.19b}$$

where H_b is the height of breaking waves just in front of the cliff, S_c is the compressive strength of the cliff material, ρ is the density of water, g is the gravitational acceleration, and A and B are dimensionless constants. Equations (7.16) through (7.19) yield

$$\frac{z_c}{h} = -\frac{1}{a_f}\left[\Gamma + \ln\left(\frac{\rho g H_b}{S_c}\right)\right] \tag{7.20}$$

where $\Gamma = \ln(A/B) =$ constant. The value of z_c obtained from the experiment (Figure 7.17) was analysed using Eq. (7.20). Figure 7.20 shows the result; the straight line can be described as

$$\frac{z_c}{h} = -0.17\left[5.8 + \ln\left(\frac{\rho g H_b}{S_c}\right)\right] \tag{7.21}$$

A comparison between Eqs (7.20) and (7.21) yields $a_f = 5.9$ and $\Gamma = 5.8$. Figure 7.20 indicates that the elevation of platforms cut in hard rocks is higher than that of platforms in soft rocks, other controlling factors being kept constant. This correlation is in harmony with the finding of Gill (1972a) and the statement of Trenhaile (1987, p. 223).

Quantitative discussion on this elevational difference will be made based on Eq. (7.21) with an attempt to apply the result to the full-scale environment. Let us consider platforms at two sites where the rock strength alone is different. From Eq. (7.20) we have

$$\frac{z_{c1}}{h} = -\frac{1}{a_f}\left[\Gamma + \ln\left(\frac{\rho g H_b}{S_{c1}}\right)\right] \quad \text{at Site 1} \tag{7.22}$$

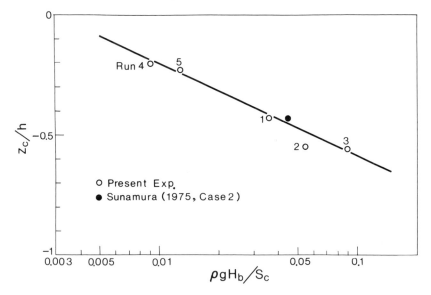

Figure 7.20 Normalized platform elevations, z_c/h, plotted against dimensionless wave-rock parameter, $\rho g H_b/S_c$. From Sunamura (1991). With permission from *J. Geol.*, copyright © 1991 the University of Chicago

$$\frac{z_{c2}}{h} = -\frac{1}{a_f}\left[\Gamma + \ln\left(\frac{\rho g H_b}{S_{c2}}\right)\right] \quad \text{at Site 2} \quad (7.23)$$

The suffixes, 1 and 2, denote quantities related to Sites 1 and 2, respectively. The elevational difference, Δz_c, is given as

$$\Delta z_c = z_{c1} - z_{c2} = -\frac{h}{a_f}\ln\left(\frac{S_{c2}}{S_{c1}}\right) \quad (S_{c1} > S_{c2}) \quad (7.24)$$

Because the height of breaking waves occurring at the cliff base is much controlled by the water depth there, it would be allowable in most cases to assume the following simple relation:

$$H_b = h \quad (7.25)$$

A further assumption is made that the degree of downward attenuation of the assailing force of waves at the cliff is the same between the small-scale experiment and the full-scale environment, i.e. that the value of a_f obtained in the laboratory can be applied to the field:

$$a_f = 5.9 \quad (7.26)$$

On these two assumptions, Eq. (7.24) reduces to

$$\Delta z_c = -0.17 H_b \ln\left(\frac{S_{c2}}{S_{c1}}\right) \quad (7.27)$$

Shore Platforms

Sample calculations were made with this equation for the three breaker heights $H_b = 3$, 5, and 7 m. The result is given in Figure 7.21. It is found that the difference in platform elevation at two sites increases as the strength ratio, S_{c2}/S_{c1}, decreases for a given breaker height. For example, the elevation of platforms created by breaking waves with a height of 5 m differs 0.6 m when the rock strength at one site is half as low as that at the other; and the elevational difference between the two sites attains 1.4 m when the strength ratio is $\frac{1}{5}$. Figure 7.21 shows that a considerable difference in platform elevation can be brought about depending on the breaker height and the strength ratio. Actually, the elevational difference is augmented or reduced by various effects of factors such as tides, weathering, and rock structure.

Equation (7.24) indicated that Δz_c is directly related to h, i.e. the water depth just in front of platforms, which is defined as the 'front depth'. This factor, having been long ignored in platform research, is indicated to be very important in explaining the local difference in platform elevation.

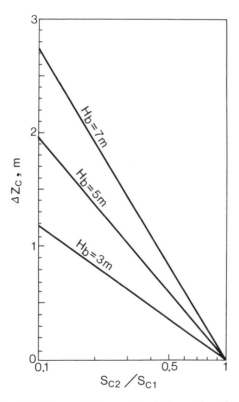

Figure 7.21 Relationship between difference in platform elevation, Δz_c, and strength ratio, S_{c2}/S_{c1}, at two sites, for three breaker heights $H_b = 3$, 5, and 7 m. From Sunamura (1991). With permission from *J. Geol.*, copyright © 1991 the University of Chicago

Sunamura (1975) conducted laboratory work in which model cliffs with the same strength were exposed to the action of two types of waves, breaking and broken waves, respectively, both waves having a similar height at the base of the cliff. The result showed that a platform was created more rapidly by breaking waves which have much larger assailing force; and its elevation was lower than that of a platform formed by broken waves.

On the coast where the front depth is smaller than the breaking depth of storm waves, the attack by breaking waves is infrequent compared with that of broken waves, as pointed out by Trenhaile and Layzell (1981). In this situation, breaking waves are more intensive but rarer, whereas broken waves are more moderate but more frequent. Which waves play an important role in determining platform elevation?

Tides control the front depth, which in turn affects the type of waves occurring in front of the cliff. This suggests the existence of tidal effects upon platform elevation. This problem also has not been deeply explored.

Except for areas that are tectonically unstable or heavily subjected to the influence of glacial isostasy, the elevation of platforms would be related to the height at which they were originally cut either by breaking or broken waves at the time of a higher stationary Holocene sea level. The original surface of platforms, the height of which is still unknown, has been lowered due to secondary processes that have varied with minor relative sea level fluctuations which would have taken place in most places after the platform initiation.

Secondary processes fall into three categories: mechanical, chemical, and biological. It is difficult to treat them in an isolated manner, because they interact with one another. Mechanical processes include (1) abrasive action of wave-moved sediment on platforms (Gill, 1976; Kirk, 1977; Emery and Kuhn, 1980; Gill and Lang, 1983); (2) potholing (Swinnerton, 1927; Wentworth, 1944); and (3) disintegration of rocks forming the platform surface due to alternation of wetting and drying (Bartrum, 1935; Wentworth, 1938; Suzuki et al., 1970, 1972; Takahashi, 1975, 1976), or due to the frost weathering and/or shore ice action (Dionne and Brodeur, 1988; and references herein). Chemical processes are characterized by solution of platform-forming rocks (Wentworth, 1939; Emery, 1946; Mii, 1962; Emery and Kuhn, 1980). Biological factors include the action of rock-boring and rock-browsing organisms (Healy, 1968b).

Measurements of the rate of secondary lowering have been attempted on some locations. Kirk (1977), applying the MEM technique to the Kaikoura coast in New Zealand, indicated that the average lowering rate is 0.15 cm/year on Tertiary mudstone and limestone platforms, where the lowering is caused mainly by abrasion. Comparing old with new photographs of the surfaces of Cretaceous sandstone platforms at La Jolla, California, Emery and Kuhn (1980) estimated the rate of lowering due to abrasion combined with solution. The result showed that it ranges from 0.003 to 0.06 cm/year. The MEM

Shore Platforms 177

measurements by Gill and Lang (1983) on the Otway coast of southeast Australia showed that the surface of greywacke platforms has been lowered by sediment-laden water mass at mean rates of 0.02 to 0.07 cm/year. These quantitative studies provide us valuable data for the discussion of platform evolution.

Platform width

The width of Type-B platforms is defined as the horizontal distance from the base of landward cliff to the top of seaward cliff (Figure 7.2b). Platform width has been considered to be determined by the relative rates of recession of both cliffs (e.g. Cotton, 1960, p. 416, 1963; Edwards 1951; Bird, 1976, p. 85; Trenhaile and Layzell, 1981; Trenhaile, 1987, p. 211). If the recession rate of landward cliff exceeds that of seaward drop, then the platform widens. If the landward cliff recedes keeping pace with the retreat of the seaward drop, then the platform narrows. Cotton (1960, p. 416) pointed out that more rapid erosion of the seaward drop than the landward cliff results in no platform development.

It is reasonable to assume, as discussed before, that no recession of the seaward drop takes place during platform development under stationary sea level conditions. Therefore, we can reckon the seaward edge of platforms as a fixed point when we discuss their width. Let us consider a simple situation in which a platform is developing under the action of breaking waves occurring immediately in front of the seaward cliff on a tideless coast composed of weathering-insusceptible rocks.

A wave breaking at the seaward edge of a platform rushes towards the landward cliff in the form of a bore (Figure 7.22), losing energy because of turbulence and bottom friction. Such a bore decreases its height with increasing travel distance. This height reduction can be assumed to be expressed in terms of an exponential function as deduced from Eq. (2.44). The assailing force of the bore, $F_W(x)$, also decreases exponentially with the distance from the seaward edge, x: $F_W(x) = F_W e^{-\alpha'_* x}$, where F_W is the assailing force of breaking waves, which can be described by Eq. (7.19a), and α'_* is an attenuation coefficient of the bore height. Then, we have

$$F_W(x) = A\rho g H_b e^{-\alpha'_* x} \qquad (7.28)$$

Because the resisting force of rocks is independent of time and space for this case, one can use again Eq. (7.19b) for F_R.

After substituting Eqs (7.19b) and (7.28) into Eq. (5.4), integration with the initial condition of $x = 0$ at $t = 0$ yields

$$x = \frac{1}{\alpha'_*}\left[\Gamma + \ln\left(\frac{\rho g H_b}{S_c}\right)\right](1 - e^{-\alpha'_* x t}) \qquad (7.29)$$

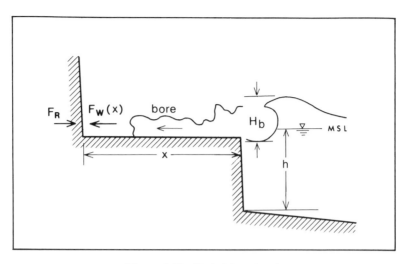

Figure 7.22 Definition sketch

It is seen that the platform width, x, increases rapidly at the initial stage and later it attains equilibrium with time. From this equation, the platform width after a certain period of time, $x_{t=t_1}$, is expressed as

$$x_{t=t_1} = G\left[\Gamma + \ln\left(\frac{\rho g H_b}{S_c}\right)\right] \quad (7.30)$$

where $G = (1 - e^{-\alpha'_* x t_1})/\alpha'_* = $ constant. Equation (7.30) indicates that platforms cut in harder rocks (i.e. larger S_c-value) have narrower widths, and coasts exposed to larger waves (larger H_b-value) have wider platforms.

As discussed before, however, the breaker height at the seaward bluff is a greatly depth-controlled quantity; so that we can assume again Eq. (7.25), i.e. $H_b = h$. It is seen from Eq. (7.30) that platform width increases with increasing front depth (water depth in front of the platform), if other factors are constant. Figure 7.23 is plotted using data obtained from a microtidal coast composed of various kinds of Tertiary volcanic ejecta in the southeastern part of Izu Peninsula, Japan, jutting out into the Pacific. This area has experienced a recent tectonic uplift with an amount of about 1.5 m, but platforms in this area had formed before the uplift (Sunamura, 1978b). Therefore, the value of front depth measured from the former sea level was used for plotting Figure 7.23. In spite of a considerable scatter in data points which may be attributed to diversity of hardness of coastal rocks, it is found that platform width is positively correlated with front depth where it is less than about 12 m. Beyond this water depth a negative correlation is suggested. This is due probably to rarity of breaking waves in such a large front depth.

Shore Platforms

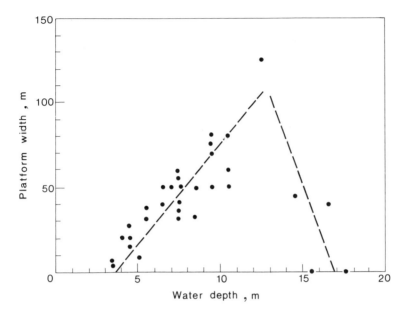

Figure 7.23   Relationship between width of Type-B platforms and water depth in front of them, on the southeastern coast of Izu Peninsula, Japan

Figure 7.23 clearly shows that the front depth is an important controlling factor for platform width. Bartrum (1935) reported from the west coast of Auckland, New Zealand that platforms on the coast exposed to the action of storm waves are narrower in some places. This may occur where the front depth is much smaller than the breaking depth of storm waves; in this situation waves break far away from the seaward drop, much reducing the assailing force. This will produce narrower platforms as compared with the coast where sizeable waves break just in front of the seaward cliff due to larger front depth.

Equation (7.30) shows that platforms become wider as the value of $[\Gamma + \ln(\rho g H_b/S_c)]$ increases. According to Eq. (7.20), as this value increases, the value of z_c/h decreases, which means that platform elevation decreases. This suggests that wider platforms have lower elevations: a negative correlation would be present between platform width and height. Duckmanton (1974) has a similar relationship on the Kaikoura Peninsula coast in New Zealand. Figure 7.24 shows a platform width vs. height relationship plotted using data of greywacke platforms developed in the microtidal environment of the Otway coast of southeast Australia (Gill and Lang, 1983, Table 1).

In the discussion of platform width, the effect of weathering should not be neglected (Abrahams and Oak, 1975). However, quantification of this effect

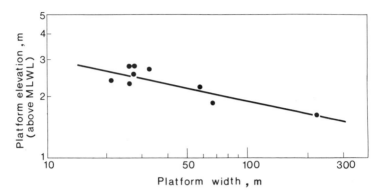

Figure 7.24 Relationship between elevation and width of Type-B platforms on the Otway coast, Victoria, Australia. Data from Gill and Lang (1983)

upon reduction in F_R (here represented by S_c) is difficult (Chapter 4); and expression in mathematical terms is still remote.

The tide controls the magnitude and the elevation of action of F_W, and affects F_R through weathering and biological processes. This factor plays a subsidiary role in platform widening. Taking account of the time that sea level occupies each elevation within the tidal range, Trenhaile and Layzell (1981) constructed a model for describing platform geometry, applied it to some coasts in Australia, Britain, Canada, and New Zealand, and concluded that the model explains the effect of tidal range on platform width.

A ROCKY COAST EVOLUTION MODEL

The preceding discussion in this chapter makes it possible to construct a comprehensive model for rocky coast evolution during a prolonged stable sea level (Figure 7.25). This model is based on the assumptions that (1) coasts made of insoluble, uniform rocks with no marked structural influence are situated in a microtidal environment, and (2) no sediment accumulation takes place in the nearshore zone. Although the model has these constraints, it may provide rudimentary modes of evolution occurring on most rocky coasts in the world.

Five kinds of coasts with different initial profiles, denoted as I through V in Figure 7.25, are assumed to be exposed to waves which have a very narrow spectrum of occurrence frequency and the same offshore properties. The assumption of considering such waves is unrealistic, but it allows us to explore the evolution of rocky coasts. Coast I is a uniformly sloping coast with a low gradient, so that waves break offshore and broken waves act on the coast. Coast II is a cliffed coast with the water depth at the cliff base, h, being zero, and the coast is exposed also to the action of broken waves. On Coast III

Shore Platforms

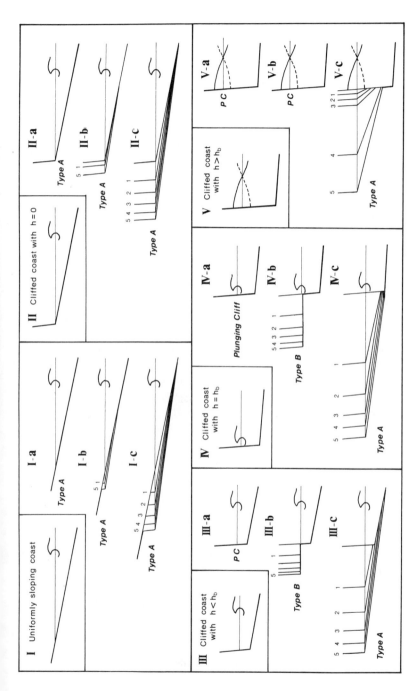

Figure 7.25 Model for rocky coast evolution beginning with five kinds of initial landform (**I** through **V**) with different degree of rock hardness (**a**, **b**, and **c**)

where $0 < h < h_b$ (= breaking depth of the incoming waves), broken waves with more energy, compared to the former two coasts, hit the cliff. Coast IV is directly subjected to breaking waves because of $h = h_b$. On Coast V ($h > h_b$), waves are always reflected from the cliff face, resulting in standing waves. If Coast I has a much steeper slope, for example, greater than 45°, then it can be regarded as one of four profiles (II–V) depending on the relative value of h to h_b.

In this model the relative strength of coast-forming rocks to waves is categorized on a qualitative basis into three: **a** (very strong and highly resistant to weathering), **b** (moderately strong and slightly susceptible to weathering), and **c** (very weak and strongly vulnerable to weathering). The mode of evolution is basically the same between Coasts I and II. Coasts I–**a** and II–**a** suffer little change due to high resistance of rocks against erosion. In the case of less resistant rocks, coasts of weaker rocks show more marked development of Type-A platforms (Compare I–**c** with I–**b**, or II–**c** with II–**b**). It should be noted that Type-B platforms and plunging cliffs never develop from Coasts I and II. On Coasts III and IV, plunging cliffs occur when rocks are very hard (III–**a** and IV–**a**), Type-B platforms develop only when rocks are of intermediate hardness (III–**b** and IV–**b**), and Type-A platforms appear when rocks are very weak (III–**c** and IV–**c**). On Coast V, plunging cliffs develop on V–**a** and V–**b**, and Type-A platforms occur only on V–**c**.

Type-A platforms occurring on Coasts I–IV grow most rapidly at the initial stage of their development, with decreasing rates with time. Such a temporal reduction in growth rate is also seen in the development of Type-B platforms (III–**b** and IV–**b**). On the other hand, a Type-A platform developed from Coast V suffers a slight change in cliff profile at the early stages when subaerial processes dominate because of impotency of standing waves in eroding the cliff. As the cliff recedes gradually, the subaqueous cliff profile becomes gentler, which finally produces the change in wave types: from standing to breaking waves. Breaking waves give rise to most rapid cliff recession at this stage; hereafter the recession rate decreases in time.

Two extreme cases are exhibited in this model. One is where the coast maintains its original profile without significant change due to the highest resistance to waves, i.e. I–**a**, II–**a**, III–**a**, IV–**a**, and V–**a**. From the purely morphological point of view, the first two, I–**a** and II–**a**, should be classified as Type-A platforms although the shoreline does not recede at all, and the remaining three, III–**a**, IV–**a**, and V–**a**, are categorized as plunging cliffs. The other is where, independent of the initial landform, the coast develops Type-A platforms due to high erodibility of materials, i.e. I–**c**, II–**c**, III–**c**, IV–**c**, and V–**c**. In the case of intermediate resistivity, Type-A platforms appear when the initial profile has already similar features to Type A, i.e. I–**b** and II–**b**; and Type-B platforms develop on the coast of III–**b** and IV–**b**, and plunging cliffs occur on V–**b**.

Edwards (1941) hypothesized that Type-B platforms (called 'storm-wave platforms' in his paper) will appear at the early stages of marine profile development and later they gradually reduce in width due to the faster recession of the seaward drop than the landward cliff, resulting in the development of Type-A platforms (called 'normal sloping wave benches'). In his hypothesis, Type-B platforms are situated as an ephemeral landform. As discussed before, the seaward drop of Type-B platforms does not recede at all, due to the abrupt downward decrease in wave assailing force along the cliff face. Therefore, Type-B platforms once formed, as seen on III–b and IV–b, never change into Type-A platforms, and should be regarded as a persistent landform over several thousand years.

An attempt will be made to quantitatively describe the three grades of relative strength of rocks to waves, i.e. **a, b,** and **c**. As shown in Figure 7.25, the development of Type-B platforms is restricted to the Coasts III–b and IV–b. This implies that examination of the occurrence condition for this type of platform enables us to evaluate the range of **b**. Figure 7.6 suggests that Type-B platforms develop only when $1.7 \times 10^{-3} \lessapprox \rho g H_1 / S_c \lessapprox 1.3 \times 10^{-2}$, which is equivalent to $77 \lessapprox S_c/\rho g H_1 \lessapprox 590$. If a parameter for the relative rock-strength can be expressed in terms of $S_c/\rho g H_1$, then we have

$$\left. \begin{array}{ll} 590 \lessapprox \dfrac{S_c}{\rho g H_1} & \text{for } \mathbf{a} \\[2ex] 77 \lessapprox \dfrac{S_c}{\rho g H_1} \lessapprox 590 & \text{for } \mathbf{b} \\[2ex] \dfrac{S_c}{\rho g H_1} \lessapprox 77 & \text{for } \mathbf{c} \end{array} \right\} \qquad (7.31)$$

where H_1 is the height of the largest waves occurring in an area under consideration. The derivation of Eq. (7.31) is based on a limited number of data of Type-B platforms (Figure 7.6). More data are necessary for a test of this relation.

Chapter Eight
Some characteristic erosional landforms

NOTCHES

A clear indicator of cliff erosion is the presence of a notch, a laterally extending hollow at the base of the cliff (Figure 7.1), its width being greater than its depth. A shallow notch is sometimes called a nip (e.g. Higgins, 1980). The notch roof, which is nearly horizontal, is termed a visor (Wentworth, 1939), and is a common feature on limestone coasts in tropical regions.

Available data concerning notch profiles measured on the coast of non-calcareous rocks are surprisingly limited in number. Figure 8.1 shows four notch profiles selected from such data. Profiles (a) and (b), taken from British coasts, show a notch cut in a London Clay cliff at Warden Point, Isle of Sheppey, and a notch scoured in a Lias shale cliff at Lingrow, northeastern Yorkshire, respectively. Profiles (c) and (d), from the Pacific coast of Japan, indicate notches developed on a Pliocene mudstone cliff at Idehama, Fukushima, and on a Miocene tuff cliff, Matsushima, Miyagi, respectively, the former located in an exposed area, and the latter in a very sheltered region.

The notches shown in Profiles (a), (b), and (c) are formed at the base of the cliff fronting Type-A platforms, and the remaining notch (d) is cut in on a plunging cliff. As shown in Profiles (a), (b), and (c), beach sediment covers the cliff base, and the deepest penetration of the notch is slightly above the cliff/beach junction. It is reported from these sites that beach material, working as an abrasive tool, plays an important role in notch development. The maximum rate of horizontal growth is estimated at 4.7 cm/year, based on the MEM measurement at Lingrow (Robinson, 1977b) and at about 1 m/year at Idehama (Aramaki, 1978). The rate depends on (1) the strength of the cliff-forming rocks, (2) the energy level of waves arriving at the cliff base, and (3) the amount of abrasive material set in motion by waves at the cliff–beach junction.

Some Characteristic Erosional Landforms 185

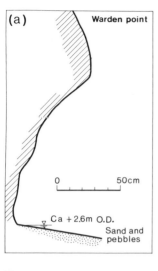
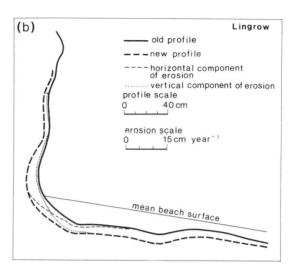
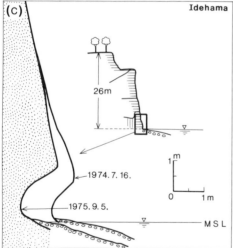
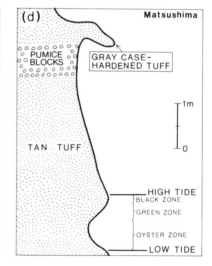

Figure 8.1 Examples of the shape of notches cut in non-calcareous rocks. (a) Notch incised in London Clay cliff at Warden Point, Isle of Sheppey. From Hutchinson (1986), by permission of National Research Council Canada. The top of the beach is located just below MHWS. (b) Notch incised in Lias shale cliff at Lingrow, Yorkshire. From Robinson (1977b), *Marine Geol.*, by permission of Elsevier Science Publishers. The top of the beach is located at about 1.6 m above O.D., between MHWS and MHWN. (c) Notch cut in mudstone cliff at Idehama, Fukushima. From Aramaki (1978), by permission. Mean tidal range is 1.3 m. (d) Notch developed on plunging cliff at Matsushima, Miyagi. From Emery and Foster (1956), *Am. J. Sci.* Reprinted by permission of *American Journal of Science*

Abrasive action is easily inferred from the fact that the surface of rocks in the vicinity of the beach is smoothed and polished. The process of notch development associated with mechanical abrasion can be observed in a laboratory experiment. When waves rushing up the beach reach the cliff, they form a vortex (with a horizontal axis) entrapping beach sand. This vortex scrapes down cliff material most efficiently to erode the notch (Sunamura, 1976, 1982a). An example of such notch development is illustrated in Figure 5.5. It is suggested that deep notching is closely associated with a sand-armed vortex as observed in the laboratory, although the observation of such a vortex in the field is not easy. Profiles (a), (b), and (c) indicate that there is a close relationship between the penetration of a notch and the elevation of the top of the beach surface. Such a relationship has been previously noted from a wave-tank experiment (Sunamura, 1982a, Figure 3).

Figure 7.17 shows the geometry of laboratory notches created by breaking waves in a laboratory experiment with no tidal effect. In this experiment no beach sediment was included, so that hydraulic action alone was involved in this notch-forming process. Three profiles out of five show that the deepest penetration of the notch is above still-water level (SWL), whereas in the remaining two it is below SWL. The discrepancy suggests that there is a considerable diversity in the elevation of breaker-cut notches in actual field situations where there is wave variability and water level fluctuations.

Profile (d) in Figure 8.1 exemplifies a notch cut in a plunging cliff attacked by standing waves. It is found that such a notch is restricted to the intertidal zone. Emery and Foster (1956) believed that the notch had been formed by the wave-induced removal of grains loosened by weathering, possibly solution and hydration.

Notches cut in cliffs composed of calcareous rocks such as limestones, aeolianites, and emerged coral reefs have been intensively studied, compared with notches on noncalcareous coasts, probably because of their peculiar shape. There are many descriptions of notch configurations, some measured and others photographed. They include intertidal notches in various parts of tropical or subtropical regions: For example, Puerto Rico (Kaye, 1959), Cayman Islands, West Indies (Woodroffe et al., 1983; Spencer, 1985b), Oahu, Hawaii (Wentworth, 1939), Ryukyu Islands, Japan (Takenaga, 1968; Kawana and Pirazzoli, 1985), Key Minor, Moluccas (Verstappen, 1960), Point Peron, Western Australia (Hodgkin, 1964), Aldabra Atoll, Indian Ocean (Trudgill, 1976), Kamaran Island, Red Sea (MacFadyen, 1930), and central and southern Greece (Higgins, 1980). However, descriptions of subtidal notches are very few. Profile measurements were made in Harrington Sound, Bermuda (Neumann, 1966) and Curacao, Netherlands Antilles (Focke, 1978a).

Notch geometry is an integration of either (1) chemical erosion—solution, (2) mechanical erosion—wave action including abrasion, or (3) biological erosion—boring and grazing by marine organisms, or a combination of these.

Trudgill (1976) investigated intertidal notches on the tropical limestone coast of Aldabra Atoll and measured their growth rate by applying the MEM technique. He found that (1) grazing is responsible for more than half the notch development occurring at a rate of 1 mm/year at a site without abrasive materials, (2) the processes of grazing and abrasion occupy each one third of the total erosion with a rate of 1.25 mm/year at a place where sand is residing at the cliff base, and (3) this abrasion-involved notch has a lower part bulging, compared with the abrasion-free notch which is symmetrical. A similar erosion rate, 1 mm/year, has been reported by Hodgkin (1964) using stainless steel pegs on an intertidal notch cut in hard aeolianite at Point Peron. Bird et al. (1979) in Western Barbados indicated that abrasion is a dominant process for cutting intertidal notches into coral rocks at annual rates of 0.2 to 2 mm/year, measured by using pegs. Tjia (1985) also reported a major role of mechanical abrasion in notching from a limestone islet in an open bay of Langkawi Island, Malaysia. In a small cove near Cabo Rojo, Puerto Rico, a solution notch is well developed on a plunging limestone cliff, but notch development is poor in the adjacent area where a coarse shingle beach lies at the foot of a cliff, abrasion by shingle obscuring the solution notching (Russell, 1963). Verstappen (1960) stressed the importance of tidal currents in notch formation on the island of Batanta, New Guinea, assuming that they brought about rapid replacement of sea water already saturated with $CaCO_3$ by unsaturated sea water. Higgins (1980) found in Greece that the solution effect of groundwater is confined to the intertidal zone and is responsible for notch formation, as already noted by Wentworth (1939) in Hawaii.

Figure 8.2 is a model presented by Focke (1978b), which shows that the profile of tropical limestone cliffs changes considerably with degree of wave exposure. This model, based on the result of field investigations in Curacao (Focke, 1978a), describes the interrelationships between wave intensity and two contrasting effects of biological activities: destructive and protective. In

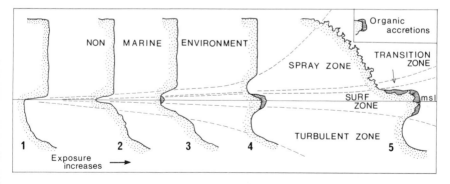

Figure 8.2 Variations in profiles of limestone cliffs with the degree of wave exposure. From Focke (1978b), Z. Geomorph., by permission of Gebrüder Borntraeger

a very sheltered area, the maximum undercutting takes place below sea level (Profile 1 in Figure 8.2) due to bioerosion; Neumann (1966) has pointed out the role of biological action in subtidal notching from his study on Bermuda. As exposure increases, a notch forms around sea level. Further increase in wave intensity activates aeration of water to promote organic accretion appearing around sea level. This increases the limestone strength by lithification, which results in a seaward protrusion (Profile 4): a profile with double notch forms, both being contemporaneous (Focke, 1978b). A platform as shown in Profile 5 has the same characteristics as the 'solution bench' that Wentworth (1939) described on Oahu (Higgins, 1980). Such platforms have been reported from many locations in the Atlantic–Caribbean, the Pacific, and the Mediterranean regions (Spencer, 1985b).

An application of Focke's model seems to be restricted to plunging limestone cliffs, because the model was constructed assuming a vertical cliff as the initial boundary condition. No studies have been made of the influence of the initial landform on morphological features of calcareous coasts: unfortunately, notch-shape models applicable to gently sloping limestone coasts are not yet available.

As suggested in Figure 8.2, the levels where notches form show considerable variation. Even for the case of a single notch, there is much disagreement. Levels of deepest notch penetration at around MSL are reported from Houtman's Abrolhos Islands (Fairbridge, 1947), Puerto Rico (Kaye, 1959), Point Peron, Australia (Hodgkin, 1964), Langkawi, Malaysia (Tjia, 1985), and Vatulele, Fiji (Nunn, 1990). On Grand Cayman Island, they vary from MSL to about 1 m above it (Woodroffe *et al.*, 1983; Spencer, 1985b). The notch floor is located slightly below HWL on Key Minor (Verstappen, 1960) and at about MSL in Western Barbados (Bird *et al.*, 1979). In sheltered regions on Bermuda (Neumann, 1966) and Curacao (Focke, 1978a), maximum undercut occurs at or below LWL.

The notch height, the vertical distance from the notch floor to the tip of the visor, increases with tidal range (Verstappen, 1960; Russell, 1963) and with exposure to storm wave action (Kaye, 1959; Russell, 1963). The notch roof tends to be horizontal on sheltered coasts, but becomes inclined in more exposed sites (Verstappen, 1960; Russell, 1963). Notch dimensions and configurations should be related to quantified wave intensity and rock strength.

SEA CAVES

A highly localized erosional feature is a sea cave (Figure 7.1), a hollow excavated in the sea cliff, with the penetration being greater than the width at the entrance. It is well recognized that the formation of sea caves is strongly controlled by geological weakness such as stratification, joints, and faults (e.g. Guilcher, 1958, pp. 62–4; Emery, 1960, p. 22; Flemming, 1965). The form of caves depends on the dip of bedding planes and the attitude of joint or fault planes (Guilcher, 1958, p. 64).

Examples of cave geometry have been documented from Santa Cruz Island, California (Emery, 1960, Figure 24), Cape Gilyanly on the Caspian, the former USSR (Zenkovich, 1967, Figure 47), and the San-In coast, Japan (Toyoshima, 1967). Figure 8.3, from the San-In coast, shows a cave developed in a plunging cliff composed of tuff breccia. This cave extends in about 60 m from the entrance along a vertical fault with N–S orientation, and at the end it branches at right angles along an E–W fault. A blowhole has formed at the intersection of the two faults.

The existence of such geological weakness does not always guarantee the development of caves. According to the work of Davies and Williams (1985) on Lower Lias cliffs (limestones and shales) on the Welsh coast, there are few caves in a sector directly exposed to incoming waves, in spite of the presence of joint systems favourable for cave development. This is because of the rapid recession of cliffs. Along a sheltered sector, however, where cliff recession has been slower, caves are well developed along major joints. This study indicates that caves are better developed and more persistent where the rock is relatively resistant but has many fractures that the sea can penetrate.

Some ephemeral caves have been reported from the rapidly eroding coast of Holderness, England by Hutchinson (1986). They form at the base of till cliffs fronted by Type-A platforms. Hutchinson described the formation and failure of caves. One of the caves, which is fairly stable, is 3.7 m long with an

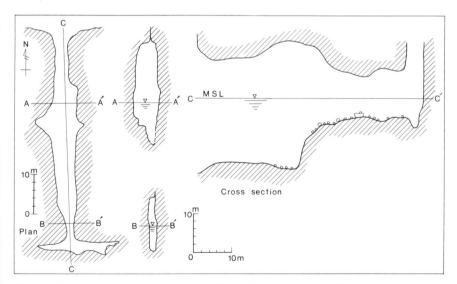

Figure 8.3 Geometry of a cave called 'Tsutendo' (a hollow leading to the sky) on the San-In coast of Japan. From Toyoshima (1967), by permission

oval cross-section about 1.7 m wide and 0.9 m high. The floor of the cave is inclined seawards with an angle of $10°-15°$. Seaward dipping floors can be seen also in caves in Chalk in Kent (Figure 7.13) and in notches produced in laboratory tests (Figure 5.5).

Wave-induced erosion of weathered cliff material along joints or faults triggers the development of sea caves. A small opening forms along the weakness, and grows as the forces produced by the hydraulic action of waves and the pneumatic action of trapped air, plus the mechanical action of abrasive tools if these are present, increase their intensity. This widens and deepens the opening, within which intensified hydraulic, pneumatic, and/or mechanical forces lead to cave development. This is a positive feedback process, during which a cave may develop more rapidly than in the initial stage; but cave development ceases when a balance is established between the wave assailing force and the resistance of the rock.

Evidence of abrasive action may be found inside caves wherever there is sand and/or gravel to act as an abrasive. The sides of caves are much scoured, smoothed, and polished around the elevation where the sediment resides, so that an oval cross-section frequently appears.

A blowhole is formed by puncturing the weaker part of the roof of a sea cave by the hydraulic and pneumatic action of waves. Tides and storm surges control the effectiveness of such wave action. Enlargement of a blowhole may lead to collapse of the cave. Australian and British examples of blowholes are described in Bird (1976, p. 63); two Hawaiian examples, one of which is on Oahu and the other on Kauai, are illustrated in Wentworth (1938) and in Macdonald and Abbott (1970, Figure 133), respectively; and a Madagascan example is shown in Battistini and Le Bourdiec (1985, Pl. 92–5).

SEA ARCHES AND STACKS

Where waves attacking either one or both sides of a promontory have succeeded in tunnelling a hollow completely through it, a sea arch (Figure 7.1) is produced. Its formation is usually associated with a geological weakness. Collapse of the vault of the arch leads to the formation of a stack (Figure 7.1).

Spectacular sea arches have attracted the attention not only of geomorphologists and geologists but also of tourists. Aside from guidebooks or pamphlets for sight-seeing, textbooks and papers have also provided photographs of arches in various places in the world, such as Cape Breton Island, Nova Scotia (Johnson, 1925, Figure 152), Arecibo, Puerto Rico (Snead, 1982, Pl. 103), Oregon (Byrne, 1963), California (Mather, 1964, Figure 55; Shepard and Kuhn, 1983), Hawaiian Islands (Snead, 1982, Pl. 104; Shepard and Kuhn, 1983), Victoria (Baker, 1943; Bird and Rosengren, 1986), Beirut, Lebanon (Snead, 1982, Pl. 105), Black Sea coast (Zenkovich, 1967, Figure 48), Island of Malta (Mather, 1964, Figure 54), English Channel coast of France

(Prêcheur, 1960; Mather, 1964, p. 139), and England and Wales (Steers, 1960, Photos 39, 85, 132, and 142).

Sea arches are ephemeral landforms especially when they occur in weak rocks. Rapid change of arch morphology has been documented in many locations. The first study on this subject was made probably by Johnson (1925, pp. 324–5). He described the history of development of an arch carved in a promontory of weak Carboniferous sedimentary rocks at Table Head, Glace Bay, Cape Breton Island, Nova Scotia, using photographs taken successively from 1900 to 1923. This arch lasted 13 years. According to the work of Byrne (1963), who used three successive photographs of a Miocene sandstone promontory called Jump-off-Joe, near Newport, Oregon, an arch had already been produced in the promontory when photographed first in 1880. This arch suffered little change from 1880 to 1930, when the last photograph was taken, whereas the promontory suffered considerable erosion and the arch collapsed in 1935. A very short life of an arch, only 1.1 years, has been documented by Bird and Rosengren (1986) who studied the evolution of a narrow promontory composed of Tertiary clayey sand at Black Rock Point near Melbourne, Victoria. Through comparison of old and recent photographs, Baker (1943) has described the collapse of an arch, formerly the 'trunk' of Elephant Rock, near Port Campbell, 200 km southwest of Melbourne. In a similar way, Shepard and Kuhn (1983) studied the history of sea arches and remnant stacks on the Californian and some other coasts. A vivid report of arch collapse was recently given with photographs by Bird (1990a,b) from the coast near Port Campbell, a well-known tourist spot with a limestone promontory that had a double arch structure, called London Bridge. The landward arch, having been a narrow neck connecting the promontory with the mainland, collapsed suddenly on 15 January 1990, leaving an islet with the other arch.

Application of the term 'sea tunnel' would be appropriate when an arch is considerably longer than the width at the entrance. A typical example of sea tunnels is Merlin's Cave at Tintagel, north Cornwall, which runs nearly straight about 100 m following a fault zone in phyllites (Wilson, 1952). A photograph of a sea tunnel southeast of Hobart, Tasmania is provided in Snead (1982, Pl. 101); the nearly rectangular cross section of the tunnel is determined by the dip of bedding and fault planes.

Cathedral Arch off La Jolla, California, formed in a promontory of massive Cretaceous sandstone, collapsed shortly after 1900 (Shepard and Grant, 1947). Until 1921 both buttresses of the arch were left standing. One of them was eroded away some time between 1921 and 1937. The remaining stack underwent slight change by 1947, but it disappeared by 1968 (Shepard and Kuhn, 1983).

Stacks are not always produced from sea arches. Figure 8.4 shows an example of stack development not following the stage of arch occurrence on a Pliocene mudstone coast of Japan. A highly fractured zone associated with

192   Geomorphology of Rocky Coasts

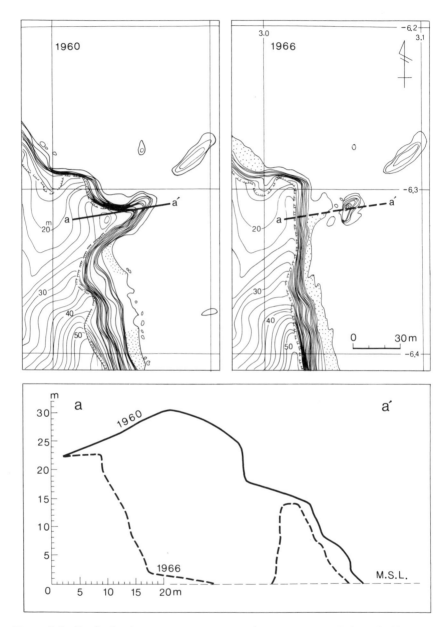

Figure 8.4   Stack development seen on a mudstone coast at Taito-misaki, Japan, facing the Pacific. From Sunamura (1973)

a steeply dipping fault between the stack and the mainland extended to the top of the cliff at the stage of a promontory before the stack formed. The presence of such a fragile material in the upper part of the cliff would have hindered the formation of an arch. May and Heeps (1985) illustrated the development of stacks on a chalk coast at Ballard Down, southern England using maps of four different editions during almost one hundred years (1887–1982). Erosion along vertical joints is responsible for stack formation in this area.

The Needles, a series of aligned stacks at the western end of the Isle of Wight, England are cut in steeply dipping, resistant chalk. Steers (1960, Photo 21), Bird (1976, Pl. 7), Twidale (1976, Pl. 18.13d), and Goudie and Gardner (1985, p. 128) have presented photographs of these stacks from different angles. Another well-known British example is the Old Man of Hoy, Orkney Island, northern Scotland, a 140 m high stack of Old Red Sandstone with prominent horizontal bedding (Twidale, 1976, Pl. 18.13b; Holmes and Holmes, 1978, Figure 23.26). A French example can be found along the Normandy coast where retreating cliffs of horizontal chalk have left spectacular stacks near Etretat and Benouville 25 km north of Le Havre (Prêcheur, 1960). On the Victorian coast in Australia, a group of stacks near Port Campbell is called the Twelve Apostles, cut in flat-lying Miocene limestone (Bird, 1976, Pl. 6; Twidale, 1976, Pl. 18.13a).

RAMPARTS

It is often seen that Type-B platforms have ridges rising a metre or more, depending on locations, above the general level of platforms at their seaward margin. Such ridges were first called 'ramparts' by Wentworth (1938) in his platform study in Hawaii, although Bartrum (1935) had noted the presence of a similar morphology in New Zealand. Because the seaward edges of the platform are kept constantly wet by spray and splash, rocks there are immune from wet/dry weathering, and are more durable than those behind them; this results in the formation of ramparts (e.g. Bartrum, 1935; Wentworth, 1938; Hills, 1949, 1971). This wet-edge hypothesis is not always adequate, because (1) ramparts are not generally continuous and not always present on all platforms (Jutson, 1949; Edwards, 1951; Cotton, 1963) and (2) ramparts, where they exist, are not always formed at the seaward edge (Edwards, 1951; Gill, 1972c). From his study of platforms cut into Jurassic (now dated Lower Cretaceous) arkoses and mudstones along the southern coast of Victoria, Australia, Edwards (1951) concluded that the most likely additional cause is a favourable combination of (1) inherent hardness of rocks, (2) a landward dip, and (3) a morphological setting that enables scouring to occur on the inner side of the platform. Baker (1947) had already noticed that rocks are slightly more resistant on ramparts near Moonlight Head on the same coast. Gill (1972c) recognized two types of rampart on the Otway coast of Victoria: (1) those due to higher resistance of massive arkose, free of joints, faults, or bedding plants, and (2) those due to ferruginous induration of the rocks preceding the Holocene transgression. Figure 8.5 shows a surveyed profile illustrating the former type of ramparts; three rows of ramparts develop although the inner rampart is less prominent. It is seen that they are closely associated with outcrops of massive arkose. According to Mii (1962), some shore platforms incised into Miocene sedimentary rocks on the western coast of Wakayama, Japan, have ramparts at their seaward edges. Such ramparts are related to the presence of harder rocks, but others have no lithological difference.

Ramparts develop at their seaward edges of some limestone platforms, although they are less marked than those formed in other rock types. Such ramparts are the result of induration due to encrusting by algae that grow in a zone that is always wet, irrespective of tidal stage. Ramparts of this kind are often found on shore platforms cut into Pleistocene dune limestone (i.e. aeolianite) on the southern and western coasts of Australia (Bird, 1976, pp. 87–8).

The processes of rampart formation may be classified into two modes. One is related to inherent rock hardness, as seen on the Otway coast. In this case the difference in strength of rocks that form ramparts had existed before the platform developed. It should be noted that a difference in mechanical strength alone does not always account for rampart formation. Suzuki *et al.* (1972) investigated the formation of ridges and furrows on a platform cut into an alternation of steeply dipping Miocene mudstone and tuff layers. The mechanical strength of tuff that forms ridges is always lower than that of mudstone forming furrows: their values are, respectively, 170 and 235 kg/cm^2 in compressive, 15 and 45 kg/cm^2 in tensile, and 30 and 74 kg/cm^2 in shear strengths. Laboratory experiments indicated that mudstone is highly susceptible to wet/dry weathering, and disintegrates into many minor flakes easily removed by waves washing across the platform. The resistance of mudstone to the waves was found to be smaller because of weathering processes than that of tuff, in spite of the inherent hardness.

The other mode of rampart formation is related to temporal increase in relative strength of rocks at the seaward edge with increasing platform width. In locations where the wet-edge hypothesis can be applied, rocks forming the landward portion of the platform deteriorate due to wet/dry weathering, resulting in reduction of strength. On algae-encrusted coasts, rocks at the seaward margin are strengthened by lithification. Under these circumstances, platform width and elevation strongly control the formation of ramparts and their morphology.

The study of ramparts should be directed not only to lithology but also to waves. Where the platform is narrow, the flow across it is bidirectional, i.e. swash and backwash. As the platform becomes wider, the flow pattern tends to be unidirectional: there is no significant backwash because wave-brought

Some Characteristic Erosional Landforms 195

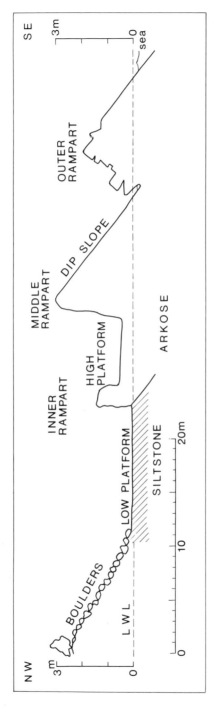

Figure 8.5 Cross section of shore platform south of George River, Otway coast, Victoria. Tidal range is 1.5 m. From Gill (1972c), by permission of Tsukiji Shokan

water masses return to the sea via 'wave furrows' that are channels developed on the platform along joints and faults. It is anticipated that the temporal change in flow pattern plus availability of abrasive tools which are mainly supplied from the landward cliff, further lower the platform surface near the base of the cliff, thus making the rampart more prominent.

RAMPS

At the base of a cliff, there are often seaward-dipping slopes which are steeper than the rest of the shore platform. These slopes, which constitute a distinct morphological element, are called ramps (Hills, 1949, Robinson, 1977a; Trenhaile, 1978) genetic terminology such as abrasion ramps (Wentworth, 1938; Jutson, 1949; Healy, 1968a; Bird, 1976, p. 76), erosion ramps (Short, 1982a), and wave-cut ramps (Fairbridge, 1968) has also been applied. Ramps may occur either at the back of Type-A or Type-B platforms. Hills (1972) actually used the term 'wave ramp' to denote a Type-A platform.

The cliff–ramp junction is usually clear, with a marked break of slope, whereas the ramp–platform junction is sometimes obscure, especially if the ramp has a concave profile. The surface of a ramp is generally smoothed and scoured, but occasionally there are structural ramps, corresponding with the outcrop of a resistant seaward-dipping rock formation.

Ramps have been reported from several places: Gaspé, Québec (Trenhaile, 1978); Oahu, Hawaii (Wentworth, 1938); Whangaparaoa, Auckland (Healy, 1968a); Lorne, Victoria (Jutson, 1949); Port Phillip Bay, Victoria (Hills, 1949); Point Peron, Western Australia (Fairbridge, 1950) and northeast Yorkshire, England (Robinson, 1977a). Only Robinson (1977a) treated ramps behind Type-A platforms; the others described them on Type-B platforms. All these studies have recognized that abrasion is involved in ramp formation. Robinson (1977a) investigated the development of ramps on Lias shale platforms along the northeastern Yorkshire coast, using MEM measurements. He found that (1) the amount and the mobility of sediment residing on the ramp greatly control its growth and extinction, and (2) there are three modes of ramp development (Figure 7.12): a concave ramp appears when a suitable amount of movable material is available (Figure 7.12a); a ramp disappears when the amount of beach material on the ramp is reduced (Figure 7.12b); and it becomes longer and linear when large immovable material is present (Figure 7.12c).

POTHOLES

Marine potholes are approximately cylindrical or bowl-shaped depressions formed on shore platforms primarily by grinding action of sand, gravel, pebbles and boulders moved or rotated under the energy of waves. Figure 8.6

Some Characteristic Erosional Landforms

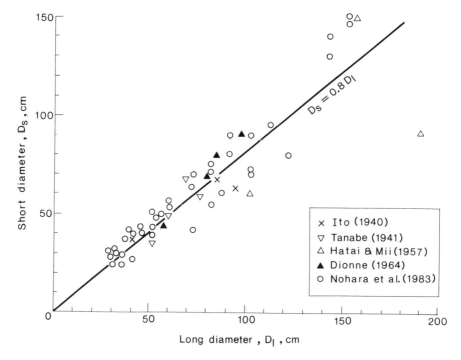

Figure 8.6 Relationship between long and short diameters of shore potholes

shows the relationship between the long and the short diameters of potholes. It is seen from this figure that most potholes are roughly circular in outline. Some potholes have an irregular outline reflecting (1) coalescence of neighbouring depressions due to their enlargement and (2) variations related to rock hardness (Bird, 1970; Trenhaile, 1987, p. 26).

When the depth of potholes is plotted against the long diameter (Figure 8.7), almost all the data points are located where depth/diameter ≤ 1. This indicates that most potholes are shallower compared with their diameter. As potholes grow, they attain a depth to diameter ratio of approximately unity, which is much smaller than for fluvial potholes, e.g. about 6 for a pothole near Namforsen in Sweden (Ängeby, 1951), 5.0 for one in the Interstate Park at Taylor's Falls in Minnesota (Alexander, 1932), and 5.3 for one at Yakama in the Kurokawa River, Japan (Sato *et al.*, 1987). A laboratory experiment on fluvial potholing by Alexander (1932) demonstrated that the vortex energy available for bottom erosion decreases abruptly with increasing depth for a constant diameter. It is probable that a similar energy reduction will occur in marine potholes, but the nature of flow in these is more complicated because of the oscillatory water motion.

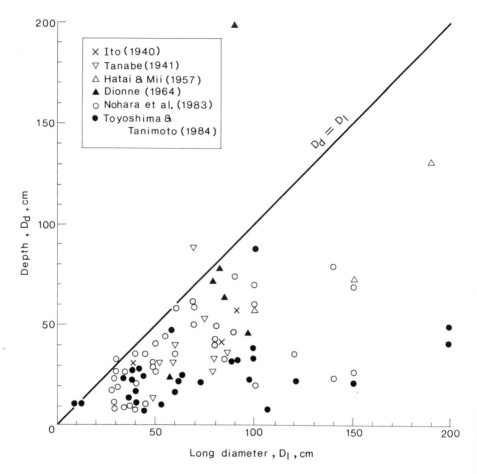

Figure 8.7 Relationship between long diameter and depth of shore potholes

Whether the abrasive material in a pothole can be moved by the energy of rotating flows is another problem. The mobility depends on (1) the size and amount of the material and (2) the flow velocity. For the vertical development of potholes, abrasive tools must be moved on the bottom, acting as a downward force. Even if no vertical erosion takes place, excavation may occur on the side wall, which results in lateral growth of potholes, due to abrasive action of pebbles and cobbles induced by the centrifugal motion of water.

It is frequently found in the field that the side wall has spiral grooves (e.g. Mii, 1962) or overhangs. This is evidence of lateral excavation due to abrasion. It is almost impossible to observe the behaviour of the abrasive material in a pothole without disturbing the flow field. Observations through a glass cylinder modelling a pothole, as used in the experiment by Alexander (1932), would increase our understanding of formative processes of marine potholes, if such experiments were conducted under oscillatory flows in the laboratory.

Potholes are found in almost all kinds of rocks (Dionne, 1964). Even if the abrasive material is of the same lithology as the pothole, repeated collisions gradually diminish the size of the material by attrition as the pothole grows. This is seen on Eno-shima Island, Japan, where a block bounded by vertical columnar joints on a shore platform cut in Miocene sandstone was detached from the bedrock by the action of waves and repeatedly moved *in situ*, abrading the bedrock and eventually scouring a pothole. Potholes thus formed range from 1 to 2 m in diameter and 1.5 to 2 m in depth, each containing a skittle-like boulder (Tanabe, 1941). When the abrasive material is harder than the bedrock, scouring is more effective and the rate of pothole growth increases, although the calibre of the abrasive material is another controlling factor. Swinnerton (1927) reported from the Moss Beach shore of California that potholes developed on platforms cut in Pliocene shale are formed by wave-induced churning of gneiss boulders supplied from adjacent outcrops. Wentworth (1944) observed that potholes had developed on a reef limestone platform on the Hawaiian shore, scoured by hard basaltic pebbles and cobbles, or hard coral gravel. Hatai *et al.* (1957) and Mii (1962) ascribed the formation of some potholes on Miocene sandstone platforms in the western part of the Wakayama coast of Japan to abrasion by calcareous concretions that were released from the host rock but retained and repeatedly moved by the action of waves; and Henkel (1906) also considered the presence of concretions working as an abrasive tool to be the most potent factor in the development of potholes on a sandstone coast west of Vancouver Island, British Columbia. Bird (1970) explained that potholes excavated into a dune calcarenite (i.e. aeolianite) platform in Victoria, Australia, were formed by the grinding action of pebbles and cobbles formed from the more resistant calcrete layers. Coasts with such a strong contrast in hardness between the abrasive tool and the bedrock, as seen in these locations, could be used for quantitative studies of temporal change in pothole geometry.

Pothole development may be initiated by the presence of depressions which can serve as lodgment sites for the abrasive material. Such depressions could be formed by solution (Elston, 1917, 1918; Wentworth, 1944; Emery, 1946) or by the extraction of concretions (Hatai and Mii, 1957; Hatai *et al.*, 1957; Mii, 1962; Gill, 1967) and erratics (Hutchinson, 1986). Where potholes are associated with joint or fault patterns (Tanabe, 1941; Mii, 1962; Nohara *et al.*, 1983; Toyoshima and Tanimoto, 1984), geological structure may have been responsible for the formation of initial depressions. In any case, the initial stages of pothole development require a subtle balance between the intensity of wave action and the mobility of the abrasive tool, as Swinnerton (1927) pointed out: waves must move the abrasives in a shallow depression without jostling them out of it.

Even if there is evidence that abrasive tool material in a pothole is being moved by waves, resulting in pothole excavation and widening, this does not necessarily mean that the pothole originated under present conditions. Some potholes are clearly of recent origin if they develop at the base of rapidly receding cliffs, as in the case of the Holderness coast of England where potholes are being formed on till platforms produced by cliff recession occurring at an average rate of more than 1 m/year (Hutchinson, 1986). A good example of inheritance from pre-existing conditions is seen on the dune calcarenite coast of Victoria, Australia. Some potholes developed on calcrete ledges were derived from the washing-out of soil from solutional pipes that had developed beneath an old (Pleistocene) soil horizon which had been exhumed by cliff recession and excavated by erosion (Bird, 1970).

SOLUTION POOLS

Solution pools are shallow, flat-bottomed depressions frequently found on shore platforms cut in calcareous rocks; they are also called pans (Wentworth, 1944), solution basins (Emery, 1946), or flat-bottomed pools (Guilcher, 1958, p. 65). The presence of solution pools has been reported from various places: California (Emery, 1946; Emery and Kuhn, 1980), Hawaii (Wentworth, 1944; Guilcher, 1953), Japan (Mii, 1962; Sunamura, 1978b), Australia (Hills, 1971), Morocco (Guilcher, 1953), and Eire (Trudgill et al., 1987).

Solution pools develop from small pits or holes with diameters of a few millimetres to several centimetres, due essentially to biochemical processes (Emery, 1946) which are characterized by (1) dissolution of calcite cement from the bedrock by sea water during nocturnal development of low pH by respiration of plants and animals living in these depressions, and then (2) removal of the disintegrated rock material by waves and grazing organisms. The pools thus extend laterally in all directions from the original depression, keeping roughly circular outlines and having the side wall sometimes vertically fluted by snails (Emery, 1946) or undercut by sea urchins (Trudgill et al., 1987). As they grow and coalesce, their outlines become irregular.

Some solution pools have elevated rims. According to Emery (1946), the origin of such pools is explained as follows. Because of the interaction between calcareous rocks and biochemical processes of living organisms in the pools, calcium carbonate is deposited on the fringes of the pools as an additional cement, so that an elevated rim is left by the reduction of less resistant surrounding rocks.

Emery (1946) investigated solution pools on shore platforms cut in Cretaceous calcareous sandstone along the coast at La Jolla, California, counted and measured the pools in a cross-sectional strip on a generally seaward-inclined platform with a width of 30 m from the cliff base to the mean water line. It was found that the pools are located between MSL and MHWL (mean tidal range: 1.1 m) and distributed across the platform, but most numerous in the middle part of the platform, which is slightly higher (at around MHWL) than the rest. Average pool diameter, 25 cm near the cliff base, increased toward the seaward margin where it attained 50 cm. Average depth, measured vertically from the top of its surrounding wall, also increased seawards from 3 to 20 cm.

Solution pools distributed along the Pacific coast (mean tidal range: 1 m) of western Wakayama, Japan, were examined by Mii (1962). He selected as a measurement site a nearly horizontal shore platform cut in Miocene sandstone. The platform, with a width of 35 m, had a rampart at the seaward edge, slightly higher than the general level of the platform located around MHWL. The landward quarter of the platform was covered with beach sand, and the pools were restricted to the seaward half, most abundant on the horizontal portion just landward of the rampart and gradually decreasing both landwards and seawards. In contrast with La Jolla, average diameter of these pools generally increased landwards from 20 to 40 cm, and the average depth exhibits a similar trend, increasing seawards from 2 to 4 cm. The depth of the pools at the seaward margin (4 cm) was much smaller than that at La Jolla (20 cm). Considering that tidal conditions and wave climate were similar for both areas, this difference could be ascribed to a difference in (1) lithology such as the amount of calcareous material or the mechanical strength of the bedrock and/or (2) biological factors such as the numbers and species of organisms living in the pool.

Pools with elevated rims are found both at La Jolla and in western Wakayama. The height of rims measured from the bottom of the pool is 2 to 8 cm and the height measured from the platform level outside the pool is 1 to 5 cm at La Jolla (Emery, 1946); the former is 0.5 to 5 cm and the latter is 0.5 to 10 cm in western Wakayama (Mii, 1962). No data on lateral dimensions were available.

In both areas, most of landward pools were nearly circular in outline, but many of those located farther seaward were of very irregular shape, suggestive of the coalescence of several smaller pools. Emery and Kuhn (1980) provided evidence of coalescence by rephotographing an area having solution pools that were described by Emery (1946). They also presented recent photographs showing the complete disappearance of rimmed pools after the rim was breached in 1944 when Emery took rock samples.

Other depressions caused by solution are solution pits, especially marked on limestone coasts. Solution pits are small cavities, usually having a diameter of a few millimetres to a centimetre or more, and can be found in the spray zone on limestone coasts in cold temperate regions (Guilcher, 1953). They also develop in shales and porphyrines, not exclusively in these regions (Guilcher,

1958, p. 67). Pinnacles, sometimes called coastal lapies, are column or spikes of rock persisting when the surrounding limestone has been dissolved away (Guilcher, 1958, Pl. IB; Bird, 1976, Pl. 28; Tricart, 1962, Photo 12; Twidale, 1976, Pl. 18.9; Trudgill, 1985, Figure 9.2). They are found in the lower intertidal zone of cool temperate limestone coasts, and in the spray zone of warm temperate or tropical regions (Guilcher, 1953).

TAFONI AND HONEYCOMBS

Tafoni (Italian; *sing.* tafone) is a generic term for features of cavernous weathering characterized by the existence of hollows or cavities on the rock surface: their diameter and depth range from a few centimetres to several metres or more. Tafoni develop not only in coastal regions but also in other environments such as inland deserts and cold regions including Antarctica (Jennings, 1968; Mustoe, 1982b). Honeycombs (i.e. alveoli or alveoles, the term used by French geomorphologists) are tafoni of smaller scale (Martini, 1978; Mustoe, 1982c; Sparks, 1986, p. 30), but no explicit distinction has been made between the two (Sparks, 1986, p. 30; Trenhaile, 1987, p. 49). As suggested by the term 'honeycomb', this should be applied only to the cavernous rock surface assuming a cell-like structure, irrespective of the size of cavities, although this application involves a subjective judgement.

Coastal tafoni and honeycombs have been reported from many locations in the world (Table 8.1). It is found that these cavernous features develop on a variety of rocks, but arkose and sandstone are probably favourable, especially for the formation of honeycombs, suggesting that the occurrence of such features is restricted to this type of rock within a certain range of physicochemical properties.

Honeycombs develop on horizontal, inclined, vertical, or overhanging rock faces; they generally form in the spray zone above HWL. On the other hand, tafoni never develop on horizontal surfaces but only on steeply inclined, vertical, or overhanging surfaces located generally, but not always, higher than the spray zone. Tafoni and honeycombs frequently coexist.

As shown in Table 8.1, quantitative data on these cavernous forms are limited, partly because of their intricate patterns. According to Mii (1962), honeycombs developed on a shore platform cut in Miocene sandstone on the Shirahama coast of Japan are of smaller dimensions than those on a platform cut in Miocene conglomerate: the former have a mean longest diameter, D_l, of 2.0 cm, a shortest diameter, D_s, of 1.5 cm, and a depth, D_d, of 1.1 cm; and the latter have $D_l = 4.1$ cm, $D_s = 3.1$ cm, and $D_d = 1.9$ cm (Mii, 1962, Figure 35). The value of D_d/D_s is 0.73 for honeycombs on sandstone and 0.61 for those on conglomerate. The value of D_s/D_l is 0.75 for the former and 0.76 for the latter. In contrast with these honeycombs on shore platforms, which are little affected by the lithological structure, honeycombs occurring on the face of cliffs are greatly controlled by the presence of weak bedding planes. Their dimensions are: $D_l = 3.5$ cm, $D_s = 1.4$ cm, and $D_d = 2.0$ cm on a sandstone cliff, and $D_l = 6.2$ cm, $D_s = 4.4$ cm, and $D_d = 7.4$ cm on a conglomerate cliff (Mii, 1962, Figure 35). The value of D_d/D_s is 1.43 for the sandstone case and 1.68 for the conglomerate; and the value of D_s/D_l is 0.40 for the former and 0.71 for the latter.

Comparison of D_d/D_s-values between structure-free and structure-controlled honeycombs shows that the former have 'relatively shallower' cavities, irrespective of rock types. As to the value of D_s/D_l, the structure-free honeycombs have similar values closer to unity, suggesting that their outlines are more circular. The structure-controlled honeycombs are more elongated, especially on sandstones ($D_s/D_l = 0.4$). Coalescence occurs where the axis of elongation is parallel to bedding planes (Mii, 1962). Similar elongation associated with bedding has been reported from a coast in northern Puget Sound, Washington, where honeycombs with an average diameter of 5 to 10 cm develop on arkosic sandstone (Mustoe, 1982c).

The honeycomb study of Mii (1962) showed that cavities with larger diameters generally have larger depths. Matsukura and Matsuoka (1991) found that this relationship also holds for tafoni on the Nojima-zaki coast, Japan, where they occur on the faces of cliffs on three uplifted shore platforms cut in Miocene tuffaceous conglomerate, with different altitudes and with different ages of emergence. The result of measurements of tafoni showed that the ratio of depth to the shortest diameter, D_d/D_s, ranged from 0.4 to 1.0, and larger tafoni (with the largest being $D_l = 160$ cm, $D_s = 60$ cm, and $D_d = 26$ cm) have formed on the higher two platforms.

Tafoni increase their diameters abruptly when coalescence occurs, but their depth increases more gradually. Suggesting that depth is an appropriate measure of the rate of tafoni growth, Matsukura and Matsuoka (1991) examined the relationship between tafoni depth and time, which is expressed by the following exponential equation:

$$D_d = 20.3(1 - e^{-0.005 t}) \qquad (8.1)$$

where D_d is the mean of the ten largest depths of tafoni (in centimetres) and t is the time (in years). This equation indicates that the growth of tafoni is rapid at the initial stages and becomes slower with time. Rapid tafoni formation has been reported from several sites. According to Grisez (1960), the surface of blocks of crystalline schists in a sea wall at Gourmalon on the French Atlantic coast has been honeycombed to a maximum depth of 66 mm in 62 years. This gives a maximum growth rate of 1.1 mm/year, the same order of magnitude as the maximum rate of tafoni formation at Nojima-zaki, 1.7 mm/year (Matsukura and Matsuoka, 1991). Gill (1981) observed on the Otway coast, Victoria, Australia, that fresh blocks of greywacke placed in a sea wall in 1943 and 1949 had been honeycombed. Mustoe (1982c) reported

Table 8.1 Tafoni and honeycombs in the coastal environment

Location	Lithology	Morphology	Size	Place of occurrence	Reference
Angola					
Serra do Sombreiro near Benguela	Calcareous marls	Tafoni	—	Sea cliffs 30–50 m high	Guilcher (1985)
Atlantic Ocean					
Tenerife, Canary Islands	Tertiary basaltic rocks and agglomerates	Tafoni	Diameter: <5 m	Cliff face (~100 m above sea level)	Höllermann (1975, Photo 2)
Tenerife, Canary Islands	Pumice-lapilli formation	Tafoni	—	Cliff face (~60 m above sea level)	Höllermann (1975, Photo 3)
Tenerife, Canary Islands	Pozzolan	Tafoni	—	Cliff face (~30 m above sea level)	Höllermann (1975, Photo 4)
Australia					
Southern Australia	Dune calcarenite	Honeycombs	—	Higher level in sea cliffs	Bird (1976, pp. 92–3)
Newcastle, New South Wales	Upper Permian sandstone	Honeycombs	—	Blocks in sea wall	Bartrum (1936, Fig. 4)
Southern coast of Victoria	Arkose	Honeycombs	—	Top and inner side of rampart	Edwards (1951, Pl. V., Fig. 2)
Cape Paterson, Victoria	Arkosic sandstone	Honeycombs	—	Projecting portion of shore platform	Hills (1971, Photo 12)
San Remo and Lorne, Victoria	Jurassic sandstone and mudstone	Honeycombs	—	—	Bartrum (1936)
Pt Sturt, Lorne, Victoria	Massive arkose	Tafoni and honeycombs	—	Landward side of rampart	Gill (1972c, Pl. 3, Fig. 2)

Some Characteristic Erosional Landforms   205

Location	Rock type	Landform	Dimensions	Position	Reference
Northeastern side of Pt Sturt, Lorne, Victoria	—	Tafoni	—	—	Gill (1973, Fig. 14)
Southwest of St George River, Lorne, Victoria	Lower Cretaceous greywacke	Honeycombs	—	Blocks in splash zone of sea wall	Gill (1981)
Pt Grey, Lorne, Victoria	—	Honeycombs	—	Shore platform	Gill (1972a, Fig. 3)
Artillery Rocks, Otway coast, Victoria	Lower Cretaceous greywacke	Honeycombs	—	Pedestal in splash zone	Gill et al. (1977, Pl. 9, Fig. 1; 1981, Pl. 12)
Artillery Rocks, Otway coast, Victoria	Arkose	Honeycombs	—	—	Jennings (1968, Fig. 2); Gill (1973, Figs. 13 and 15); Trenhaile (1987, Fig. 2.2)
France					
Agay, Provence	Porphyry and sandstone	Honeycombs	—	—	Boucart (1930)
Gourmalon, Clion-sur-Mer, Loire-Atlantique	Crystalline schist	Honeycombs	Max. depth: 6.6 cm	Blocks in sea wall (near MHWS)	Grisez (1960)
Italy					
Western half of Isle of Elba, Tuscany	Granodiorite	Tafoni	Diameter: several tens centimetres to several metres	Boulders on steep slopes	Martini (1978)
Japan					
Hatake Is., Shirahama, Wakayama	Miocene sandstone	Tafoni	—	Sea cliff	Mii (1962, Pl. 4, Fig. 4)
near Seto Marine Biological Lab., Shirahama, Wakayama	Miocene sandstone and conglomerate	Honeycombs	Depth: 1.1–7.4 cm Diameter: 2.0–6.2 cm	Sea cliff and shore platform	Mii (1962)
Nojima-zaki, Boso Peninsula	Miocene tuffaceous conglomerate	Tafoni and honeycombs	Mean depth: 6.5–20 cm	Cliff faces on uplifted shore platforms	Matsukura and Matsuoka (1991)

(continued)

Table 8.1 Continued

Location	Lithology	Morphology	Size	Place of occurrence	Reference
Liberia					
Mamba Point, Monrovia	Dolerite	Honeycombs	—	Blocks	Tricart (1962, Photo 9)
New Zealand					
Northeastern shore of South Island near Auckland City	Sandstone	Tafoni	Diameter: <15 cm	—	Trenhaile (1987, Fig. 2.1)
	Tertiary argillaceous sandstone	Tafoni (?)	—	Sea cliff	Bartrum (1936)
Whangarei Heads, Auckland	Dacite lava	Honeycombs	Max. depth: 5 cm	Landward margin of shore platform ab. 1 m above HWL	Bartrum (1936)
Waiheke Is., Auckland near Manukau Harbour, Auckland West Coast	Greywacke Pyroclastic rocks	Honeycombs Tafoni (?)	— Diameter: 0.3–3 m Depth: ~1 m	— Sea cliff 45 m high	Bartrum (1936, Fig. 3) Bartrum (1936, Fig. 1)
Muriwai, Auckland West Coast	Tuffaceous sandstone	Honeycombs	—	—	Bartrum (1936)
UK					
Ballycastle, Northern Ireland	Carboniferous sandstone	Honeycombs	—	Cliff face, blocks in sea wall, and boulders on shore platform	McGreevy (1985)
Isle of Arran, Western Scotland	Old Red sandstone	Honeycombs	—	Upper part of shore platform	Kelletat (1980, Photo 1)
Isle of Skye, Western Scotland	Cretaceous calcareous sandstone	Honeycombs	—	Notch surface	Kelletat (1980, Photo 2)

Location	Rock type	Landform	Dimensions	Position	Reference
USA					
Northern Puget Sound, Washington	Arkosic sandstone	Honeycombs	Av. width: 5–10 cm Depth: <10 cm	Outcrops up to 3 m above HWL	Mustoe (1982c)
near Bellingham, Washington	Arkose	Honeycombs	Width and depth: several centimetres	Blocks used for railway construction	Mustoe (1982c)
near Bellingham, Washington	Arkose	Honeycombs	Av. width: 5 cm Av. depth: 5 cm	Cliff face nearest HWL	Mustoe (1982c)
near Bellingham, Washington	Metavolcanic greenstone	Honeycombs	Av. width: 5 cm Av. depth: 5 cm	Cliff face nearest HWL	Mustoe (1982c)
Cape Arago, Oregon	Massive sandstone	Honeycombs	—	Cliff face above HWL	Johannessen et al. (1982)
Bean Gulch, south of San Francisco, California	Arkose	Honeycombs	—	—	Twidale (1976, Pl. 18.10d)
near Scripps Pier, La Jolla, California	Eocene sandstone	Honeycombs	—	Boulders	Emery (1960, Fig. 9)
USSR (former)					
near Baku, West Coast of Caspian Sea	Sandstone	Honeycombs	—	—	Mustoe (1982a, Fig. 2)

that blocks of arkose and boulders of metavolcanic greenstone, both used for railway construction near Bellingham, Washington, in the early twentieth century, had cavities as large as several centimetres in depth and width. Comparing old (1903) and recent (1981) (G. E. Mustoe, 1991, personal commun.) photographs of the face of sea cliffs composed of arkose, Mustoe (1982c) found that honeycombs have developed close to high-tide level, on surfaces that were planar when the first photographs were taken, and have now average dimensions 5 cm in width and depth. These data enable us to estimate an average rate of honeycombing of about 0.6 mm/year (5 cm/78 years). This is comparable to the mean rate of tafoni deepening, 1 mm/year (6.5 cm/66 years) obtained using the mean value of the ten largest depths of tafoni on the face of cliff just above HWL at Nojima-zaki (Matsukura and Matsuoka, 1991).

Various hypotheses proposed for the origin of cavernous features include wind erosion, frost shattering, thermal weathering, and salt weathering (e.g. Mustoe, 1982c; Twidale, 1982, pp. 280–300; Trenhaile, 1987, pp. 44–51). According to Mustoe (1982c), the suggestion that salt weathering (Chapter 4) is responsible for cavity formation was first presented by Hume (1925, pp. 213–14) and has since become popular for explaining coastal tafoni and honeycombs (Bourcart, 1930; Bartrum, 1936; Cailleux, 1953). Recent studies which applied chemical analyses (Höllermann, 1975; Mustoe, 1982b), scanning electron microscopy and/or X-ray diffraction analyses (Gill *et al.*, 1981; McGreevy, 1985; Matsukura and Matsuoka, 1991) have detected salts in the hollowed rock surface. As McGreevy (1985) stated, however, the presence of salts does not necessarily indicate that salt weathering is the cause of tafoni development. The process of rock disintegration requires further research.

Chapter Nine
Effects of human activity on rocky coasts

INTRODUCTION

There are various human impacts on rocky coasts due to increased use of the coastal zone. Most of them are associated with engineering works characterized by the construction of sea walls, groynes, and breakwaters. These engineering structures disturb natural processes of geomorphological development. Unintentional consequences of engineering works may accelerate cliff erosion or induce mass movement. Beach mining or filling also affects rocky coasts. The most indirect effects of human activities will be where future rocky coasts are placed under conditions of a sea level rise induced by a man-induced intensification of the greenhouse effect.

EFFECTIVENESS OF ENGINEERING STRUCTURES

Various types of engineering structures have been built to protect receding rocky coasts. They are sea walls, bulkheads, revetments, detached breakwaters, and groynes. The first three are structures constructed continuously along the coastline with a purpose of directly protecting the shore from wave attack. Although the massiveness and size of the structure decrease in the order of sea walls, bulkheads, and revetments, there is little functional difference between the three (Coastal Engineering Research Center, 1984, p. 6:1). Because usage of the term 'sea walls' is generally accepted to represent these structures, this terminology will be employed here, unless otherwise stated.

Sea walls can be classified into two types depending on whether they are impermeable or not: rigid and mound types (Figure 9.1). Rigid-type sea walls are made of concrete, timber, or steep sheet piles. Concrete structures are usually constructed at some distance from the cliff (Figure 9.1, I-a). They have vertical, uniformly sloping, curved, or stepped faces. Figure 9.2 shows an example of a combination curved- and sloping-face sea wall which was erected

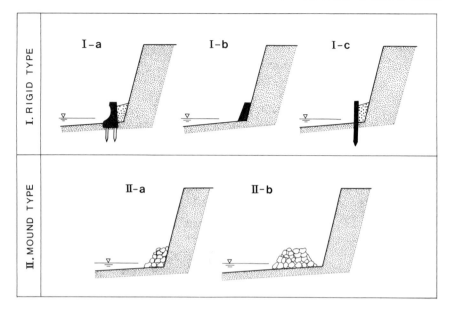

Figure 9.1 Schematic diagram showing two types of sea walls. After Sunamura and Horikawa (1972)

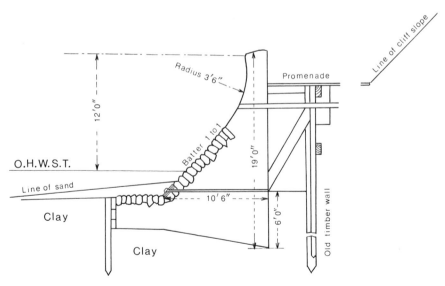

Figure 9.2 Sea wall built to protect London Clay cliffs on the Clacton coast in Essex, England. After Wheeler (1902, Figure 14)

on the Clacton coast in Essex, England, to protect London Clay cliffs. A concrete sea wall may be directly attached to the cliff (Figure 9.1, I-b). Figure 9.3 exemplifies this type, built to prevent recession of the cliff cut into alternating limestone and shale beds at Lyme Regis in Dorset, England (Fox, 1924, pp. 166–7). Vertical or steeply inclined sea walls made of steel sheet pilings or timber (Figure 9.1, I-c) are constructed in some places. Figure 9.4 shows an erosion-preventive measure of this type which was set up on the Lake Erie till (glacial drift) shore.

According to Clayton (1989), 70% of the 33 km coastline of glacial-drift cliffs in northeastern Norfolk, England, has been protected by rigid-type sea walls or revetments (less massive structures than sea walls) combined with the groyne system, most of which were constructed after World War II, especially after the storm surge of 1953. Using erosion data of the recent 10 years, Clayton examined the effectiveness of these shore protection structures. Unprotected coastal sectors receded at an average rate of 0.7 m/year, whereas the sectors protected by sea walls cut back only at a rate of less than 0.1 m/years and those fronted by revetments retreated at a rate of 0.3 m/years.

Mound-type sea walls are piles of stones or concrete blocks, thus being permeable. They are divided into two categories: one has the structure sloping up against the cliff (Figure 9.1, II-a), and the other has the structure of a self-supporting trapezoidal mound, detached some distance from the cliff base (Figure 9.1, II-b). The former structure is called a riprap sea wall when stones are used, and the latter is a rubble-mound sea wall irrespective of the kind of

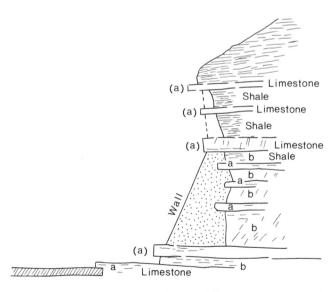

Figure 9.3 A concrete wall built on a hard layer of limestone to protect the lower part of a cliff at Lyme Regis, Dorset, England. From Fox (1924, Figure 17), by permission of John Murray (Publishers) Ltd

212   Geomorphology of Rocky Coasts

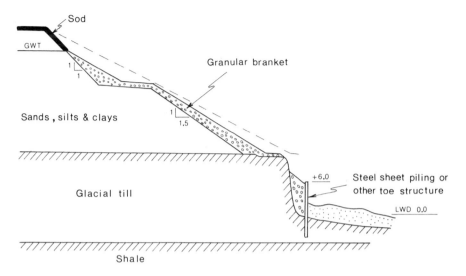

Figure 9.4   Slope protection and cliff erosion preventive measure using sheet pilings on the Lake Erie shore at Perry Township Park near Painesville, Ohio. From Chieruzzi and Baker (1959), by permission of American Society of Civil Engineers

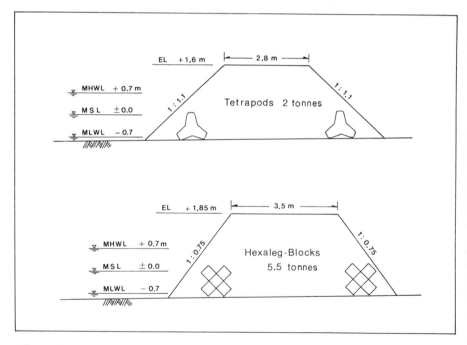

Figure 9.5   Sea walls of concrete amour units, constructed to protect the Byobugaura mudstone cliff in Japan. From Sunamura and Horikawa (1972)

materials used. Riprap sea walls are widely employed to protect receding cliffs of the Central California coast (Fulton-Bennett and Griggs, 1986). Cliffs cut into Neogene sedimentary rocks in the Santa Cruz area on this coast receded at average rates of 0.2 to 0.4 m/year (Sorensen, 1968) before riprap sea walls were constructed in the early 1960s. A follow-up study by Magoon et al. (1988) on the effectiveness of these sea walls indicated that they have exhibited satisfactory performance although bedrock scouring at the toe of the structure occurred in some locations to cause stone to shift.

Rubble-mound sea walls are usually constructed using quarry stones where they are available inexpensively; in other places this type of structure is built using concrete blocks or other artificial materials. Figure 9.5 shows an example of rubble-mound sea walls using concrete armour units, which were installed in front of an actively eroding mudstone cliff on the Byobugaura coast, Japan (Sunamura and Horikawa, 1972). The protection works began from both ends of a nearly straight, 9 km long coastline towards the centre. The effect of the protective measure was examined using large-scale photogrammetric maps of 1960, 1967, and 1970. Figure 9.6 shows seven selected cliff profiles: four from the protected section where the sea wall construction was completed by 1967, and three from the unprotected region. It is found that the sea walls are effective in protecting the cliff base from wave erosion. Figure 9.7 illustrates the alongshore variation in erosion rate during three years from 1967 to 1970. This figure shows that there is considerable scatter in data points for the unprotected area, which is inherent in short-term cliff erosion (Chapter 5), and that the average rate of erosion in this area is 1 m/year. In contrast, the erosion rates in the protected sections are only 0.05 and 0.3 m/year.

A detached breakwater is a structure constructed parallel to and a certain distance from the shoreline, which attenuates the incoming wave energy and protects the shore. Detached breakwaters have become popular for protecting sandy beaches. Use of this type of structure, however, is not common for the protection of eroding cliffed coasts. An example has been reported by Dean et al. (1986) from the US East Coast. A detached breakwater system combined with beach fill was constructed in 1982 at two sites along Colonial Beach, Virginia, a low wave-energy environment in the lowermost reach of Potomac River, in order to protect severely receding low bluffs (with an erosion rate of 0.5 m/year) cut into Pleistocene marine and fluvial deposits. At one site, four segmented detached breakwaters of rubble-mound type were built 35–70 m from the bluff, and at the other site, three breakwaters of the same type were installed about 50 m apart from the bluff. After the breakwater building, beach fill was placed in front of the bluff at both sites. Three-year investigations after the construction indicate that this project using detached breakwaters and beach fill has succeeded in protecting the cohesive bluff and providing an artificial beach now as a recreational area.

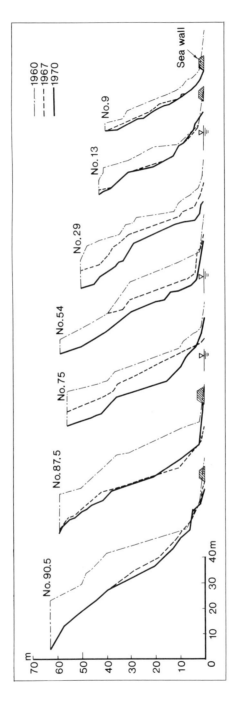

Figure 9.6 Cliff profile changes on the Byobugaura coast. In the protected section (Profile Nos. 9, 13, 87.5, and 90.5), the sea wall was built by 1967. From Sunamura and Horikawa (1972)

Effects of Human Activity on Rocky Coasts

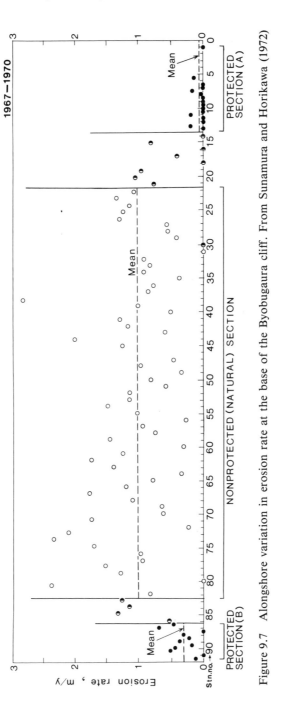

Figure 9.7 Alongshore variation in erosion rate at the base of the Byobugaura cliff. From Sunamura and Horikawa (1972)

A groyne is a structure extending perpendicular to the shoreline, designed to trap littoral drift and prevent erosion of the shore. On a cliffed coast where littoral drift dominates in one direction, large quantities of sediment accumulate on the updrift side of a groyne, resulting in formation of a wide beach on which storm wave energy is expended, and the cliff behind is protected. The downdrift area, on the other hand, suffers starvation, which allows waves to penetrate landwards, and the cliff is subjected to wave attack.

At Clacton-on-Sea, a holiday resort on the East Coast of England, an incessant struggle with waves has taken place during the last 100 years (Fryde, 1968; Harris and Ralph, 1980). Sea walls and groynes had been constructed to protect the cliff cut into Pleistocene gravels overlying London Clay. After World War II, two longer groynes were built in an area of pre-existing shorter groynes. These two groynes were successful in trapping large amounts of sediment on the updrift side, which in turn caused depletion of the beach material on the downdrift portion, exposed London Clay on the foreshore, and eventually led to the failure of the adjacent sea walls and cliff. Remedial measures, such as reshaping the groyne system or placing beach fill, have been reasonably successful (Fryde, 1968).

A series of studies by Carter and his coworkers (Carter et al., 1981; 1982; 1986) discussed the influence of shore protection structures on geomorphological changes along the 300 km Ohio shore of Lake Erie, which is characterized by cliffs cut into glacial deposits. Changes in cliff line associated with the groyne construction are most typically demonstrated on a till coast near Painesville-on-the-Lake, 40 km east of Cleveland (Carter et al., 1987, p. 43). This is shown in Figure 9.8; nearly parallel retreat of the cliff took place from 1876 to 1937 at an average rate of about 0.6 m/year. During the 1937–73 period the erosion decelerated with a recession rate of less than 0.3 m/year along the updrift area of the groynes where intercepted sand made the beach wider. In contrast to this, the erosion accelerated on the downdrift side due to depletion of the beach, a maximum recession rate attaining 3 m/year. As illustrated by this case, well-functioning groynes exacerbate the downdrift situation. It is possible, however, to retard erosion of this area with appropriate engineering works, although they much depend on the cost–benefit relationship.

The shore protection scheme to use artificial headlands (groynes of large scale) has been undertaken at Barton-on-Sea in South England to halt recession of clay cliffs. The scheme is based on the idea that a stable bay-shaped beach to be formed between the headlands can retain beach material, which in turn protects the underlying clay from wave erosion (Fleming and Summers, 1986). The headlands-and-bay concept was originally proposed by Silvester (1960) for shoreline stabilization of eroding sandy beaches. Application of this concept to rocky coasts has been investigated to stabilize the coastline of Holderness, England, a 60 km long, severely eroding cliff cut into

Effects of Human Activity on Rocky Coasts 217

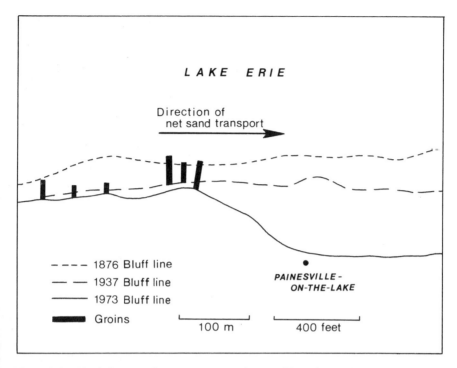

Figure 9.8 The influence of groyne construction on till erosion at the Lake Erie shore in Ohio. From Carter *et al.* (1987, Figure 3.2), *Living with the Lake Erie Shore*, by permission of Duke University Press

glacial drift; artificial offshore banks or reefs are considered for creating fixed points to be worked as headlands (Fleming, 1986).

Modern technology has enabled us to protect any type of rocky coasts from wave erosion. However, shore protection structures conflict with the conservation of natural environments of the coast. If the coast is a feature of scenic and/or scientific interest, it may be preferable to leave it as it is, as suggested by Bird and Rosengren (1987), for those who wish to enjoy studying nature as well as to conduct further research of rocky coast geomorphology.

BEDROCK SCOURING IN FRONT OF SEA WALLS

As discussed in Chapter 6, the downward erosion of sea floor is most active at the water's edge on natural coasts. Sea walls, erected at the cliff base to protect the shore, are most subjected to scouring at the toe of the structure. Bedrock lowering becomes more salient when sea walls are of rigid type, compared with mound type, due to higher reflectivity of wave energy.

To protect a low cliff near the Byobugaura coast in Japan, a vertical concrete sea wall was erected in 1936 on the bedrock of Pliocene mudstone, which is the same as that forming the Byobugaura cliff. The bedrock surface at the toe of the structure, located at MLWL at the time of erection, was lowered approximately 1 m by 1952 when the foot protection works were needed (Horikawa and Sunamura, 1970). The average scouring rate is estimated at 6.3 cm/year, which is found almost four times as high as the lowering rate of the mudstone in the natural section of the Byobugaura cliff, 1.65 cm/year (Figure 6.7).

According to Zenkovich (1967, pp. 166–8), the sea floor of flysch in front of a sea wall (the type of structure unknown) near Sochi in the former USSR was eroded down by the action of reflected waves, with an average rate of 15 cm/year at the water's edge and 6 cm/year at a depth of 2 m. The lowering process ceased at a depth of 3 m where the rate of destruction of the flysch was not more than 1 cm/year. It is noted that the influence of sea walls is restricted to a very shallow region up to a water depth of a few metres.

Scouring at the toe of the structure is one of typical causes of failure for sea walls, which reduces the life of engineering structures. Davidson-Arnott and Keizer (1982) reported this type of damage from a cohesive shore at Stoney Creek, western Lake Ontario, Canada. Sea walls and groynes, or a combination of these have been constructed to protect till bluffs (10 km long) receding at average rates exceeding 1 m/year. The protected section occupies 30% of the whole shoreline in 1934, but it rose to 64% in 1979. Sea walls constructed prior to 1972 were of concrete structure, most being vertical but some being stepped. According to Davidson-Arnott and Keizer, only 18% of the sea walls could survive in an undamaged state in 10 years after construction, and the proportion of survival became null in 30 years. The main cause of this low durability is considered to be collapse resulting from rapid scouring of till floor at the toe of the structure. Although this rapid scouring was not quantified, it is anticipated that the rate must have been extremely high, because the lowering rate of till floor shallower than 1 m in water depth is estimated at more than 10 cm/year even in the unprotected section near Fifty Mile Point located at the eastern end of the study area (Davidson-Arnott, 1986a).

Incidentally, Fulton-Bennett and Griggs (1986) illustrated causes of damage for various types of sea walls erected on the central California coast. Most of the sea walls described in their report are protection measures for eroding rocky coasts but some are for receding dune fields.

CLIFF EROSION CONTROLLED BY BEACH SEDIMENT VOLUME

It is well known that a wide beach in front of the cliff is the best natural wave-energy absorber to protect the shore. A reduction in the volume of beach sediment decreases beach width and lowers beach level, which allows storm waves to reach the cliff and to have intensified force of assault when sand is mobilized and hurled on the cliff face: severe cliff erosion is likely to occur. Human activities such as harbour works, dam construction on rivers, and beach mining, may bring about reduction in beach volume, producing unintentional results of rapid cliff recession in some areas. An intentional activity to halt cliff erosion is the building of a wide, artificially nourished beach in front of the cliff—beach nourishment works.

A breakwater is a structure protecting harbour, anchorage, or basin from waves. Where a harbour is located at the river side, a jetty is built, extending into the sea, to prevent shoaling of a channel by littoral materials and to facilitate entry to the harbour. On coasts with prevailing littoral drift, these harbour structures may exert serious influences on the neighbouring area: the updrift side of these structures has sediment accumulation, whereas the downdrift side suffers depletion of beach material.

A typical example of the influence of harbour works on downdrift rocky coasts can be found on the coast near Newhaven in Sussex, England (Figure 9.9). A breakwater constructed for Newhaven Harbour interrupted eastward longshore drift and caused shingle accumulation, resulting in depletion of beaches at Seaford a few kilometres east of the breakwater; erosion of Chalk cliffs at Seaford Head, located further downdrift, was accelerated (Bird, 1976, p. 73; 1985a). Seaford beach was artificially restored in 1988.

At West Bay in Dorset, England, a jetty was built for the harbour, and had little effect on the neighbouring shoreline by 1860. Gradual accumulation of littoral sediment took place updrift of the jetty, whereas depletion of a beach on the downdrift side had already started in 1900, decreasing the beach width in front of a cliff some distance away from the structure. By 1976 beach erosion was aggravated and the cliff had to be protected by a sea wall (Goudie, 1981, pp. 229-30).

Augmented cliff erosion by similar processes, i.e. associated with depletion of littoral sediments induced by engineering works, has been reported from various locations: Portballintrae, County Antrim, Northern Ireland (Carter et al., 1983), the Black Sea coast of the former USSR (Shuisky and Schwartz, 1988), the Southern California coast (Shepard and Grant, 1947), and the Iwaki coast, Fukushima, Japan (Aramaki, 1978).

On the north central shore of Lake Erie, however, the situation is not so simple as the locations mentioned above. According to Philpott (1986), a claim that the breakwater at Port Burwell Harbour on this coast had caused an increase in the ongoing erosion of till bluffs located east (downdrift) of the harbour was recently filed to the Federal Court of Canada by people who owned property at or near the Lake shore east of Port Burwell, with the claim area extending to more than 20 km far from the harbour. A report of scientific investigations associated with this litigation stated that the coastal development process of cohesive shores on the Great Lakes is different from that of

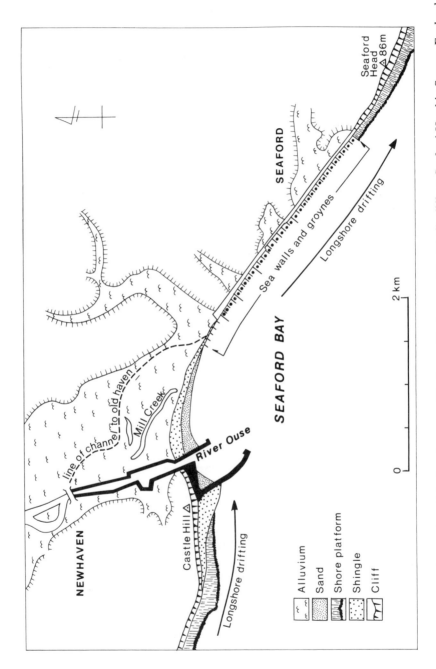

Figure 9.9 Geographical setting of the breakwater at Newhaven Harbour and the Chalk cliffs at Seaford Head in Sussex, England. From Bird (1985a), *The World's Coastline*, by permission of Van Nostrand Reinhold

oceanic beaches, and emphasized that a comprehensive research including field surveys, archive searches, geotechnical tests, and hydraulic laboratory experimentsis required for the elucidation of causality (Philpott, 1986).

Suppose a rocky coast that is fronted with a wide beach composed of sediment mainly supplied from a neighbouring river. The construction of dams or reservoirs on the river certainly reduces the sediment supply to the coast, resulting in the gradual decrease in beach volume, which leads to the progressive exposure of the cliff to wave assault. If the coast consists of soft rocks, erosion may occur.

A classical example of erosion induced by beach mining is the event of Hallsands in Devon, England (see also Chapter 5). Large quantities of shingle were removed in the late 1890s from the foreshore of a micaschist cliff on which the fishing village of Hallsands stood, to obtain material for harbour works at Plymouth. This shingle removal triggered rapid erosion of the cliff, leading to the ruin of the fishing village (Robinson, 1961). Other examples of accelerated cliff erosion by the beach material extraction are found in some places in England: West Bay in Dorset, and Gunwalloe near Helston in Cornwall (Bird, 1985b, p. 74).

Beach nourishment (or replenishment) has been a widely used scheme on the sandy coast to build up the beach by the artificial addition of appropriate sediments. This has recently been applied to rocky coasts to hinder cliff erosion. According to Pinchin and Nairn (1986), to protect clay bluffs on the north shore of Lake Ontario, beach fill was placed between 'armoured hard points' (some sort of groynes) constructed alongshore at a certain distance. This artificial-beach scheme achieved reasonable success in Toronto, the beach now being used as a water-front recreational space.

Compared with conventional engineering structures, artificial beaches provide much easier access to the water and have no problem of sea-floor scouring. Pinchin and Nairn (1986) recommended beach nourishment, when it is cost-effective, as a shore-protection alternative for eroding rocky coasts. Occasional replenishment would be required to maintain the function of protective beaches, because alongshore and/or offshore leakage of the material are likely to occur.

HUMAN ACTIVITIES AND MASS MOVEMENT

There are various types of coastal mass movement (Figure 5.20), one of which is the deep-seated rotational landslide. Landslides of this type occurred at Folkestone Warren in Kent, England, with a close connection with harbour works (Hutchinson *et al.*, 1980). Since 1716 The Warren (Figure 9.10) had experienced ten large rotational or multiple rotational landslidings until 1915 when the major one happened. The geological sequence of this area is Chalk, underlain by Gault (clayey facies) and Folkestone Beds (sandy facies)

222   Geomorphology of Rocky Coa.

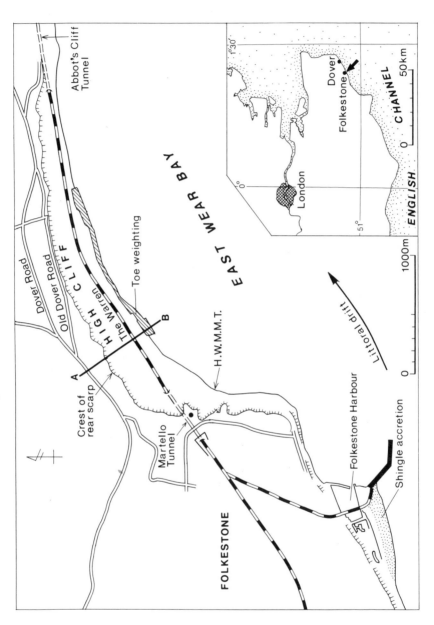

Figure 9.10 Map showing geographical relationship between The Warren and Folkestone Harbour. After Hutchinson *et al.* (1980)

(Figure 9.11), and the main slip surface is located on the boundary between Gault and Folkestone Beds. At Folkestone Harbour, located about 3 km up drift of The Warren (Figure 9.10), progressive extensions of the breakwater, completed in 1905, gave rise to shingle accumulation against the breakwater, which resulted in depletion of beach material in front of The Warren. This depletion led to unloading at the toe of The Warren and caused a series of landslidings on the pre-existing slip surface, culminating in the great slide of 1915 (Figure 9.11), which derailed a train on the railway connecting Folkestone and Dover. The Warren landslides have exhibited no major movement since 1950 when remedial measures were completed, which included a massive toe weighting structure (Figure 9.10), sea defence works (a sea wall and groynes), and drainage works to reduce the ground-water pressure (Hutchinson, 1980).

As a slope stabilization measure, artificial cutting of the cliff slope to make it gentler has been adopted at some coastal locations. They include clayey sandstone cliffs near Black Rock Point, Melbourne, Australia (Bird and Rosengren, 1987), London Clay cliffs at Clacton-on-Sea, Essex, England (Harris and Ralph, 1980), and limestone cliffs at Llantwit Major, Wales (Williams and Davies, 1980), the last case having used a unique technique, blasting, to produce a stable slope. Slope degradation by artificial excavation is commonly considered to be a safe way of improving stability. The slope flattening indeed assists to halt the shallow landslide, but, as Hutchinson (1984b) pointed out, for the area having experienced deep-seated landslides, such slope cutting means unloading at the toe of the overall landslide mass and may provoke an unanticipated large-scale landslide. This occurred in 1962 at Lyme Regis, Dorset, England. This site, composed of Liassic clays, suffered a renewal of a pre-existing landslide a few days after the completion of slope flattening in the front area (Hutchinson, 1984b).

Increase in the slope stress caused by automobile traffic along a road located at the edge of cliffs adds some contribution to the cliff instability, as found in the Santa Cruz area in California, where cliffs develop in Neogene sedimentary rock (Griggs and Johnson, 1979). A similar effect by construction of dwellings near the cliff edge has been reported from the Seaton–Downderry area, south Cornwall, England, where a cliff cut in Pleistocene solifluction deposits stands on a Devonian slate platform (Sims and Ternan, 1988). The saturation of the cliff material by storm-drain discharge is pointed out as a contributing factor to slope failure at the location.

At Black Rock Point, Melbourne, Australia, cliffs are cut in soft Tertiary clayey sands overlying hard ferruginous sandstone, which forms shore platforms. The cliffs had been shaped partly by basal undercutting by storm waves, and partly by gully erosion during heavy rainfall due to runoff from a seaward-sloping ramp at the cliff-crest cascading down the cliff face (Bird and Rosengren, 1986). In order to halt the run-off, railway

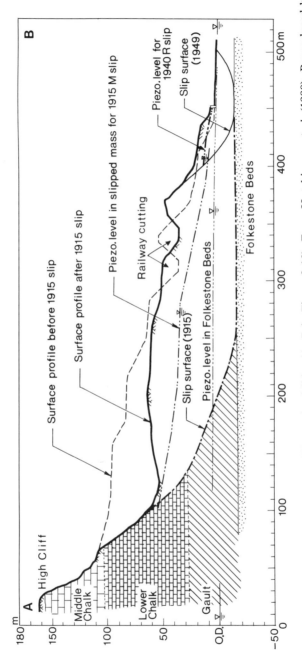

Figure 9.11 Cross section of Folkestone Warren landslides (A–B in Figure 9.10). From Hutchinson *et al.* (1980). Reproduced by permission of the Geological Society from *Additional observations on the Folkestone Warren landslides* by Hutchinson, J. N., Bromhead, E. N. and Lupini, J. F. in *Q. J. Eng. London*, **13**, 1980. *M*slip = massive multiple rotational slip and *R*slip = rotational slip (Hutchinson, 1969)

sleepers were placed in 1981 to form a bank and drain along the top of the cliff, so that the flow during heavy rains was intercepted and gulleying has been much reduced. This resulted in a smoother cliff face. Continued basal erosion by storm waves has also steepened the overall cliff profile (Bird and Rosengren, 1987).

Hengistbury Head in Dorset, England, a recreational area with maximum daily visitors of 3000, has a cliff cut in Eocene sands and clays (May, 1977). Trampling occurs when visitors climb up or down the slope of the cliff to the beach below. Damage to the vegetation cover has allowed the underlying material to be displaced downslope by trampling, so that cliff-face erosion occurs that is directly induced by man. Cottonaro (1975) and Lee (1980) reported similar erosion from the Southern California coast, to which Lee added carving of graffitti and excavation of caves.

FUTURE SEA LEVEL RISE AND CLIFF EROSION

It is widely accepted that a world-wide sea level rise is likely to occur during the next century due to global warming induced by human activities. The first attempt to estimate future sea level rise through the year 2100 was made by the US Environmental Protection Agency (EPA) in the early 1980s under various scenarios incorporating the major contributing factors to the rise in sea level (Hoffman, 1984). As shown in Figure 9.12, the EPA's estimates are in a wide range with the low scenario producing a rise of 50 cm by the year 2100, whereas the high scenario yields a 350 cm rise. The result of sea level rise projection by Thomas (1986) gives considerably lower values, with a narrow range of estimates, 90–170 cm by the year 2100 (Figure 9.12). The updated result of projections made for 'business-as-usual' scenario by the Intergovernmental Panel on Climate Change (IPCC) shows more reduced values (Figure 9.12): 30–110 cm over the 1990 level through the year 2100 (Warrick and Oerlemans, 1990). Figure 9.12 shows that both Thomas' and the IPCC's projections generally fall in an area between the EPA's estimates based on 'mid-range low' and 'low' scenarios, suggesting that a possible sea level rise by 2100 will be in a range between 0.5 and 1 m.

Even a sea level rise of this order greatly influences rocky coasts. Much attention has been focused on beaches and coastal lowlands (e.g. Titus, 1986; Leatherman, 1989, Dubois, 1990), but little on rocky coasts. On hard-rock coasts (largely characterized by Type-B shore platforms or plunging cliffs—Chapter 7), a rise in sea level may bring about submergence of existing platforms, resulting in a temporary landward shift of the shoreline, or the sea may simply move up the cliff face. On soft-rock coasts (where Type-A platforms develop), on the other hand, the submergence is likely to accelerate cliff erosion (Bird, 1986), as is clearly shown by erosion studies along the Great Lakes which have indicated that when lake levels rise bluff recession

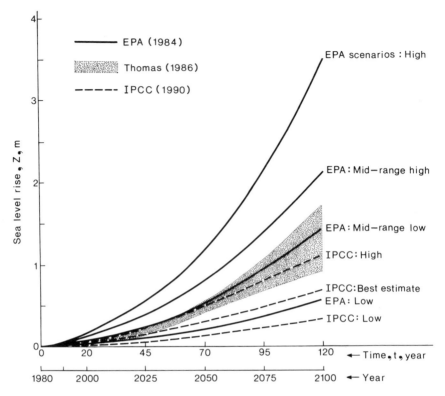

Figure 9.12 Projections of sea level rise through the year 2100

accelerates (Birkemeier, 1980; Quigley and Zeman, 1980; Carter and Guy, 1988).

The most probable rate of sea level rise in the next century is 0.5 to 1 cm/year. This is considerably higher than the average rate of global sea level rise thought to have occurred over the last 100 years, 0.1 to 0.2 cm/year (Warrick and Oerlemans, 1990), but it should be recalled that the world's coasts have already experienced a sea level rise of up to 1 cm/year during the Holocene marine transgression.

In an attempt to hindcast cliff erosion during the sea level rise following the last glacial maximum, Sunamura (1978a) constructed a simple mathematical model which is described by:

$$\frac{dX}{dt} = \frac{dx}{dt} + \frac{dZ}{dt}\bigg|I \qquad (9.1)$$

where dX/dt is the cliff erosion rate under a rising sea level, dx/dt is the cliff recession rate with a stationary sea level (x = the recession distance), dZ/dt is

Effects of Human Activity on Rocky Coasts

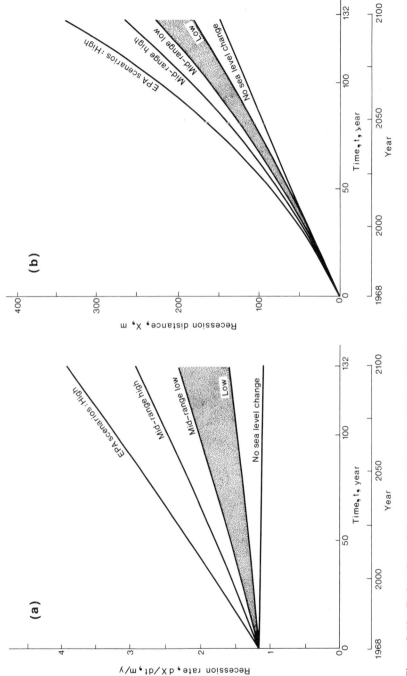

Figure 9.13 Estimation of future recession of a cliff on Nii-jima Island, Japan: (a) recession rate–time curves, and (b) recession distance–time curves. In both diagrams the lowest curve indicates the case of stable sea level. After Sunamura (1988b)

the rising speed of sea level, and I is the platform gradient given by $I = h_a/(x + W_p)$ in which W_p is the width of the present platform, defined as the horizontal distance from the cliff base to the offshore point providing h_a, the wave base (Chapter 6). Using Eq. (9.1) Sunamura (1988b) predicted future erosion of a cliff on a small volcanic island in the Pacific Ocean. The cliff, cut in poorly consolidated deposits, is receding at an average rate of 1.2 m/year. The prediction was made to the year 2100; calculation of dx/dt in Eq. (9.1) was based on Eq. (7.11) (Sunamura, 1987), and dZ/dt was based on the EPA's predictions (Figure 9.12).

Results are plotted in Figure 9.13, which shows that recession rates and distances will be greatly influenced by rates of rising sea level. Figure 9.13a indicates that the recession rates are almost linear in relation to time. Because the future sea level rise is likely to be between the mid-range low and the low scenarios as mentioned before, the most probable recession rate by the year 2100 is from 1.6 to 2.3 m/year, whereas the erosion rate if no sea level change occurred would be 1.1 m/year to 2100, slightly lower than the initial rate (1.2 m/year) because of the increasing width of the underwater platform. As shown in Figure 9.13b, the stationary sea level case produces an almost linear trend between recession distance and time, but with a rising sea level there will be accelerated recession. According to the two likely scenarios, the cliff would retreat 180 to 230 m by the year 2100, which is 1.2 to 1.5 times greater than the recession distance of 150 m with a stable sea level. Thus a rise in sea level of only 0.5 to 1.5 m will substantially accelerate erosion.

The assumption of an unchanging wave climate in the future was made in Sunamura's (1988b) projection. If global warming occurs with marked regional differences in air temperature patterns as predicted by Manabe and Stouffer (1980) and Hansen *et al.* (1984), the new wind systems could produce changes in local patterns of wave climate. Studies by Nicholls (1984) and Emanuel (1987) suggest that a rise in sea surface temperature will increase the intensity and frequency of tropical storms. This means that more severe storm waves will occur more frequently. The accuracy of predictions of future cliff erosion will depend on improved predictions of changes in wave climate as well as in sea level.

References

Abrahams, A. D. and Oak, H. L. (1975) Shore platform widths between Port Kembla and Durras Lake, New South Wales, *Australian Geogr. Studies*, 13, 190–4.
Adey, W. H. and MacIntyre, I. G. (1973) Crustose coralline algae: a re-evaluation in the geological sciences, *Geol. Soc. Am. Bull.*, 84, 883–904.
Agar, R. (1960) Postglacial erosion of the north Yorkshire coast from Tees estuary to Ravenscar, *Proc. Yorkshire Geol. Soc.*, 32 408–25.
Ahr, W. M. and Stanton, R. J., Jr. (1973) The sedimentologic and paleoecologic significance of *Lithotrya*, a rock-boring barnacle, *Jour. Sedimentary Petrology*, 43, 20–3.
Aida, T. and Okamoto, T. (1960) On the cutting mechanism of coal, *Jour. Mining Metalurgical Inst. Japan*, 76, 167–72 (in Japanese with English abstract).
Airy, G. B. (1845) Tides and waves, *Encyclopedia Metropolitana*, Vol. 5, pp. 241–396.
Alexander, H. S. (1932) Pothole erosion, *Jour. Geology*, 4, 305–37.
Allen, J. R. L. (1982) *Sedimentary Structures—Their Character and Physical Basis*, Vol. 1, Elsevier, Amsterdam, 593pp.
Allison, R. J. (1989) Rates and mechanisms of change in hard rock coastal cliffs, *Zeit Geomorphologie N.F.*, Suppl. Bd. 73, 125–38.
Andersen, K. H. (1976) Behaviour of clay subjected to undrained cyclic loading, *Proc. Int. Conference on Behaviour of Offshore Structures*, Trondheim, 1, pp. 392–403.
Ängeby, O. (1951) Pothole erosion in recent water-falls, Lund Studies in Geography, Ser. A, *Physical Geography*, 2, 1–34.
Aramaki, M. (1978) Erosion of coastal cliff at Iwaki coast in Fukushima prefecture, Japan, *Bull. Assoc. Natural Science, Senshu University*, 11, 5–36 (in Japanese with English abstract).
Askin, R. W. and Davidson-Arnott, R. G. D. (1981) Micro-erosion meter modified for use underwater, *Marine Geology*, 40, M45–8.
Bagnold, R. A. (1939) Interim report on wave pressure research, *Jour. Inst. Civil Engr.*, 12, 202–26.
Baker, G. (1943) Features of a Victorian limestone coastline, *Jour. Geology*, 51, 359–86.
Baker, G. (1947) Geology and physiography of the Moonlight Head district, Victoria, *Proc. Roy. Soc. Victoria*, 60, 17–43.
Baker, G. (1958) Stripped zones at cliff edges along a high wave energy coast, Port Campbell, Victoria, *Proc. Roy. Soc. Victoria*, 71, 175–9.
Barnes, H. L. (1956) Cavitation as a geological agent, *Am. Jour. Science*, 254, 493–505.

Barrell, J. (1920) The piedmont terraces of the northern Appalachians, *Am. Jour. Science*, **49**, 327–62.

Bartrum, J. A. (1916) High-water rock platforms: a phase of shoreline erosion, *Trans. Proc. New Zealand Inst.*, **48**, 132–4.

Bartrum, J. A. (1924) The shore-platform of the west coast near Auckland: its storm-wave origin, *Rept. Australian New Zealand Assoc. Advancement Science*, **16**, 493–5.

Bartrum, J. A. (1926) 'Abnormal' shore platforms, *Jour. Geology*, **34**, 793–806.

Bartrum, J. A. (1935) Shore-platforms, *Rept. Australian New Zealand Assoc. Advancement Science*, **22**, 135–43.

Bartrum, J. A. (1936) Honeycomb weathering of rocks near the shoreline, New Zealand Jour. *Science Technology*, **18**, 593–600.

Bascom, W. (1980) *Waves and Beaches*, Anchor Press/Doubleday, Garden City, New York, 366 pp.

Battistini, R. and Le Bourdiec, P. (1985) Madagascar, in E. C. F. Bird and M. L. Schwartz, *The World's Coastline*, Van Nostrand Reinhold, New York, pp. 679–89.

Battjes, J. A. (1974) Surf similarity, *Proc. 14th Coastal Eng. Conf.*, Am. Soc. Civil Engr., pp. 569–87.

Battjes, J. A. and Janssen, J. P. F. M. (1978) Energy loss and set-up due to breaking of random waves, *Proc. 16th Coastal Eng. Conf.*, Am. Soc. Civil Engr., pp. 569–587.

Baulig, H. (1930) Le littoral Dalmate, *Ann. Géographie*, **39**, 305–10.

Belikov, B. P., Zalesskii, B. V., Rozanov, Y. A., Sanina, E. A., and Timchenko, I. P. (1967) Methods of studying the physicomechanical properties of rocks, in B. V. Zalesskii, *Physical and Mechanical Properties of Rocks*, Israel Program for Scientific Translations, Jerusalem, pp. 1–58.

Bell, F. G. (1983) *Engineering Properties of Soils and Rocks*, Butterworths, London, 149pp.

Bernaix, J. (1967) La mesure de la résistance des roches, *Proc. Geotechnical Conf.*, Oslo, pp. 245–57.

Bieniawski, Z. T. (1975) The point-load test in geotechnical practice, *Eng. Geology*, **9**, 1–11.

Biésel, F. (1952) Equations générales au second ordre de la houle irrégulière, *La Houille Blanche*, **3**, 372–6.

Bird, E. C. F. (1970) Shore potholes at Diamond Bay, Victoria, *Victorian Naturalist*, **87**, 312–18.

Bird, E. C. F. (1974) Pitted rocks at Jubilee Point, Victoria, *Victorian Naturalist*, **91**, 60–5.

Bird, E. C. F. (1976) *Coasts*, 2nd edn, Australian National University Press, Canberra, 282pp.

Bird, E. C. F. (1985a) England and Wales, in E. C. F. Bird and M. L. Schwartz, *The World's Coastline*, Van Nostrand Reinhold, New York, pp. 359–69.

Bird, E. C. F. (1985b) *Coastline Changes—A Global Review*, Wiley, Chichester, 219pp.

Bird, E. C. F. (1986) Potential effects of sea level rise on the coasts of Australia, Africa and Asia, in J. G. Titus, *Effects of Changes in Stratospheric Ozone and Global Climate*, Vol. 4: Sea Level Rise, Proc. Int. Conf. Health Environmental Effects of Ozone Modification and Climatic Change, U.N. Environment Programme and US Environmental Protection Agency, pp. 83–98.

Bird, E. C. F. (1988) The tubeworm *Galeolaria caespitosa* as an indicator of sea level rise, *Victorian Naturalist*, **105**, 98–104.

Bird, E. C. F. (1990a) Photographs of London Bridge and their explanation, *Jour. Coastal Res.*, **6** (3) (preliminaries).

Bird, E. C. F. (1990b) Cliff instability on the Victorian coast, *Victorian Naturalist*, **107**, 86–98.
Bird, E. C. F. and Dent, O. F. (1966) Shore platforms on the South Coast of New South Wales, *Australian Geographer*, **10**, 71–80.
Bird, E. C. F. and Rosengren, N. J. (1984) The changing coastline of the Krakatau Islands, Indonesia, *Zeit. Geomorphologie N.F.*, **28**, 347–66.
Bird, E. C. F. and Rosengren, N. J. (1986) Changes in cliff morphology at Black Rock Point 1973–1986, *Victoria Naturalist*, **103**, 106–13.
Bird, E. C. F. and Rosengren, N. J. (1987) Coastal cliff management: an example from Black Rock Point, Melbourne, Australia, *Jour. Shoreline Management*, **3**, 39–51.
Bird, J. B., Richards, A., and Wong, P. P. (1979) Coastal subsystems of Western Barbados, West Indies, *Geografiska Ann.*, **61A**, 221–36.
Birkeland, P. W. and Larson, E. E. (1978) *Putnam's Geology*, Oxford University Press, New York, 659pp.
Birkemeier, W. A. (1980) The effect of structures and lake level on bluff and shore erosion in Berrien County, Michigan, 1970–74, US Army Coastal Eng. Res. Center, Misc. Rept. No. 80-2, 74pp.
Birkemeier, W. A. (1981) Coastal changes, Eastern Lake Michigan, 1970–74, US Army Coastal Eng. Res. Center, Misc. Rept. No. 81-2, 89pp.
Bishop, A. W. (1966) Soils and soft rocks as engineering materials, Inaugural lecture, Imperial College, 17 May, 1966. Cited in Marsland (1972).
Blackmore, P. A. and Hewson, P. J. (1984) Experiments on full-scale wave impact pressures, *Coastal Eng.*, **8**, 331–46.
Bokuniewicz, H. and Tanski, J. (1980) Managing localized erosion of coastal bluffs, *Proc. Coastal Zone '80*, Am. Soc. Civil Engr., **3**, pp. 1883–98.
Bosence, D. W. J. (1983) Coralline algal reef frameworks, *Jour. Geol. Soc. London*, **140**, 365–76.
Boulden, R. S. (ed.) (1975) Canada/Ontario Great Lakes Shore Damage Survey, Tech. Rep., Environment Canada and Ontario, Ministry of Natural Resources, Ottawa, 97pp.
Bourcart, J. (1930) Le problème de 'taffoni' de Corse et l'érosion alvéolaire, *Rev. Geographie Physique et Géologie Dynamique*, **111**, 5–15.
Bowen, A. J., Inman, D. L., and Simmons, V. P. (1968) Wave 'set-down' and set-up, *Jour. Geophysical Res.*, **73**, 2569–77.
Bradley, W. C. (1958) Submarine abrasion and wave-cut platforms, *Geol. Soc. Am. Bull.*, **69**, 967–74.
Bradley, W. C. and Griggs, G. B. (1976) Form, genesis, and deformation of central California wave-cut platforms, *Geol. Soc. Am. Bull.*, **87**, 433–49.
Bretschneider, C. L. (1954) Generation of wind waves over a shallow bottom, US Army Beach Erosion Board, Tech. Memo. No. 51, 24pp.
Bretschneider, C. L. (1958) Revisions in wave forecasting: deep and shallow water, *Proc. 6th Conf. Coastal Eng.*, Am. Soc. Civil Engr., pp. 30–67.
Bretschneider, C. L. (1966a) Wave generation by wind, deep and shallow water, in A. T. Ippen, *Estuary and Coastline Hydrodynamics*, McGraw-Hill, New York, pp. 133–96.
Bretschneider, C. L. (1966b) Engineering aspects of hurricane surge, in A. T. Ippen, *Estuary and Coastline Hydrodynamics*, McGraw-Hill, New York, pp. 231–56.
Bretschneider, C. L. (1968) Decay of wind generated waves to ocean swell by significant wave method, Ocean Industry, 3(3), 36–9 and 51.
Brighenti, G. (1979) Mechanical behaviour of rocks under fatigue, *Proc. 4th Cong. Int. Soc. Rock Mechanics*, Montreux, **1**, pp. 65–70.

Broch, E. and Franklin, J. A. (1972) The point-load strength test, *Int. Jour. Rock Mechanics Mining Sciences*, **9**, 669–97.

Bromhead, E. N. (1986) *The Stability of Slopes*, Surrey University Press, London, 373pp.

Brunsden, D. (1974) The degradation of a coastal slope, Dorset, England, in E. H. Brown and R. S. Waters, *Progress in Geomorphology*, Inst. British Geographers, Special Publ. No. 7, pp. 79–98.

Brunsden, D. (1976) Weathering, in C. Embleton and J. Thornes, *Process in Geomorphology*, Edward Arnold, London, pp. 73–129.

Brunsden, D. and Jones, D. K. C. (1980) Relative time scales and formative events in coastal landslide systems, *Zeit. Geomorphologie N.F.*, Suppl. Bd. **34**, 1–19.

Bryan, R. B. and Price, A. G. (1980) Recession of the Scarborough bluffs, Ontario, Canada, *Zeit. Geomorphologie N.F.*, Suppl. Bd. **34**, 48–62.

Buckler, W. R. and Winters, H. A. (1983) Lake Michigan bluff recession, *Ann. Assoc. Am. Geographers*, **73**, 89–110.

Burshtein, L. S. (1969) Effect of moisture on the strength and deformability of sandstone, *Soviet Mining Science*, **5**, 573–6.

Buttle, J. M. and von Bulow, P. (1986) Crest retreat along the Bluffer's park section of the Scarborough bluffs, Ontario, Canada, *Proc. Symp. Cohesive Shores*, National Research Council, Canada, pp. 87–102.

Byrne, J. V. (1963) Coastal erosion, northern Oregon, in T. Clements, *Essays in Marine Geology in Honor of K. O. Emery*, University of Southern California Press, Los Angeles, pp. 11–33.

Byrne, J. V. (1964) An erosional classification for the northern Oregon coast, *Ann. Assoc. Am. Geographers*, **54**, 329–35.

Cailleux, A. (1953) Taffonis et érosion alvéolaire, *Cahier Géologique de Thoiry*, **16–17**, 130–3.

Cambers, G. (1976) Temporal scales in coastal erosion systems, *Trans. Inst. British Geographers*, New Ser., **1**, 246–56.

Camfield, F. E. (1991) Wave forces on wall, *Jour. Waterway, Port, Coastal, and Ocean Eng.*, **117**, 76–9.

Camfield, F. E. and Street, R. L. (1969) Shoaling of solitary waves on small slopes, *Jour. Waterway Harbors Div., Proc. Am. Soc. Civil Engr.*, **95**, 1–22.

Carr, A. P. and Graff, J. (1982) The tidal immersion factor and shore platform development: discussion, *Trans. Inst. British Geographers*, New Ser., **7**, 240–5.

Carson, M. A. (1971) *The Mechanics of Erosion*, Pion, London, 174pp.

Carter, C. H. (1976) Lake Erie shore erosion, Lake County, Ohio: setting, processes, and recession rates from 1876 to 1973, Div. Geol. Survey, Dept. Natural Resources, Ohio, Invest. Rept. No. 99, 105pp.

Carter, C. H. and Guy, D. E., Jr. (1988) Coastal erosion: processes, timing and magnitudes at the bluff toe, *Marine Geology*, **84**, 1–17.

Carter, C. H., Benson, D. J., and Guy, D. E., Jr. (1981) Shore protection structures: effects on recession rates and beaches from the 1870s to the 1970s along the Ohio shore of Lake Erie, *Environmental Geology*, **3**, 353–62.

Carter, C. H., Benson, D. J., and Guy, D. E., Jr. (1982) Man-made structures and geomorphic changes since 1876 along the Ohio shore of Lake Erie, in R. G. Craig and J. L. Craft, *Applied Geomorphology*, George Allen & Unwin, London, pp. 148–64.

Carter, C. H., Monroe, C. B. and Guy, D. E., Jr. (1986) Lake Erie shore erosion: the effect of beach width and shore protection structures, *Jour. Coastal Res.*, **2**, 17–23.

Carter, C. H., Neal, W. J., Haras, W. S. and Pilkey, O. H., Jr. (1987) *Living with the Lake Erie Shore*, Duke University Press, Durham, North Carolina, 263pp.

References

Carter, R. W. G. (1988) *Coastal Environments*, Academic Press, London, 617pp.
Carter, R. W. G., Lowry, P., and Shaw, J. (1983) An eighty year history of erosion in a small Irish bay, *Shore and Beach*, **51**(3), 34–9.
Cernia, J. N. (1982) *Geotechnical Engineering*, Holt, Rinehart and Winston, New York, 488pp.
Challinor, J. (1949) A principle in coastal geomorphology, *Geography*, **34**, 213–15.
Chandler, R. J. (1972) Lias clay: weathering processes and their effect on shear strength, *Géotechnique*, **22**, 403–31.
Chandler, R. J. and Apted, J. P. (1988) The effect of weathering on the strength of London Clay, *Quart. Jour. Eng. Geology*, **21**, 59–68.
Chappell, J. and Shackleton, N. J. (1986) Oxygen isotopes and sea level, *Nature*, **324**, 137–40.
Chien, N. (1956) Sediment motion at the vicinity of littoral barrier, *US Army Beach Erosion Board, Bull.*, **10**, 21–31.
Chieruzzi, R. and Baker, R. F. (1959) Investigation of bluff recession along Lake Erie, *Jour. Waterway Harbor Div., Proc. Am. Soc. Civil Engr.*, **85**, 109–32.
Clague, J. J. and Bornhold, B. D. (1980) Morphology and littoral processes of the Pacific coast of Canada, in S. B. McCann, *The Coastline of Canada, Geol. Survey Canada*, Paper 80–10, pp. 339–80.
Clayton, K. M. (1989) Sediment input from the Norfolk cliffs, eastern England—A century of coast protection and its effect, *Jour. Coastal Res.*, **5**, 433–42.
Coakley, J. P. and Boyd, G. L. (1979) Fifty Mile Point case history part 1: Long-term recession and sediment sources, National Water Research Institute, Environment Canada, Burlington, Ontario, Unpub. Rept., 25pp.
Coakley, J. P. and Cho, H. K. (1972) Shore erosion in western Lake Erie, *Proc. 15th Conf. Great Lakes Res.*, pp. 344–60.
Coakley, J. P., Rukavina, N. A., and Zeman, A. (1986) Wave-induced subaqueous erosion of cohesive tills: preliminary results, *Proc. Symp. Cohesive Shores*, National Research Council, Canada, pp. 120–36.
Coakley, J. P., Skafel, M. G., Davidson-Arnott, R. G. D., Zeman, A. J., and Rukavina, N. A. (1988) Computer simulation of near shore profile evolution in cohesive materials, *Proc. IAHR Symp. Mathematical Modelling of Sediment Transport in the Coastal Zone*, pp. 290–9.
Coastal Engineering Research Center (1984) *Shore Protection Manual*, US Army Corps of Engineers, US Govt. Printing Office, Washington, D.C., 3 Volumes.
Colback, P. S. B. and Wild, B. L. (1965) The influence of moisture content on the compressive strength of rocks, *Proc. Rock Mechanics Symp., 3rd Canadian Symp.*, Toronto, pp. 65–83.
Coleman, J. M., Gagliano, S. M., and Smith, W. G. (1966) Chemical and physical weathering on saline high tidal flats, N. Queensland, Australia, *Geol. Soc. Am. Bull.*, **77**, 205–6.
Cooke, R. U. and Smally, I. J. (1968) Salt weathering in deserts, *Nature*, **220**, 1226–7.
Cotton, C. A. (1949) Plunging cliffs, Lyttelton Harbour, *New Zealand Geographer*, **5**, 130–6.
Cotton, C. A. (1951) Sea cliffs of Banks Peninsula and Wellington: some criteria for coastal classification, *New Zealand Geographer*, **7**, 103–20.
Cotton, C. A. (1952a) Cyclic resection of headlands by marine erosion, *Geological Mag.*, **89**, 221–5.
Cotton, C. A. (1952b) The Wellington coast: an essay in coastal classification, *New Zealand Geographer*, **8**, 48–62.

Cotton, C. A. (1960) *Geomorphology*, 7th edn, Whitcombe and Tombs, Christchurch, 505pp.
Cotton, C. A. (1963) Levels of plantation of marine benches, *Zeit. Geomorphologie N.F.*, **7**, 97–111.
Cotton, C. A. (1967) Plunging cliffs and Pleistocene coastal cliffing in the southern hemisphere, in J. A. Sporck, *Mélanges de Géographie offerts à M. Omer Tulippe*, J. Duculot, S. A., Gembloux, Vol. 1, pp. 37–59.
Cotton, C. A. (1968) Plunging cliffs, in R. W. Fairbridge, *The Encyclopedia of Geomorphology*, Reinhold, New York, pp. 872–3.
Cottonaro, W. F. (1975) *Sea Cliff Erosion*, Isla Vista, California, California Geology, June, pp. 140–3.
Craig, A. K., Dobkin, S., Grimm, R. B., and Davidson, J. B. (1969) The gastropod *Siphonaria pectinata*: a factor in destruction of beach rock, *Am. Zoologist*, **9**, 895–901.
Croad, R. N. (1981) Physics of erosion of cohesive soils, Ph.D. Thesis, Dept. Civil Eng., University of Auckland, Rept. No. 247, 331pp. Cited in Nairn *et al*. (1986).
Curtis, C. D. (1976) Chemistry of rock weathering: fundamental reactions and controls, in E. Derbyshire, *Geomorphology and Climate*, Wiley, Chichester, pp. 25–57.
Dally, W. R., Dean, R. G., and Dalrymple, R. A. (1984) A model for breaker decay on beaches, *Proc. 19th Coastal Eng. Conf.*, Am. Soc. Civil Engr., pp. 82–98.
Dalrymple, R. A., Biggs, R. B., Dean, R. G., and Wang, H. (1986) Bluff recession rates in Chesapeake Bay, *Jour. Waterway Port Coastal Ocean Eng.*, **112**, 164–8.
Daly, R. A. (1927) The geology of Saint Helena Island, *Proc. Am. Acad. Arts Sci.*, **62** (2) 31–92.
Dana, J. D. (1849) *Geology*, Report of the US Exploring Expedition during the Years 1938 to 1942, Vol. 10, p. 109. Cited in Bartrum (1916).
Dana, J. D. (1875) *Manual of Geology*, 2nd edn, Ivison, Blakemane, Taylor and Co., New York, 828pp.
Davidson-Arnott, R. G. D. (1986a) Erosion of the nearshore profile in till: rates, controls, and implications for shoreline protection, *Proc. Symp. Cohesive Shores*, National Research Council, Canada, pp. 137–49.
Davidson-Arnott, R. G. D. (1986b) Rates of erosion of till in the nearshore zone, *Earth Surface Processes Landforms*, **11**, 53–8.
Davidson-Arnott, R. G. D. and Keizer, H. I. (1982) Shore protection in the town of Stoney Creek, southwest Lake Ontario, 1934–1979. Historical changes and durability of structures, *Jour. Great Lakes Res.*, **8**, 635–47.
Davies, D. S., Axelrod, E. W., and O'Conner, J. S. (1972) Erosion of the north shore of Long Island, Marine Sciences Res. Center, State University of New York, Stony Brook, Tech. Rept. 18, 101pp.
Davies, J. L. (1972) *Geographical Variation in Coastal Development*, Oliver and Boyd, Edinburgh, 204pp.
Davies, P. and Williams, A. T. (1985) Cave development in Lower Lias coastal cliffs, the Glamorgan Heritage Coast, Wales, UK, *Proc. Iceland Coastal and River Symp.*, Reykjavik, Iceland, pp. 75–92.
Davies, P., Williams, A. T., and Bomboe, P. (1991) Numerical modelling of Lower Lias rock failures in the coastal cliffs of South Wales, *Proc. Coastal Sediments '91*, Am. Soc. Civil Engr., pp. 1599–612.
Davis, R. A., Jr. (1976) Coastal changes, Eastern Lake Michigan 1970–73, US Army Coastal Eng. Res. Center Tech. Paper 76–16, 64pp.
Davis, W. M. (1909) The outline of Cape Cod, in W. M. Davis, *Geographical Essays*, Ginn, Boston, pp. 690–724.

Davis, W. M. (1928) *The Coral Reef Problem*, American Geographical Society, New York, 596pp.
De Boer, G. (1977) Coastal erosion, in D. Brunsden, J. C. Doornkamp and D. Ingle-Smith, *The Unquiet Landscape*, David & Charles, Newton Abbot, pp. 73–8.
Dean, J. L., Pope, J. and Fulford, E. (1986) The use of segmented detached breakwaters to protect cohesive shores: Colonial Beach, VA, *Proc. Symp. Cohesive Shores*, National Research Council, Canada, pp. 292–308.
Dean, R. G. (1966) Tides and harmonic analysis, in A. P. Ippen, *Estuary and Coastline Hydrodynamics*, McGraw-Hill, New York, pp. 197–230.
Dean, R. G. (1977) Equilibrium beach profiles: US Atlantic and Gulf coast, Ocean Eng. Rept., Dept. Civil Eng., University of Delaware, No. 12, 45pp.
Dean, R. G. and Eagleson, P. S. (1966) Finite amplitude waves, in A. T. Ippen, *Estuary and Coastline Hydrodynamics*, McGraw-Hill, New York, pp. 93–132.
Deere, D. U. (1964) Technical description of cores for engineering purposes, *Rock Mechanics Eng. Geology*, **1**, 16–22.
Deere, D. U. and Miller, R. P. (1966) Engineering classification and index properties for intact rock, Air Force Weapons Laboratory, Tech. Rept. No. AFWL-TR-65-116, Kirkland Air Force Base, New Mexico, 300pp.
Deere, D. U., Hendron, A. J., Jr., Patton, F. D., and Cording, E. J. (1966) Design of surface and near-surface construction in rock, in C. Fairhurst, *Failure and Breakage of Rock, Proc. 8th Symp. Rock Mechanics*, Am. Inst. Mining, Metallurgical, Petroleum Engr., pp. 237–302.
Denny, D. F. (1951) Further experiments on wave pressures, *Jour. Inst. Civil Eng.*, **35**, 330–45.
Dick, T. M. and Zeman, A. J. (1983) Coastal processes on soft shores, *Proc. Canadian Coastal Conf.*, pp. 19–35.
Dietz, R. S. (1952) Geomorphic evolution of continental terrace (continental shelf and slope), *Am. Assoc. Petroleum Geologists Bull.*, **36**, 1802–19.
Dietz, R. S. (1963) Wave-base, marine profile of equilibrium, and wave-built terraces: a critical appraisal, *Bull. Geol. Soc. Am.*, **74**, 971–90.
Dietz, R. S. and Fairbridge, R. W. (1968) Wave base, in R. W. Fairbridge, *The Encyclopedia of Geomorphology*, Reinhold, New York, pp. 1224–8.
Dietz, R. S. and Menard, H. (1951) Origin of abrupt change in slope at continental-shelf margin, *Am. Assoc. Petroleum Geologists Bull.*, **35**, 1994–2016.
Dionne, J. C. (1964) Notes sur les marmites littorales, *Revue Géographie Montreal*, **18**, 249–77.
Dionne, J. C. and Brodeur, D. (1988) Frost weathering and ice action in shore platform development with particular reference to Quebec, Canada, *Zeit. Geomorphologie N.F.*, Suppl. Bd. **71**, 117–30.
Doornkamp, J. C. and King, C. A. M. (1971) *Numerical Analysis in Geomorphology*, an Introduction, Edward Arnold, London, 372pp.
Dubois, R. N. (1990) Barrier-beach erosion and rising sea level, *Geology*, **18**, 1150–2.
Duckmanton, N. K. (1974) The shore platforms of the Kaikoura Peninsula, Unpub. M.A. Thesis, University of Canterbury. Cited in Trenhaile and Layzell (1981).
Duncan, N. (1967) Rock mechanics and earthwork engineering, *Muck Shifter*, January, 35–40.
Eagleson, P. S. and Dean, R. G. (1966) Small amplitude wave theory, in A. T. Ippen, *Estuary and Coastine Hydrodynamics*, McGraw-Hill, New York, pp. 1–92.
Edil, T. B. and Haas, B. J. (1980) Proposed criteria for interpreting stability of lakeshore bluffs, *Eng. Geology*, **16**, 97–110.

Edil, T. B. and Vallejo, L. E. (1980) Mechanics of coastal landslides and the influence of slope parameters, *Eng. Geology*, **16**, 83–96.

Edwards, A. B. (1941) Storm-wave platforms, *Jour. Geomorphology*, **4**, 223–36.

Edwards, A. B. (1951) Wave action in shore platform formation, *Geological Mag.*, **88**, 41–9.

Edwards, A. B. (1958) Wave-cut platforms at Yampi Sound in the Buccaneer Archipelago, W. A., *Jour. Roy. Soc. Western Australia*, **41**, 17–21.

Elston, E. D. (1917) Potholes: their variety, origin and significance, *Scientific Monthly*, **5**, 554–67.

Elston, E. D. (1918) Potholes: their variety, origin and significance. II, *Scientific Monthly*, **6**, 37–51.

Emanuel, K. A. (1987) The dependence of hurricane intensity on climate, *Nature*, **326**, 483–5.

Emery, K. O. (1941) Rate of surface retreat of sea cliffs based on dated inscriptions, *Science*, **93**, 617–18.

Emery, K. O. (1946) Marine solution basins, *Jour. Geology*, **54**, 209–28.

Emery, K. O. (1960) *The Sea off Southern California*, Wiley, New York, 366pp.

Emery, K. O. and Foster, H. L. (1956) Shoreline nips in tuff at Matsushima, Japan, *Am. Jour. Science*, **254**, 380–5.

Emery, K. O. and Kuhn, G. G. (1980) Erosion of rock shores at La Jolla, California, *Marine Geology*, **37**, 197–208.

Emery, K. O. and Kuhn, G. G. (1982) Sea cliffs: their processes, profiles, and classifications, *Geol. Soc. Am. Bull.*, **93**, 644–54.

Emmons, W. H., Thiel, G. A., Stauffer, C. R., and Allison, I. S. (1955) *Geology*, McGraw-Hill, New York, 638pp.

Evans, I. and Pomeroy, C. D. (1958) The strength of cubes of coal in uniaxial compression, in W. H. Walton, *Mechanical Properties of Non-Metallic Brittle Materials*, Butterworths, London, pp. 1–28.

Evans, I. and Pomeroy, C. D. (1966) *The Strength, Fracture and Workability of Coal*, Pergamon Press, Oxford, 277pp.

Evans, J. W. (1968) The role of *Penitella penita* (Conarad 1837) (Family Pholadidae) as eroders along the Pacific Coast of North America, *Ecology*, **49**, 156–9.

Everts, C. H. (1991) Seacliff retreat and coarse sediment yields in southern California, *Proc. Coastal Sediments '91*, Am. Soc. Civil Engr., pp. 1586–98.

Eyles, N., Buergin, R., and Hincenbergs, A. (1986) Sedimentological controls on piping structures and the development of scalloped slopes along an eroding shoreline, Scarborough bluffs, Ontario, *Proc. Symp. Cohesive Shores*, National Research Council, Canada, pp. 69–86.

Fairbridge, R. W. (1947) Notes on the geomorphology of the Pelsart Group of the Houtman's Abrolhos Islands, *Jour. Roy. Soc. Western Australia*, **33**, 1–43.

Fairbridge, R. W. (1950) The geology and geomorphology of Point Person, Western Australia, *Jour. Roy. Soc. Western Australia*, **34**, 35–72.

Fairbridge, R. W. (1952) Marine erosion, *Proc. 7th Pacific Science Cong.*, 1949, New Zealand, **3**, pp. 347–59.

Fairbridge, R. W. (1968) Platforms—wave-cut, in R. W. Fairbridge, *The Encyclopedia of Geomorphology*, Reinhold, New York, pp. 859–65.

Farmer, I. (1983) *Engineering Behaviour of Rocks*, 2nd edn, Chapman and Hall, London, 208pp.

Feda, J. (1966) The influence of water content on the behaviour of subsoil, formed by highly weathered rocks, *Proc. 1st Cong. Int. Soc. Rock Mechanics*, Lisbon, **1**, pp. 283–8.

Fenneman, N. M. (1902) Development of the profile of equilibrium of the subaqueous shore terrace, *Jour. Geology*, **10**, 1–32.
Flaxman, E. M. (1963) Channel stability in undisturbed cohesive soils, *Jour. Hydraulics Div., Proc. Am. Soc. Civil Engr.*, **89**, 87–96.
Fleming, C. A. (1986) Holderness coast protection project, *Proc. Symp. Cohesive Shores*, National Research Council, Canada, pp. 394–420.
Fleming, C. A. and Summers, L. (1986) Artificial headlands on a clay cliff shoreline, *Proc. Symp. Cohesive Shores*, National Research Council, Canada, pp. 262–76.
Flemming, N. C. (1965) Form and relation to present sea level of Pleistocene marine erosion features, *Jour. Geology*, **73**, 799–811.
Focke, J. W. (1978a) Limestone cliff morphology and organism distribution on Curaçao (Netherlands Antilles), *Leidse Geol. Meded.*, **51**, 131–50.
Focke, J. W. (1978b) Limestone cliff morphology on Curaçao (Netherlands Antilles), with special attention to the origin of notches and vermetid/coralline algal surf benches ('cornices', 'trottoirs'). *Zeit. Geomorphologie N.F.*, **22**, 329–49.
Fox, F. (1924) *Sixty-Three Years of Engineering*, John Murray, London, 338pp.
Friedman, G. M. and Sanders, J. E. (1978) *Principles of Sedimentology*, Wiley, New York. 792pp.
Fryde, W. T. (1968) Cliff drainage and beach distribution, *Proc. 11th Coastal Eng. Conf.*, Am. Soc. Civil Engr., pp. 644–52.
Fulton-Bennett, K. and Griggs, G. B. (1986) *Coastal Protection Structures and Their Effectiveness*, Joint Publication of Department of Boating, State of California and Waterways and Marine Science Institute, University of California (Santa Cruz), 48pp.
Galvin, C. J., Jr. (1968) Breaker type classification on three laboratory beaches, *Jour. Geophysical Res.*, **73**, 3651–59.
Galvin, C. J., Jr. (1969) Breaker travel and choice of design wave height, *Jour. Waterways Harbors Div., Proc. Am. Soc. Civil Engr.*, **95**, 175–200.
Galvin, C. J., Jr. (1972) Wave breaking in shallow water, in R. E. Meyer, *Waves on Beaches and Resulting Sediment Transport*, Academic Press, New York, pp. 413–56.
Garner, H. F. (1974) *The Origin of Landscapes, A Synthesis of Geomorphology*, Oxford University Press, New York, 734pp.
Garrels, R. M. (1951) *A Textbook of Geology*, Harper, New York, 511pp.
Gary, M., McAfee, R., Jr., and Wolf, C. L. (1974) *Glossary of Geology*, American Geological Institute, Washington, D.C., 805pp.
Gatto, L. W. (1978) Shoreline changes along the outer shore of Cape Cod from Long Point to Monomoy Point, CRREL Rep. 78-17, 49pp.
Gaughan, M. K. (1978) Depth of disturbance of sand in surf zones, *Proc. 16th Coastal Eng. Conf.*, Am. Soc. Civil Engr., pp. 1513–30.
Gelinas, P. J. and Quigley, R. M. (1973) The influence of geology on erosion rates along the north shore of Lake Erie, *Proc. 16th Conf. Great Lakes Res.*, pp. 421–30.
Gibb, J. G. (1978) Rates of coastal erosion and accretion in New Zealand, *New Zealand Jour. Marine Freshwater Res.*, **12**, 429–50.
Giese, G. S. and Aubrey, D. G. (1987) Bluff erosion on Outer Cape Cod, *Proc. Coastal Sediments '87*, Am. Soc. Civil Engr., pp. 1871–6.
Gill, E. D. (1950) Some unusual shore platforms near Gisborne, North Island, New Zealand, *Trans. Proc. Roy. Soc. New Zealand*, **78**, 64–8.
Gill, E. D. (1967) The dynamics of shore platform process and its relation to changes in sea-level, *Proc. Roy. Soc. Victoria*, **80**, 183–92.
Gill, E. D. (1971) Rocks plucked by the sea, *Victorian Naturalist*, **88**, 287–90.

Gill, E. D. (1972a) The relationship of present shore platforms to past sea levels, *Boreas*, **1**, 1–25.
Gill, E. D. (1972b) Sanders' wave tank experiments and shore platforms, *Papers Proc. Roy. Soc. Tasmania*, **106**, 17–20.
Gill, E. D. (1972c) Ramparts on shore platforms, *Pacific Geology*, **4**, 121–33.
Gill, E. D. (1973) Rate and mode of retrogradation on rocky coasts in Victoria, Australia, and their relationship to sea level changes, *Boreas*, **2**, 143–71.
Gill, E. D. (1976) Large waves at Lorne, Victoria, *Victorian Naturalist*, **93**, 92–5.
Gill, E. D. (1981) Rapid honeycomb weathering (tafoni formation) in greywacke, S. E. Australia, *Earth Surface Processes Landforms*, **6**, 81–3.
Gill, E. D. and Lang, J. G. (1983) Micro-erosion meter measurements of rock wear on the Otway coast of southeast Australia, *Marine Geology*, **52**, 141–56.
Gill, E. D., Segnit, E. R. and McNeill, N. H. (1977) Concretions in Otway group sediments, South-east Australia, *Proc. Roy. Soc. Victoria*, **89**, 51–5.
Gill, E. D., Segnit, E. R., and McNeil, N. H. (1981) Rate of formation of honeycomb weathering features (small scale tafoni) on the Otway coast, S.E. Australia, *Proc. Roy. Soc. Victoria*, **92**, 149–54.
Gilluly, J., Waters, A. C., and Woodford, A. O. (1959) *Principles of Geology*, 2nd edn, Freeman, San Francisco, 534pp.
Goda, Y. (1970) A synthesis of breaker indices, *Trans. Japan Soc. Civil Engr.*, **2** (Part 2), 227–30.
Goda, Y. (1985) *Random Seas and Design of Maritime Structures*, University of Tokyo Press, Tokyo, 323pp.
Goudie, A. (1981) *The Human Impact—Man's Role in Environmental Charge*, Basil Blackwell, Oxford, 326pp.
Goudie, A. and Gardner, R. (1985) *Discovering Landscape in England and Wales*, Allen & Unwin, London, 177pp.
Grainger, P. and Kalaugher, P. G. (1987) Intermittent surging movements of a coastal landslide, *Earth Surface Processes Landforms*, **12**, 597–603.
Greenwood, B. and Hale, P. B. (1980) Depth of activity, sediment flux and morphological change in a barred beach environment, in S. B. McCann, *The Coastline of Canada*, Geol. Survey Canada, Paper 80–10, pp. 89–109.
Griggs, G. B. and Johnson, R. E. (1979) Coastal erosion, Santa Cruz County, *California Geology*, April, 67–76.
Grisez, L. (1960) Alvéolisation littorale de schistes metamorphiques, *Rev. Géomorphologie Dynamique*, **11**, 164–7.
Groen, P. and Groves, G. W. (1962) Surges, in M. N. Hill, *The Sea*, Vol. 1, Interscience, Wiley, New York, pp. 611–46.
Grosvenor, N. E. (1963) Specimen proportion—Key to better compressive strength tests, *Mining Eng.*, **15**, 31–3.
Guidicini, G., Nieble, C. M., and Cornides, A. T. (1973) Analysis of point load test as a method for preliminary geotechnical classification of rocks, *Bull. Int. Assoc. Eng. Geology*, **7**, 37–52.
Guilcher, A. (1953) Essai sur la zonation et la distribution des formes littorales de dissolution du calcaire, *Ann. Géographie*, **62**, 161–79.
Guilcher, A. (1957) Formes de corrosion littorale du calcaire sur les côtes du Portugal, in *The Earth, its Crust and Atmosphere*, Brill, Leiden, pp. 49–55. Cited in Hills (1972).
Guilcher, A. (1958) *Coastal and Submarine Morphology*, Methuen, London, 274pp.
Guilcher, A. (1985) Angola, in E. C. F. Bird and M. L. Schwartz, *The World's Coastline*, Van Nostrand Reinhold, New York, pp. 639–43.

Gulliver, F. (1899) Shoreline topography, *Proc. Am. Acad. Arts Sciences*, **34**, 151–258.
Guza, R. T. and Thornton, E. B. (1981) Wave set-up on a natural beach, *Jour. Geophysical Res.*, **86**, 4133–7.
Hails, J. R. (1975) Sediment distribution and Quaternary history, *Jour. Geol. Soc. London*, **131**, 19–35.
Hails, J. R. (1977) Applied geomorphology in coastal-zone planning and management, in J. R. Hails, *Applied Geomorphology*, Elsevier, Amsterdam, pp. 317–62.
Hansen, J., Lacis, A., Rind, D., Russell, G., Stone, P., Fung, I., Ruedy, R., and Lerner, J. (1984) climate sensitivity: analysis of feedback mechanism, in J. E. Hansen and T. Takahashi, *Climate Sensitivity*, American Geophysical Union, Washington, D.C., pp. 130–63.
Harris, D. L. (1963) Characteristics of the hurricane storm surge, US Weather Bureau, Tech. Paper, No. 48, 139pp.
Harris, W. B. and Ralph, K. J. (1980) Coastal engineering problems at Clacton-on-Sea, Essex, *Quart. Jour. Eng. Geology London*, **13**, 97–104.
Hatai, K. and Mii, H. (1957) Observations along the southern coast of Izu peninsula, Shizuoka prefecture, Japan, *Record Oceanographic Works Japan*, **4**, 23–9.
Hatai, K., Funayama, Y. and Mii, H. (1957) A note on the development of certain marine pot-holes along the west coast of Wakayama prefecture, Japan, *Record Oceanographic Works Japan*, **4**, 45–8.
Hattori, M. (1982) Field study on onshore-offshore sediment transport, *Proc. 18th Coastal Eng. Conf.*, Am. Soc. Civil Engr., pp. 923–40.
Hawkes, I. and Mellor, M. (1970) Uniaxial testing in rock mechanics laboratories, *Eng. Geology*, **4**, 177–285.
Hayami, S. (1958) Type of breakers, wave steepness and beach slope, *Coastal Eng. Japan*, **1**, 21–3.
Hayashi, M. (1966) Strength and dilatancy of brittle jointed mass—The extreme value stochastic and anisotropic failure mechanism, *Proc. 1st Cong. Int. Soc. Rock Mechanics*, Lisbon, **1**, pp. 275–302.
Hayes, M. O. (1972) Forms of sediment accumulation in the beach zone, in R. E. Meyer, *Waves on Beaches and Resulting Sediment Transport*, Academic Press, New York, pp. 297–356.
Healy, T. R. (1968a) Shore platform morphology on the Whangaparaoa Peninsula, Auckland, *Conference Series, New Zealand Geogr. Soc.*, **5**, 163–8.
Healy, T. R. (1968b) Bioerosion on shore platforms in the Waitemata Formation, Auckland, *Earth Science Jour.*, **2**, 26–37.
Healy, T. R. and Wefer, G. (1980) The efficacy of submarine abrasion vs cliff retreat as a supplier of marine sediment in the Kieler Bucht, Western Baltic, *Meyniana*, **32**, 89–96.
Healy, T. R., Sneyd, A. D., and Werner, F. (1987) First approximation sea-level dependent mathematical model for volume eroded and submarine profile development in a semienclosed sea: Kiel Bay, Wertern Baltic, *Math. Geology*, **19**, 41–56.
Henkel, I. (1906) A study of tide pools on the west coast of Vancouver Island, Postelsia, in *The Year Book of the Minnesota Seaside Station*, pp. 277–304. Cited in Elston (1917).
High, C. and Hanna, F. K. (1970) A method for the direct measurement of erosion on rock surfaces, *British Geomorph. Res. Group, Tech. Bull.*, No. 5, 24pp.
Higgins, C. G. (1980) Nips, notches, and the solution of coastal limestone: an overview of the problem with examples from Greece, *Estuarine Coastal Marine Science*, **10**, 15–30.

Hills, E. S. (1949) Shore platforms, *Geological Mag.*, **86**, 137–52.
Hills, E. S. (1971) A study of cliffy coastal profiles based on examples in Victoria, Australia, *Zeit. Geomorphologie N.F.*, **15**, 137–80.
Hills, E. S. (1972) Shore platforms and wave ramps, *Geological Mag.*, **109**, 81–8.
Hinds, N. E. A. (1930) The geology of Kauai and Niihau, *Bernice P. Bishop Museum, Bull.*, No. 71, 103pp.
Hobbs, D. W. (1964) Rock compressive strength, *Colliery Engineering*, **41**, 287–92.
Hobbs, N. B. (1974) Factors affecting the prediction of settlement of structures on rocks: with particular reference to the Chalk and Trias, in British Geotechnical Society, *Settlement of Structures*, Pentech Press, London, pp. 579–610.
Hodgkin, E. P. (1964) Rate of erosion of intertidal limestone, *Zeit. Geomorphologie N.F.*, **8**, 385–92.
Hoek, E. and Bray, J. W. (1981) *Rock Slope Engineering*, 3rd edn, Institute of Mining and Metallurgy, London, 358pp.
Hoffman, J. S. (1984) Estimates of future sea level rise, in M. C. Barth and J. G. Titus, *Greenhouse Effect and Sea Level Rise*, Van Nostrand Reinhold, New York, pp. 79–103.
Höllermann, P. (1975) Formen kavernöser verwitterung ('tafoni') auf Teneriffa, *Catena*, **2**, 385–410.
Holmes, A. and Holmes, D. L. (1978) *Holmes Principles of Physical Geology*, 3rd edn, Thames Nelson and Sons, Middlesex, 730pp.
Hom-ma, M. and Horikawa, K. (1964) Wave force against sea wall, *Proc. 9th Coastal Eng. Conf.*, Am. Soc. Civil Engr., pp. 490–503.
Hom-ma, M., Horikawa, K. and Hase, N. (1962) On wave forces against sea wall, *Proc. 9th Japan. Conf. Coastal Eng.*, Japan Soc. Civil Engr., pp. 133–7 (in Japanese).
Hondros, G. (1959) The evaluation of Poisson's ratio and the modulus of materials of a low tensile resistance by the Brazilian (indirect tensile) test with particular reference to concrete, *Australian Jour. Applied Science*, **10**, 243–68.
Horibe, T., Kobayashi, R., and Ikemi, Y. (1970) Fatigue test of rocks, *Proc. 3rd Japan. Symp. Rock Mechanics*, pp. 29–34 (in Japanese with English Abstract).
Horikawa, K. (1978) *Coastal Engineering*, University of Tokyo Press, Tokyo, 402pp.
Horikawa, K. and Sunamura, T. (1967) A study on erosion of coastal cliffs by using aerial photographs, *Coastal Eng. Japan*, **10**, 67–83.
Horikawa, K. and Sunamura, T. (1968) An experimental study on erosion of coastal cliffs due to wave action, *Coastal Eng. Japan*, **11**, 131–47.
Horikawa, K. and Sunamura, T. (1970) A study on erosion of coastal cliffs and of submarine bedrocks, *Coastal Eng. Japan*, **13**, 127–39.
Hoshino, M. (1975) *Eustacy in Relation to Orogenic Stage*, Tokai University Press, Tokyo, 397pp.
Houpert, R. (1970) La résistance à la rupture des roches en compression simple, *Proc. 2nd Cong. Int. Soc. Rock Mechanics*, Belgrade, **2**, pp. 49–55.
Houston, W. N. and Herrmann, H. G. (1980) Undrained cyclic strength of marine soils, *Jour. Geotechnical Eng. Div., Proc. Am. Soc. Civil Engr.*, **106**, 691–712.
Hudson, R. Y. (1952) Wave forces on breakwaters, *Proc. Am. Soc. Civil Engr.*, **78**, 1–22.
Hume, W. F. (1925) *Geology of Egypt*, Vol. 1, *The Surface Features of Egypt, their Determining Causes and Relation to Geological Structure*, Government Press, Cairo, 380pp.
Hutchinson, J. N. (1969) A reconsideration of the coastal landslides at Folkestone Warren, Kent, *Géotechnique*, **19**, 6–38.

Hutchinson, J. N. (1970) A coastal mudflow on the London Clay cliffs at Beltinge, north Kent, *Géotechnique*, **20**, 412–38.

Hutchinson, J. N. (1972) Field and laboratory studies of a fall in Upper Chalk cliffs at Joss Bay, Isle of Thanet, in R. H. G. Parry, *Stress–Strain Behaviour of Soils, Proc. Roscoe Memorial Symposium*, G. T. Foulis & Co. Ltd, Henley-on-Thames, Oxfordshire, pp. 692–706.

Hutchinson, J. N. (1973) The response of London Clay cliffs to differing rates of toe erosion, *Geologia Applicata e Idrogeologia*, **8**, 221–39.

Hutchinson, J. N. (1980) Various forms of cliff instability arising from coastal erosion in south-east England, in A. M Heltzen, K. Garshol, and A. Mowinckel-Amundsen, *Fjellspnengringsteknikk, Bergmekanikk, Geoteknikk 1979*, Tapir for Norsk Jord-og Fjellteknisk Forbund tilkuyttet NIF, Trondheim, pp. 19.1–19.32.

Hutchinson, J. N. (1983) A pattern in the incidence of major coastal mudslides, *Earth Surface Processes Landforms*, **8**, 391–7.

Hutchinson, J. N. (1984a) Landslides in Britain and their countermeasures, *Jour. Japan Landslide Soc.*, **21**, 1–24.

Hutchinson, J. N. (1984b) An influence line approach to the stabilization of slopes by cuts and fills, *Canadian Geotechnical Jour.*, **21**, 363–70.

Hutchinson, J. N. (1986) Cliffs and shores in cohesive materials: geotechnical and engineering geological aspects, *Proc. Symp. Cohesive Shores*, National Research Council, Canada, pp. 1–44.

Hutchinson, J. N., Bromhead, E. N. and Lupini, J. F. (1980) Additional observations on the Folkestone Warren landslides, *Quart. Jour. Geol. Soc. London*, **13**, 1–31.

Ijima, T., Takahashi, T., and Nakamura, K. (1956) A study of wave properties in the surfzone by using stereo-camera system, *Proc. 3rd Japan. Conf. Constal Eng.*, Japan Soc. Civil Engr., pp. 99–116 (in Japanese).

Ikeda, K. (1979) The property and the strength of fissured rock masses, *Jour. Japan Soc. Eng. Geology*, **20**, 158–70 (in Japanese with English abstract).

Imanaga, I. (1975) On the coastal erosive features of Enoshima island, *Bull. Kanagawa Prefectural Museum*, **8**, 79–90 (in Japanese with English abstract).

Inman, D. L. and Filloux, J. (1960) Beach cycles related to tide and local wind regine, *Jour. Geology*, **68**, 225–31.

Inman, D. L. Zampol, J. A., White, T. E., Hanes, D. M., Waldorf, B. W., and Kastens, K. A. (1980) Field measurements of sand motion in the surf zone, *Proc. 17th Coastal Eng. Conf.*, Am. Soc. Civil Engr., pp. 1215–34.

Inoue, M. and Omi, M. (1971) Relations among compressional wave velocity, apparent density and uniaxial compressivstrength of common sedimentary and igneous rocks, *Jour. Japan Soc. Eng. Geology*, **12**, 1–9 (in Japanese with English abstract).

Inozemtsev, Yu. P., Pashkov, N. N., Pshenitsyn, P. A., Rosanov. N. P., Sakharov, V. I., and Shalnev, K. K. (1965) Cavitational-erosion resistance of hydrotechnical concretes on cement and polymer binders, *Proc. 11th Cong. Int. Assoc. Hydraulic Res.*, Paper 1.48, pp. 1–17.

International Bureau for Rock Mechanics (1961) Group 3, Strength Research, Appendix H, Prague, March 21–25, 1961. Cited in Vutukuri *et al.* (1974, p. 69).

Ippen, A. T. and Kulin, G. (1954) The shoaling and breaking of the solitary wave, *Proc. 5th Coastal Eng. Conf.*, Am. Soc. Civil Engr., pp. 27–49.

Irfan, T. Y. and Dearman, W. R. (1978) Engineering classification and index properties of a weathered granite, *Bull. Int. Assoc. Eng. Geology*, **17**, 79–90.

Ito, R. (1940) New examples of pothole erosion at the seashore of Inbo, Tiba Prefecture, and in the Valley of the Hida-gawa, *Geogr. Rev. Japan*, **16**, 73–93 (in Japanese with English abstract).

Iversen, H. W. (1952) Waves and breakers in shoaling water, *Proc. 3rd Coastal Eng. Conf.*, Am. Soc. Civil Engr., pp. 1–12.
Jaeger, C. (1972) *Rock Mechanics and Engineering*, Cambridge University Press, Cambridge, 417pp.
Jehu, T. J. (1918) Rock-boring organisms as agents in coastal erosion, *Scottish Geogr. Mag.*, **34**, 1–11.
Jennings, J. N. (1968) Tafoni, in R. W. Fairbridge, *The Encyclopedia of Geomorphology*, Reinhold, New York, pp. 1103–4.
Johannessen, C. L., Feiereisen, J. J. and Wells, A. N. (1982) Weathering of ocean cliffs by salt expansion in a mid-latitude coastal environment, *Shore and Beach*, **50**, 26–34.
Johnson, D. W. (1919) *Shore Processes and Shoreline Development*, Hafner, New York, 584pp.
Johnson, D. W. (1925) *The New England Acadian Shoreline*, Wiley, New York, 608pp.
Johnson, D. W. (1931) Supposed two-metre eustatic bench of the Pacific shores, *Int. Geog. Cong.*, Paris, **13**, 158–63.
Johnson, D. W. (1938) Shore platforms, discussion, *Jour. Geomorphology*, **1**, 268–72.
Jones, D. G. and Williams, A. T. (1991) Statistical analysis of factors influencing cliff erosion along a section of the West Wales coast, UK, *Earth Surface Processes Landforms*, **16**, 95–111.
Jones, J. R., Fisher, J. J., and Reigler, P. (1985) Shoreline change at Thompson Island, Boston Harbor, Massachusetts, 1938–1977, *Physical Geography*, **5**, 198–206.
Jonsson, I. G. (1966) Wave boundary layers and friction factors, *Proc. 10th Coastal Eng. Conf.*, Am. Soc. Civil Engr., pp. 127–48.
Jutson, J. T. (1939) Shore platforms near Sydney, New South Wales, *Jour. Geomorphology*, **2**, 237–50.
Jutson, J. T. (1949) The shore platforms of Lorne, Victoria, *Proc. Roy. Soc. Victoria*, **61**, 43–59.
Kaizuka, S. (1955) On the age of submarine shelves of southern Kanto, *Geogr. Rev. Japan*, **28**, 15–24 (in Japanese with English abstract).
Kamphuis, J. W. (1975) Friction factor under oscillatory waves, *Jour. Waterways Harbors Coastal Eng.*, Proc. Am. Soc. Civil Engr., **101**, 135–45.
Kamphuis, J. W. (1987) Recession rate of glacial till bluffs, *Jour. Waterway Port Coastal Ocean Eng.*, **113**, 60–73.
Kamphuis, J. W. and Hall, K. R. (1983) Cohesive material erosion by unidirectional current, *Jour. Hydraulic Eng.*, **109**, 49–61.
Kanazaki, H. (1961) Shoreline recession on the western Kanazawa coast, Japan, in F. Tada, *Dr. Tsujimura Commemoriative Volume of Geographical Treatises*, Kokonshoin, Tokyo, pp. 145–58 (in Japanese with English abstract).
Kawana, T. and Pirazzoli, P. A. (1985) Holocene coastline changes and seismic uplift in Okinawa Island, the Ryukyus, Japan, *Zeit. Geomorphologie N.F.*, Suppl. Bd. **57**, 11–31.
Kawasaki, I. (1954) Geomorphological study of the Byobugaura sea-cliff in the vicinity of Iioka-machi, Chiba prefecture, *Geogr. Rev. Japan*, **27**, 213–17 (in Japanese with English abstract).
Kayanne, H. and Yoshikawa, T. (1986) Comparative study between present and emergent erosional landforms on the southeast coast of Boso Peninsula Central Japan, *Geogr. Rev. Japan*, **59** (Ser. A), 18–36 (in Japanese with English abstract).
Kaye, C. A. (1959) Shoreline features and Quaternary shoreline changes, Puerto Rico, US Geol. Survey Prof. Paper 317–B, pp. 49–140.

Kaye, C. A. (1967) Erosion of a sea cliff, Boston Harbor, Massachusetts, in O. C. Farquhar, *Economic Geology in Massachusetts*, Graduate School, University of Massachusetts, Amherst, Massachusetts, pp. 521–8.

Kelletat, D. (1980) Studies on the age of honeycombs and tafoni features, *Catena*, **7**, 317–25.

Kennedy, M. P. (1973) *Sea-cliff erosion at Sunset Cliffs*, San Diego, *California Geology*, February, 27–31.

Keunen, Ph. H. (1960) *Marine Geology*, Wiley, New York, 568pp.

Kézdi, Á. (1974) *Soil Physics, Handbook of Soil Mechanics*, Vol. 1, Elsevier, Amsterdam, 294pp.

King, C. A. M. (1951) Depth of disturbance of sand on sea beaches by waves, *Jour. Sedimentary Petrology*, **21**, 131–40.

King, C. A. M. (1963) Some problems concerning marine plantation and the formation of erosion surfaces, *Trans. Papers Inst. British Geographers*, **33**, 29–43.

King, C. A. M. (1972) *Beaches and Coasts*, 2nd edn, Edward Arnold, London, 570pp.

King, C. A. M. (1974) *Introduction to Marine Geology and Geomorphology*, Edward Arnold, London, 309pp.

Kino, Y. (1958) *The Geological Map of Hyuga-Aoshima and its Explanatory Text*, Geol. Survey Japan, 5pp.

Kirk, R. M. (1975) Coastal changes at Kaikoura, 1942–74, determined from air photographs, *New Zealand Jour. Geology Geophysics*, **18**, 788–801.

Kirk, R. M. (1977) Rates and forms of erosion on intertidal platforms at Kaikoura Peninsula, South Island, New Zealand, *New Zealand Jour. Geology Geophysics*, **20**, 571–613.

Kirkgoz, M. S. (1982) Shock pressure of breaking waves on vertical walls, *Jour. Waterway Port Coastal Ocean Div., Proc. Am. Soc. Civil Engr.*, **108**, 81–95.

Kobayashi, M. (1983) A study on wave-cut bench at Tateyama, Boso Peninsula, Japan, Unpub. B.S. Thesis, University of Tsukuba, 51pp. (in Japanese with English abstracts).

Kobayashi, R. (1970) On mechanical behaviours of rocks under various loading-rates, *Rock Mechanics in Japan*, **1**, 56–8.

Kobayashi, R. and Okumura, K. (1971) Study on shear test of rocks, *Jour. Mining Metalurgical Inst. Japan*, **87**, 407–12 (in Japanese with English abstract).

Kohno, F., Nagamatsu, K., and Kiyan, T. (1978) Field observation of wave transformation on a reef, *Proc. 25th Japan. Conf. Coastal Eng.*, Japan Soc. Civil Engr., pp. 146–50 (in Japanese).

Kolberg, T. O. (1974) Evaluation of water levels effect on erosion rate, *Summaries 14th Coastal Eng. Conf.*, Am. Soc. Civil Engr., pp. 413–16.

Komar, P. D. (1976) *Beach Processes and Sedimentation*, Prentice-Hall, Englewood Cliffs, New Jersey, 429pp.

Komar, P. D. (1983) Beach processes and erosion—an introduction, in P. D. Komar, *CRC Handbook of Coastal Processes and Erosion*, CRC Press, Boca Raton, Florida, pp. 1–20.

Komar, P. D. and Gaughan, M. K. (1972) Airy wave theory and breaker height prediction, *Proc. 13th Coastal Eng. Conf.*, Am. Soc. Civil Engr., pp. 405–18.

Komar, P. D. and Inman, D. L. (1970) Longshore sand transport on beaches, *Jour. Geophysical Res.*, **75**, 5914–27.

Komar, P. D. and McDougal, W. G. (1988) Coastal erosion and engineering structures: The Oregon experience, *Jour. Coastal Res.*, Special Issue, No. 4, 77–92.

Komar, P. D. and Shih, S-M. (1991) Sea-cliff erosion along the Oregon coast, *Proc. Coastal Sediments '91*, Am. Soc. Civil Engr., pp. 1558–70.

Kraus, N. C. (1985) Field experiments on vertical mixing of sand in the surf zone, *Jour. Sedimentary Petrology*, **55**, 3–14.
Kraus, N. C., Gingerich, K. J. and Rosati, J. D. (1988) Toward an improved empirical formula for longshore sand transport, *Proc. 21st Coastal Eng. Conf.*, Am. Soc. Civil Engr., pp. 1182–96.
Kuenen, Ph. H. (1960) *Marine Geology*, Wiley, New York, 568pp.
Kuhn, G. G. and Shepard, F. P. (1980) Coastal erosion in San Diego, California, *Proc. Coastal Zone '80*, Am. Soc. Civil Engr., **III**, pp. 1899–918.
Lama, R. D. and Vutukuri, V. S. (1978) *Handbook on Mechanical Properties of Rocks*, Vol. II, Trans Tech Publications, Clausthal, 481pp.
Le Méhauté, B. (1976) *An Introduction to Hydrodynamics and Water Waves*, Springer, New York, 315pp.
Le Méhauté, B. and Koh, R. C. Y. (1967) On the breaking of waves arriving at an angle to the shore, *Jour. Hydraulic Res.*, **5**, 67–88.
Leatherman, S. P. (1986) Cliff stability along western Chesapeake Bay, Maryland, *Marine Technology Soc. Jour.*, **20** (3), 28–36.
Leatherman, S. P. (1989) Impact of accelerated sea level rise on beaches and coastal wetlands, in J. C. White, *Global Climate Change Linkages*, Elsevier, Amsterdam, pp. 43–57.
Lee, K. L. and Focht, J. A., Jr. (1976) Strength of clay subjected to cyclic loading, *Marine Geotechnology*, **1**, 165–85.
Lee, L. J. (1980) Sea cliff erosion in Southern California, *Proc. Coastal Zone '80*, Am. Soc. Civil Engr., pp. 1919–38.
Lee, L. J., Pinckney, C., and Bemis, C. (1976) Sea cliff base erosion, *National Water Resources Ocean Eng. Conv.*, San Diego, California, Am. Soc. Civil Engr., Preprint 2708, pp. 1–14.
Leet, L. D. and Judson, S. (1958) *Physical Geology*, 2nd edn, Prentice Hall, Englewood Cliffs, New Jersey, 502pp.
Longuet-Higgins, M. S. (1952) On the statistical distribution of the heights of sea waves, *Jour. Marine Res.*, **11**, 245–66.
Longuet-Higgins, M. S. (1980) The unsolved problem of breaking waves, *Proc. 17th Coastal Eng. Conf.*, Am. Soc. Civil Eng., pp. 1–28.
Longuet-Higgins, M. S. and Stewart, R. W. (1963) A note on wave set-up, *Jour. Marine Res.*, **21**, 4–10.
Longuet-Higgins, M. S. and Stewart, R. W. (1964) Radiation stresses in water waves— a physical discussion with applications, *Deep-sea Res.*, **11**, 529–62.
Longwell, C. R., Flint, R. F. and Sanders, J. E. (1969) *Physical Geology*, Wiley, New York, 685pp.
Louis, H. (1968) *Allgemeine Geomorphologie*, Walter de Gruyter, Berlin, 522pp.
Lugt, H. J. (1983) *Vortex Flow in Nature and Technology*, Wiley, New York, 297pp.
Lundborg, N. (1967) The strength-size relation of granite, *Int. Jour. Rock Mechanics Mining Sciences*, **4**, 269–72.
Lundborg, N. (1968) Strength of rock-like materials, *Int. Jour. Rock Mechanics Mining Sciences*, **5**, 421–54.
McCowan, J. (1894) On the highest wave of permanent type, *Philosophical Mag.*, Ser. 5, **38**, 351–8.
Macdonald, G. A. and Abbott, A. T. (1970) *Volcanoes in the Sea: The Geology of Hawaii*, University Press of Hawaii, Honolulu, 441pp.
MacFadyen, W. A. (1930) The undercutting of coral reef limestone on the coast of some islands in the Red Sea, *Geogr. Jour.*, **75**, 27–34.

McGreal, W. S. (1979a) Factors promoting coastal slope instability in southeast County Down, N. Ireland, *Zeit. Geomorphologie N.F.*, **23**, 76–90.
McGreal, W. S. (1979b) Marine erosion of glacial sediments from a low-energy cliffline environment near Kilkeel, Northern Ireland, *Marine Geology*, **32**, 89–103.
McGreal, W. S. (1979c) Cliffline recession near Kilkeel, N. Ireland; an example of a dynamic coastal system, *Geografiska Ann.*, **61A**, 211–19.
McGreal, W. S. and Craig, D. (1977) Mass-movement activity: An illustration of differing responses to groundwater conditions from two sites in northern Ireland, *Irish Geography*, **10**, 28–35.
McGreevy, J. P. (1985) A preliminary scanning electron microscope study of honeycomb weathering of sandstone in a coastal environment, *Earth Surface Processes Landforms*, **10**, 509–18.
Machatschek, F. (1959) *Geomorphologie*, B. G. Teubner, Stuttgart, 219pp.
McLean, R. F. (1967) Measurement of beachrock erosion by some tropical marine gastropods, *Bull. Marine Science*, **17**, 551–61.
McLean, R. F. and Davidson, C. F. (1968) The role of mass-movement in shore platform development along the Gisborne coastline, New Zealand, *Earth Science Jour.*, **2**, 15–25.
Magoon, O. T., Pope, J. L., Sloan, R. L., and Treadwell, D. D. (1988) Long term experience with seawalls on an exposed coast, *Proc. 21st Coastal Eng. Conf.*, Am. Soc. Civil Engr., pp. 2455–68.
Manabe, S. and Stouffer, R. J. (1980) Sensitivity of a global climate model to an increase of CO_2 concentration in the atmosphere, *Jour. Geophysical Res.*, **85**, 5529–54.
Marsland, A. (1972) The shear strength of stiff fissured clays, in R. H. G. Parry, Stress–Strain Behaviour of Soils, *Proc. Roscoe Memorial Symposium*, G. T. Foulis, Henley-on-Thames, Oxfordshire, pp. 59–139.
Martini, I. P. (1978) Tafoni weathering, with examples from Tuscany, Italy, *Zeit. Geomorphologie N.F.*, **22**, 44–67.
Mather, K. F. (1964) *The Earth beneath Us*, Nelson, Sunbury-on-Thames, Middlesex, 320pp.
Matsukura, Y. and Yatsu, E. (1982) Wet-dry slaking of Tertiary shale and tuff, *Trans. Japan. Geomorph. Union*, **3**, 25–39.
Matsukura, Y. and Matsuoka, N. (1991) Rates of tafoni weathering on uplifted shore platforms in Nojima-zaki, Boso Peninsula, Japan, *Earth Surface Processes Landforms*, **16**, 51–6.
Matthews, E. R. (1934) *Coast Erosion and Protection*, Charles Griffin, London, 228pp.
May, V. J. (1971) The retreat of chalk cliffs, *Geogr. Jour.*, **137**, 203–6.
May, V. J. (1977) Earth cliffs, in R. S. K. Barnes, *The Coastline*, Wiley, Chichester, pp. 215–35.
May, V. J. and Heeps, C. (1985) The nature and rates of change on chalk coastlines, *Zeit. Geomorphologie N.F.*, Suppl. Bd. **57**, 81–94.
Mellor, M. and Hawkes, I. (1971) Measurement of tensile strength by diametral compression of discs and annuli, *Eng. Geology*, **5**, 173–225.
Miche, M. (1944) Mouvements ondulatoires de la mer en profondeur constante ou décroissante, *Ann. des Ponts et Chaussées*, **114**, 369–406.
Michell, J. H. (1893) The highest waves in water, *Philosophical Mag.*, Ser. 5, **36**, 430–7.
Mii, H. (1962) Coastal geology of Tanabe bay, *Sci. Rept., Tohoku University*, 2nd Ser. (Geology), **34**, 1–93.

Miller, A. A. (1939) Attainable standards of accuracy in the determination of preglacial sea levels by physiographic methods, *Jour. Geomorphology*, **2**, 95–115.
Minikin, R. R. (1963) *Winds, Waves and Maritime Structures*, 2nd ed., Griffin, London, 294pp.
Mitsuyasu, H. (1962) Experimental study on wave force against a wall, Rept. Transportation Technical Res. Inst., Ministry of Transportation, No. 47, 39pp.
Mitsuyasu, H. (1963) Wave pressure and wave runup, in S. Yokota, *Handbook of Hydraulics*, Japan Soc. Civil Engr., pp. 505–28 (in Japanese).
Mizuguchi, M., Tsujioka, K., and Horikawa, K. (1978) A model of change of wave heights in surf zone, *Proc. 25th Japan. Conf. Coastal Eng.*, Japan Soc. Civil Engr., 155–9 (in Japanese).
Mizuguchi, M., Isobe, M., and Ogawa, Y. (1988) Waves in the surfzones, in K. Horikawa, *Nearshore Dynamics and Coastal Processes—Theory, Measurement, and Predictive Models*, University of Tokyo Press, Tokyo, pp. 79–115.
Mogi, A., Tsuchide, M., and Fukushima, M. (1980) Coastal erosion of new volcanic island Nishinoshima, *Geogr. Rev. Japan*, **53**, 449–62 (in Japanese with English abstract).
Mogi, K. (1962) The influence of the dimensions of specimens on the fracture strength, *Bull. Earthquake Research Inst., University of Tokyo*, **4**, 175–85.
Mogi, K. (1966) Some precise measurements of fracture strength of rocks under uniform compressive stress, *Rock Mechanics Eng. Geology*, **4**, 41–55.
Munk, W. H. (1949) The solitary wave theory and its application to surf problems, *Ann. New York Acad. Sci.*, **51**, 376–424.
Muromachi, T. (1957) Relationship between cone-type penetration resistance in clayey soils and uniaxial compressive strength, *Jour. Japan Soc. Civil Engr.*, **42** (10), 7–12 (in Japanese).
Mustoe, G. E. (1982a) Alveolar weathering, in M. L. Schwartz, *The Encyclopedia of Beaches and Coastal Environments*, Hutchinson Ross, Stroudsburg, Pennsylvania, pp. 37–8.
Mustoe, G. E. (1982b) Tafone, in M. L. Schwartz, *The Encyclopedia of Beaches and Coastal Environments*, Hutchinson Ross, Stroudsburg, Pennsylvania, pp. 805–6.
Mustoe, G. E. (1982c) The origin of honeycomb weathering, *Geol. Soc. Am. Bull.*, **93**, 108–15.
Nagai, S. (1961) Shock pressures exerted by breaking waves on breakwaters, *Trans. Am. Soc. Civil Engr.*, **126**, 772–809.
Nagai, S. and Otsubo, M. (1968) Wave pressures acting on low-mound breakwaters, *Proc. 15th Japan. Conf. Coastal Eng.*, Japan Soc. Civil Engr., 109–14 (in Japanese).
Nairn, R. B. (1986) Physical modeling of wave erosion on cohesive profiles, *Proc. Symp. Cohesive Shores*, National Research Council, Canada, pp. 210–25.
Nairn, R. B., Pinchin, B. M. and Philpott, K. L. (1986) Cohesive profile development model, *Proc. Symp. Cohesive Shores*, National Research Council, Canada, pp. 246–61.
Nash, D. (1987) A comparative review of limit equilibrium methods of stability analysis, in M. G. Anderson and K. S. Richards, *Slope Stability*, Wiley, Chichester, pp. 11–75.
Neumann, A. C. (1966) Observations on coastal erosion in Bermuda and measurements of the boring sponge, *Cliona lampa*, *Limnology Oceanography*, **11**, 92–108.
Newell, N. D. and Imbrie, J. (1955) Biogeological reconnaissance in the Bimini area, Great Bahama Bank, *New York Acad. Sci. Trans.*, Ser. 2, **18**, 3–14.
Nicholls, N. (1984) The southern oscillation, sea-surface temperature, and interannual fluctuations in Australian tropical cyclone activity, *Jour. Climatology*, **4**, 661–70.

Noda, H. (1971) Beach processes, *Summer Seminar Text on Hydraulic Eng.*, Japan Soc. Civil Engr., (B – 5) pp. 1–27 (in Japanese).

Nohara, M., Tsuchiya, R., and Yamaji, H. (1983) Potholes on the Kiwado coast, Yamaguchi prefecture, *Abstracts Spring Meeting Assoc. Japan. Geographers*, pp. 68–9 (in Japanese).

Norris, R. M. (1968) Sea cliff retreat near Santa Barbara, California, *Mineral Information Service*, **21** (6), 87–91.

Norrman, J. O. (1980) Coastal erosion and slope development in Surtsey Island, *Zeit. Geomorphologie N.F.*, Suppl. Bd. **34**, 20–38.

North, W. J. (1954) Size distribution, erosive activities, and gross metabolic efficiency of the marine intertidal snails, *Littorina planaxis* and *L. scutulata*, *Biological Bull.*, **106**, 185–97.

Nunn, P. D. (1990) Coastal processes and landforms of Fiji: their bearing on Holocene sea-level changes in the South and West Pacific, *Jour. Coastal Res.*, **6**, 279–310.

Ohshima, H. (1974) Erosional pattern of sea cliffs, with special reference to geological aspects, *Proc. Symp. Coastal Cliff Recession*, Japan Soc. Civil Engr., pp. 9–22 (in Japanese).

Okazaki, S. and Sunamura, T. (1991) Re-examination of breaker-type classification on uniformly inclined laboratory beaches, *Jour. Coastal Res.*, **7**, 559–64.

Ollier, C. (1984) *Weathering*, 2nd edn, Longman, London, 270pp.

Ongley, M. (1940) Note on coastal benches formed by spray weathering, *New Zealand Jour. Science Technology* (Sec. B), **22**, 34–5.

Onodera, T. F. (1963) Dynamic investigation of foundation rocks *in situ*, in C. Fairburst, *Rock Mechanics, Proc. 5th Symp. Rock Mechanics*, Pergamon Press, Oxford, pp. 517–33.

Parker, G. F., Matrich, M. A. J., and Denney, B. E. (1986) Stabilization studies — South Marine Drive Sector Scarborough Bluffs, Ontario, *Proc. Symp. Cohesive Shores*, National Research Council, Canada, pp. 356–77.

Partenscky, H. W. (1988) Dynamic forces due to waves breaking at vertical coastal structures, *Proc. 21st Coastal Eng. Conf.*, Am. Soc. Civil Engr., pp. 2504–18.

Patrick, D. A. and Wiegel, R. L. (1955) Amphibian tractors in the surf, *Proc. 1st Conf. Ships and Waves*, Am. Soc. Naval Archit. Marine Engr., pp. 397–422.

Pethick, J. (1984) *An Introduction to Coastal Geomorphology*, Edward Arnold, London, 260pp.

Phillips, B. A. M. (1970) Effective levels of marine plantation on raised and present rock platforms, *Révue Géographie Montreal*, **24**, 227–40.

Philpott, K. L. (1984) Comparison of cohesive coasts and beach coasts, *Proc. Coastal Eng. Canada*, pp. 227–44.

Philpott, K. L. (1986) Coastal engineering aspects of the Port Burwell shore erosion damage litigation, *Proc. Symp. Cohesive Shores*, National Research Council, Canada, pp. 309–38.

Pierson, W. J., Neumann, G., and James, R. W. (1955) Observing and forecasting ocean waves by means of wave spectra and statistics, US Navy Hydrog. Office, Publ. No. 603, 284pp.

Pinchin, B. M. and Nairn, R. B. (1986) The use of numerical models for the design of artificial beaches to protect cohesive shores, *Proc. Symp. Cohesive Shores*, National Research Council, Canada, pp. 196–209.

Plakida, M. E. (1970) Pressure of waves against vertical walls, *Proc. 12th Coastal Eng. Conf.*, Am. Soc. Civil Engr., pp. 1451–68.

Prêcheur, C. (1960) Le littoral de la Manche, de Sainte-Adresse à Ault, Étude morphologique, Special issue, Norois, Centre National de la Recherche Scientifique ed du Conseil Général de la Seine-Maritime, Poitiers, pp. 1–138.

Pringle, A. W. (1985) Holderness coast erosion and the significance of ords, *Earth Surface Processes Landforms*, **10**, 107–24.
Prior, D. B., Stephens, N., and Archer, D. R. (1968) Composite mudflows on the Antrim coast of north-east Ireland, *Geografiska Ann.*, **50A**, 65–78.
Protodyakonov, M. M. (1960) New methods of determining mechanical properties of rocks, *Proc. 3rd Int. Conf. Strata Control*, Paris, pp. 187–91.
Protodyakonov, M. M. (1962) Mechanical properties and drillability of rocks, in C. Fairhurst, *Rock Mechanics, Proc. 5th Symp. Rock Mechanics*, Pergamon Press, Oxford, pp. 103–18.
Protodyakonov, M. M. (1969) Method of determining the strength of rocks under uniaxial compression, in M. M. Protodyakonov, M. I. Koifman, and others, *Mechanical Properties of Rocks*, Israel Program for Scientific Translations, Jerusalem, pp. 1–8.
Pugh, D. T. (1987) *Tides, Surges and Mean Sea-Level*, Wiley, Chichester, 472pp.
Quigley, R. M. and Gelinas, P. J. (1976) Soil mechanics aspect of shoreline erosion, *Geoscience Canada*, **3**, 169–73.
Quigley, R. M. and Di Nardo, L. R. (1980) Cyclic instability modes of eroding clay bluffs, Lake Erie Northshore Bluffs at Port Bruce, Ontario, Canada, *Zeit. Geomorphologie N.F.*, Suppl. Bd. **34**, 39–47.
Quigley, R. M. and Zeman, A. J. (1980) Strategy for hydraulic, geologic and geotechnical assessment of Great Lakes shoreline bluffs, in S. B. McCann, *The Coastline of Canada*, Geol. Survey Canada, Paper 80–10, pp. 397–406.
Quigley, R. M., Gelinas, P. J., Bou, W. T., and Packer, R. W. (1977) Cyclic erosion-instability relationships: Lake Erie northshore bluffs, *Canadian Geotechnical Jour.*, **14**, 310–23.
Raichlen, F. (1966) Harbor resonance, in A. T. Ippen, *Estuary and Coastline Hydrodynamics*, McGraw-Hill, New York, pp. 281–340.
Raudkivi, A. J. (1967) *Loose Boundary Hydraulic*, Pergamon Press, Oxford, 331pp.
Reffell, G. (1978) Descriptive analysis of the subaqueous extensions of subaerial rock platforms, Unpub. B.A. Thesis, University of Sydney, 144p.
Revelle, R. and Emery, K. O. (1957) Chemical erosion of beach rock and exposed reef rock, U.S. Geol. Survey Prof. Paper, 260-T, pp. 699–709.
Rich, J. L. (1951) Three critical environments of deposition, and criteria for recognition of rocks deposited in each of them, *Geol. Soc. Am. Bull.*, **62**, 1–20.
Richards, A. F. (1965) Geology of the Islas Revillagigedo, 3. Effect of erosion on Isla San Benedicto 1952–61 following the birth of Volcán Bárcena, *Bull. Volcanologique*, **28**, 381–403.
Richards, K. S. and Lorriman, N. R. (1987) Basal erosion and mass movement, in M. G. Anderson and K. S. Richards, *Slope Stability*, Wiley, Chichester, pp. 331–57.
Ritter, D. F. (1986) *Process Geomorphology*, WmC. Brown, Dubuque, Iowa, 579pp.
Robinson, A. H. W. (1961) The hydrography of Start Bay and its relationship to beach changes at Hallsands, *Geogr. Jour.*, **127**, 63–77.
Robinson, A. H. W. (1980a) Erosion and accretion along part of the Suffolk coast of East Anglia, England, *Marine Geology*, **37**, 133–46.
Robinson, A. H. W. (1980b) Coastal changes in Holderness with special reference to the influence of the offshore zone, Unpub. Rept. Cited in Fleming (1986).
Robinson, D. A. and Jerwood, L. C. (1987) Sub-aerial weathering of chalk shore platforms during harsh winters in southeast England, *Marine Geology*, **77**, 1–14.
Robinson, L. A. (1976) The micro-erosion meter technique in a littoral environment, *Marine Geology*, **22**, M51–8.

Robinson, L. A. (1977a) The morphology and development of northeast Yorkshire shore platform, *Marine Geology*, **23**, 237–55.
Robinson, L. A. (1977b) Marine erosive processes at the cliff foot, *Marine Geology*, **23**, 257–71.
Robinson, L. A. (1977c) Erosive processes on the shore platform of northeast Yorkshire, England, *Marine Geology*, **23**, 339–61.
Rode, K. (1930) Geomorphologenie des Ben Lemond (Kalifornien) eine Studie uber Terrassenbuildung durch Marine Abrasion, *Zeit. Geomorphologie*, **5**, 16–78.
Ross, C. W. (1955) Laboratory study of shock pressures of breaking waves, US Army Beach Erosion Board, Tech. Memo., No. 59, 22pp.
Rudberg, S. (1967) The cliff coast of Gotland and the rate of cliff retreat, *Geografiska Ann.*, **49A**, 283–98.
Rundgren, L. (1958) Water wave force, *Bull. Hydraulics Div.*, Roy. Inst. Tech., Stockholm, No. 54, 121pp.
Russell, R. J. (1958) Geological geomorphology, *Geol. Soc. Am. Bull.*, **69**, 1–22.
Russell, R. J. (1963) Recent recession of tropical cliffy coasts, *Science*, **139**, January, 9–15.
Russell, R. J. (1970) Oregon and northern California coastal reconnaissance, Tech. Rept., Coastal Studies Inst., Louisiana State University, No. 86, 25pp.
Russell, R. J. (1971) Water-table effects on sea coasts, *Geol. Soc. Am. Bull.*, **82**, 2343–8.
Sainflou, M., (1928) Essai sur les digues maritimes verticales, *Ann. des Ponts et Chaussées*, **98**, 5–48.
Sakamoto, N. and Ijima, T. (1963) Properties of wave caused by typhoons along the Pacific coast of Japan and their estimations by significant wave method, *Coastal Eng. Japan*, **6**, 103–14.
Sanders, N. K. (1968a) The development of Tasmanian shore platforms, Unpub. Ph.D. Thesis, University of Tasmania, 402pp.
Sanders, N. K. (1968b) Wave tank experiments on the erosion of rocky coasts, *Papers Proc. Roy. Soc. Tasmania*, **102**, 11–16.
Sanders, N. K. (1970) Production of horizontal high tidal shore platforms, *Australian Natural History*, **16**, 315–19.
Sanglerat, G. (1972) *The Penetrometer and Soil Exploration*, Elsevier, Amsterdam, 488pp.
Sasaki, M. and Saeki, H. (1974) A study of the deformation of waves after breaking (2), *Proc. 21st Japan. Conf. Coastal Eng.*, Japan Soc. Civil Engr., pp. 39–44 (in Japanese).
Sato, S., Matsuura, H., and Miyazaki, M. (1987) Potholes in Shikoku, Japan, Part 1. Potholes and their hydrodynamics in the Kurokawa river, Ehime, *Memoirs, Faculty Educ., Ehime University*, **7**, 127–90.
Sato, Sh., Ijima, T., and Tanaka, N. (1962) A study of critical depth and mode of sand movement using radioactive glass sand, *Proc. 8th Coastal Eng. Conf.*, Am. Soc. Civil Engr., pp. 304–23.
Schalk, M. (1938) A texture study of the outer beach of Cape Cod, Massachusetts, *Jour. Sedimentary Petrology*, **8**, 41–54.
Schlichting, H. (1968) *Boundary Layer Theory*, McGraw-Hill, New York, 748pp.
Schneider, J. (1976) Biological and inorganic factors in the destruction of limestone coasts, *Contributions to Sedimentology*, **6**, 1–112.
Schwartz, M. L. (1971) Shannon Point cliff recession, *Shore and Beach*, April, 45–8.
Seed, H. B. and Chan, C. K. (1966) Clay strength under earthquake loading conditions, *Jour. Soil Mechanics Foundations Div., Proc. Am. Soc. Civil Engr.*, **92**, 53–78.

Seibold, E. (1963) Geological investigation of nearshore sand-transport, examples of methods and problems from the Baltic and North Seas, in M. Sears, *Progress in Oceanography*, Pergamon Press, Oxford, pp. 1–70.

Shepard, F. P. and Grant, U. S. (1947) Wave erosion along the Southern California coast, *Bull. Geol. Soc. Am.*, **58**, 919–26.

Shepard, F. P. and Inman, D. L. (1950) Nearshore circulation, *Proc. 1st Coastal Eng. Conf.*, Council on Wave Research, Berkley, California, pp. 50–9.

Shepard, F. P. and Kuhn, G. G. (1983) History of sea arches and remnant stacks of La Jolla, California, and their bearing on similar features elsewhere, *Marine Geology*, **51**, 139–61.

Short, A. D. (1982a) Erosion ramp, wave ramp, in M. L. Schwartz, *The Encyclopedia of Beaches and Coastal Environments*, Hutchinson Ross, Stroudsburg, Pennsylvania, p. 393.

Short, A. D. (1982b) Quarrying processes, in M. L. Schwartz, *The Encyclopedia of Beaches and Coastal Environments*, Hutchinson Ross, Stroudsburg, Pennsylvania, p. 674.

Short, A. D. (1982c) Wave-cut bench and wave-cut platform, in M. L. Schwartz, *The Encyclopedia of Beaches and Coastal Environments*, Hutchinson Ross, Stroudsburg, Pennsylvania, pp. 856–7.

Shuisky, Y. D. and Schwartz, M. L. (1988) Human impact and rates of shore retreat along the Black Sea Coast, *Jour. Coastal Res.*, **4**, 405–16.

Shumway, G., Dietz, R. S., Dill, R. F., Hamilton, E. L., Menard, H. W., and Moore, D. G. (1954) Observations by diving geologists of bedrock off southern California, *Abstract, Geol. Soc. Am. Bull.*, **65**, 1304–5.

Silvester, R. (1960) Stabilization of sedimentary coastlines, *Nature*, **188**, 467–9.

Silvester, R. (1974a) *Coastal Engineering*, Vol. 1, Elsevier, Amsterdam, 457pp.

Silvester, R. (1974b) *Coastal Engineering*, Vol. 2, Elsevier, Amsterdam, 338pp.

Sims, P. and Ternan, L. (1988) Coastal erosion: protection and planning in relation to public policies—a case study from Downderry, south-east Cornwall, in J. M. Hooke, *Geomorphology in Environmental Planning*, Wiley, Chichester, pp. 231–44.

Singh, D. P. (1981) Determination of some engineering properties of weak rocks, *Proc. Int. Symp. Weak Rock*, Tokyo, pp. 315–20.

Skinner, B. J. and Porter, S. C. (1987) *Physical Geology*, Wiley, New York, 750pp.

Snead, R. E. (1982) *Coastal Landforms and Surface Features*, Hutchinson Ross, Stroudsburg, Pennsylvania, 247pp.

So, C. L. (1965) Coastal platforms of the Isle of Thanet, Kent, *Trans. Inst. British Geographers*, **37**, 147–56.

So, C. L. (1967) Some coastal changes between Whitstable and Reculver, Kent, *Proc. Geologists' Assoc.*, **77**, 475–90.

Sorensen, R. M. (1968) Recession of marine terraces—with special reference to the coastal area north of Santa Cruz, California, *Proc. 11th Coastal Eng. Conf.*, Am. Soc. Civil Engr., pp. 653–70.

Sparks, B. W. (1986) *Geomorphology*, 3rd edn, Longman, London, 561pp.

Spencer, T. (1985a) Weathering rates on a Caribbean reef limestone: results and implications, *Marine Geology*, **69**, 195–201.

Spencer, T. (1985b) Marine erosion rates and coastal morphology of reef limestones on Grand Cayman Island, West Indies, *Coral Reefs*, **4**, 59–70.

Spencer, T. (1988) Coastal biogeomorphology, in H. A. Viles, *Biogeomorphology*, Basil Blackwell, Oxford, pp. 255–318.

Stearns, H. T. (1935) Shore benches on the island of Oahu, Hawaii, *Geol. Soc. Am. Bull.*, **46**, 1467–82.

Steers, J. A. (1951) Notes on erosion along the coast of Suffolk, *Geological Mag.*, **88**, 435–9.
Steers, J. A. (1960) *The Coast of England & Wales in Pictures*, Cambridge University Press, Cambridge, 146pp.
Strahler, A. N. 1963 *The Earth Sciences*, Harper and Row, New York, 681pp.
Sturm, M. and Kolberg, M. (1986) Protecting public lands in St Catharines: A case study, *Proc. Symp. Cohesive Shores*, National Research Council, Canada, pp. 378–93.
Summers, D. (1978) *The East Coast Floods*, David & Charles, Newton Abbot, 176pp.
Sunamura, T. (1972) A study on the formation of continental shelves, *Geogr. Rev. Japan*, **45**, 813–23 (in Japanese with English abstract).
Sunamura, T. (1973) Coastal cliff erosion due to waves—field investigations and laboratory experiments, *Jour. Faculty Eng., University of Tokyo*, **32**, 1–86.
Sunamura, T. (1975) A laboratory study of wave-cut platform formation, *Jour. Geology*, **83**, 389–97.
Sunamura, T. (1976) Feedback relationship in wave erosion of laboratory rocky coast, *Jour. Geology*, **84**, 427–37.
Sunamura, T. (1977) A relationship between wave-induced cliff erosion and erosive force of waves, *Jour. Geology*, **85**, 613–18.
Sunamura, T. (1978a) A model of the development of continental shelves having erosional origin, *Geol. Soc. Am. Bull.*, **89**, 504–10.
Sunamura, T. (1978b) Mechanisms of shore platform formation on the southeastern coast of the Izu Peninsula, Japan, *Jour. Geology*, **86**, 211–22.
Sunamura, T. (1978c) A mathematical model of submarine platform development, *Math. Geology*, **10**, 53–8.
Sunamura, T. (1982a) A wave tank experiment on the erosional mechanism at a cliff base, *Earth Surface Processes Landforms*, **7**, 333–43.
Sunamura, T. (1982b) A predictive model for wave-induced cliff erosion, with application to Pacific coasts of Japan, *Jour. Geology*, **90**, 167–78.
Sunamura, T. (1982c) Determination of breaker height and depth in the field, *Ann. Rept., Inst. Geosci., University of Tsukuba*, **8**, 53–4.
Sunamura, T. (1983) Processes of sea cliff and platform erosion, in P. D. Komar, *CRC Handbook of Coastal Processes and Erosion*, CRC Press, Boca Raton, Florida, pp. 233–65.
Sunamura, T. (1984) Onshore–offshore sediment transport rate in the swash zone of laboratory beaches, *Coastal Eng. Japan*, **27**, 205–12.
Sunamura, T. (1985) A simple relationship for predicting wave height in the surf zone with a uniformly sloping bottom, *Trans. Japan. Geomorph. Union*, **6**, 361–4.
Sunamura, T. (1987) Coastal cliff erosion in Nii-jima Island, Japan: present, past, and future—an application of mathematical model, in V. Gardiner, *International Geomorphology* 1986, Part I, Wiley, Chichester, pp. 1199–1212.
Sunamura, T. (1988a) Parameters for evaluating rock mass quality from visibly measured discontinuity frequency, *Ann. Rept., Inst. Geosci., University of Tsukuba*, **14**, 17–19.
Sunamura, T. (1988b) Projection of future coastal cliff recession under sea level rise induced by the greenhouse effect: Nii-jima Island, Japan, *Trans. Japan. Geomorph. Union*, **9**, 17–33.
Sunamura, T. (1989) Sandy beach geomorphology elucidated by laboratory modelling, in V. C. Lakhan and A. S. Trenhaile, *Applications in Coastal Modeling*, Elsevier, Amsterdam, pp. 159–213.
Sunamura, T. (1990a) A dynamical approach to wave-base problems, *Trans. Japan. Geomorph. Union*, **11**, 41–8.

Sunamura, T. (1990b) A wave-flume study on the elevation of shore platforms, *Ann. Rept., Inst. Geosci., University of Tsukuba*, 16, 36–8.

Sunamura, T. (1991) The elevation of shore platforms: a laboratory approach to the unsolved problem, *Jour. Geology*, **99**, 761–6.

Sunamura, T. and K. Horikawa (1969) A study on erosion of coastal cliffs by using aerial photographs (Report No. 2), *Coastal Eng. Japan*, **12**, 99–120.

Sunamura, T. and Horikawa, K. (1971) A quantitative study on the effect of beach deposits upon cliff erosion, *Coastal Eng. Japan*, **14**, 97–106.

Sunamura, T. and Horikawa, K. (1972) A study using aerial photographs of the effect of protective structures on coastal cliff erosion, *Coastal Eng. Japan*, **15**, 105–11.

Sunamura, T. and Horikawa, K. (1974) Two-dimensional beach transformation due to waves, *Proc. 14th Coastal Eng. Conf.*, Am. Soc. Civil Engr., pp. 920–38.

Sunamura, T. and Kraus, N. C. (1985) Prediction of average mixing depth of sediment in the surf zone, *Marine Geology*, **62**, 1–12.

Sunamura, T. and Takeda, I. (1984) Landward migration of inner bars, *Marine Geology*, **60**, 63–78.

Sunamura, T. and Tsujimoto, H. (1982) Long-term recession rate of Taitomisaki coastal cliff, *Ann. Rept., Inst. Geosci., University of Tsukuba*, 8, 55–6.

Suwardi, A. and Rosengren, N. J. (1983) Coastal changes on Anak Krakatau and Sertung Island, in E. C. F. Bird and A. Soegiarto, *Proc. Workshop Krakatau and Sunda Strait*. Cited in Bird and Rosengren (1984).

Suzuki, T. (1982) Rate of lateral plantation by Iwaki River, Japan, *Trans. Japan. Geomorph. Union*, 3, 1–24.

Suzuki, T., Takahashi, K., Sunamura, T., and Terada, M. (1970) Rock mechanics on the formation of washboard-like relief on wave-cut benches at Arasaki, Miura Peninsula, Japan, *Geogr. Review Japan*, **43**, 211–22 (in Japanese with English abstract).

Suzuki, T., Takahashi, K., and Sunamura, T. (1972) Rock control in coastal erosion at Arasaki, Miura peninsula, Japan, *Abstracts, 22nd Int. Geogr. Cong.*, Canada, pp. 66–8.

Suzuki, T., Hirano, M., Takahashi, K., and Yatsu, E. (1977) The interaction between the weathering processes of granites and the evolution of landforms in the Rokko Mountains, Japan—Part I. Vertical changes in physical, mechanical, mineral and chemical properties of the weathered Rokko granite, *Bull. Faculty Sci. Eng., Chuo University*, **20**, 343–89 (in Japanese with English abstract).

Svendsen, I. A. (1984) Wave attenuation and set-up on a beach, *Proc. 19th Coastal Eng. Conf.*, Am. Soc. Civil Engr., pp. 54–69.

Sverdrup, H. U. and Munk, W. H. (1947) Wind, sea, and swell theory of relationships in forecasting, U.S. Navy Hydrog. Office, Publ. No. 601, 44pp.

Swinnerton, A. C. (1927) Observations on some details of wave erosion: wave furrows and shore potholes, *Jour. Geology*, **15**, 171–9.

Swinchatt, J. P. (1969) Algal boring: a possible depth indicator in carbonate rocks and sediments, *Geol. Soc. Am. Bull.*, **80**, 1391–6.

Takahashi K. (1975) Differential erosion originating washboard-like relief on wave-cut bench at Aoshima Island, Kyushu, Japan, *Geogr. Rev. Japan*, **48**, 43–62 (in Japanese with English abstract).

Takahashi, K. (1976) Differential erosion on wave-cut bench, *Bull. Faculty Sci. Eng.*, Chuo University, **19**, 253–316 (in Japanese with English abstract).

Takahashi, T. (1974) Geomorphological study of shore platforms—analytical and genetical, *Sci. Rept., Tohoku University*, 7th Ser. (Geography), **24**, 115–63.

Takenaga, K. (1968) The classification of notch profiles and the origin of notches, *Jour. Geography*, **77**, 329–41 (in Japanese with English abstract).
Tanabe, K. (1941) Several marine pot-holes, *Geogr. Rev. Japan*, **17**, 938–44 (in Japanese).
Tayama, R. (1950) The submarine configuration off Sikoku, especially the continental slope, *Bull. Japan Hydrographic Office*, Special Publ. No. 7, 54–82 (in Japanese with English abstract).
Taylor, D. W. (1948) *Fundamentals of Soil Mechanics*, Wiley, New York, 700pp.
Thomas, R. H. (1986) Future sea level rise and its early detection by satellite remote sensing, in J. G. Titus, *Effects of Changes in Stratospheric Ozone and Global Climate*, Vol. 4: Sea Level Rise, *Proc. Int. Conf. Health Environmental Effects of Ozone Modification and Climatic Change*, U.N. Environment Programme and U.S. Environmental Protection Agency, pp. 19–36.
Thornbury, W. D. (1960) *Principles of Geomorphology*, Wiley, New York, 618pp.
Titus, J. G. (1986) *Effects of Changes in Stratospheric Ozone and Global Climate*, Vol. 4: Sea Level Rise, *Proc. Int. Conf. Health Environmental Effects of Ozone Modification and Climatic Change*, U.N. Environment Programme and U.S. Environmental Protection Agency, 193pp.
Tjia, H. D. (1985) Notching by abrasion on a limestone coast, *Zeit. Geomorphologie N. F.*, **29**, 367–72.
Toyoshima, O. (1974) Sea cliff erosion and countermeasures, *Proc. Symp. Coastal Cliff Recession*, Japan Soc. Civil Engr., pp. 1–8 (in Japanese).
Toyoshima, Y. (1956) On wave-cut benches along the south coast of the Miura Peninsula, Kanagawa Prefecture, *Geogr. Rev. Japan*, **29**, 240–52 (in Japanese with English abstract).
Toyoshima, Y. (1967) A study on marine erosive features along San-In coast, *Sci. Rept., Faculty Educ., Tottori University*, **18**, 64–98 (in Japanese with English abstract)
Toyoshima, Y. (1968) Abrasion platform and wave-cut bench at Kushimoto-machi, Wakayama prefecture, *Jour. Faculty Educ., Tottori University*, **19**, 41–8. (in Japanese with English abstract).
Toyoshima, Y. (1973) Coastal geomorphology—with special reference to coastal profiles and coastal changes, *Marine Sci. Monthly*, **5**, 839–43 (in Japanese).
Toyoshima, Y. and Tanimoto, I. (1984) On some marine potholes along tajima coast, *Sci. Rept., Faculty Educ., Tottori University*, **33**, 71–9 (in Japanese with English abstract).
Trenhaile, A. S. (1971) Lithological control of high-water rock ledges in the Vale of Glamorgan, Wales, *Geografiska Ann.*, **53A**, 59–69.
Trenhaile, A. S. (1972) The shore platforms of the Vale of Glamorgan, Wales, *Trans. Inst. British Geographers*, **56**, 127–44.
Trenhaile, A. S. (1974) The morphology and classification of shore platforms in England and Wales, *Geografiska Ann.*, **56A**, 103–10.
Trenhaile, A. S. (1978) The shore platforms of Gaspé, Québec, *Ann. Assoc. Am. Geographers*, **68**, 95–114.
Trenhaile, A. S. (1980) Shore platforms: a neglected coastal feature, *Progress Physical Geography*, **4**, 1–23.
Trenhaile, A. S. (1987) *The Geomorphology of Rock Coasts*, Oxford University Press, Oxford, 384pp.
Trenhaile, A. S. and Layzell, M. G. J. (1981) Shore platform morphology and the tidal duration factor, *Trans. Inst. British Geographers*, **6**, 82–102.
Trenhaile, A. S. and Mercan, D. W. (1984) Frost weathering and the saturation of coastal rocks, *Earth Surface Processes Landforms*, **9**, 321–31.

Tricart, J. (1959) Problèmes géomorphologiques du littoral oriental du Bresil, *Cahiers Océanogr.*, **11**, 276–308.
Tricart, J. (1962) Observations de géomorphologie littorale a Mamba Point, *Erdkunde*, **16**, 49–57.
Trudgill, S. T. (1976) The marine erosion of limestones on Aldabra Atoll, Indian Ocean, *Zeit. Geomorphologie N.F.*, Suppl. Bd. **26**, 164–200.
Trudgill, S. T. (1985) *Limestone Geomorphology*, Longman, London, 196pp.
Trudgill, S. T. and Crabtree, R. W. (1987) Bioerosion of intertidal limestone, Co. Clare, Eire—2: *Hialella arctica*, *Marine Geology*, **74**, 99–109.
Trudgill, S. T., High, C. J., and Hanna, F. K. (1981) Improvements to the micro-erosion meter, British Geomorph. Res. Group, Tech. Bull., No. 29, 17pp.
Trudgill, S. T., Smart, P. L., Friederich, H., and Crabtree, R. W. (1987) Bioerosion of intertidal limestone, Co. Clare, Eire—1: *Paracentrotus lividus*, *Marine Geology*, **74**, 85–98.
Tsujimoto, H. (1985) Types of rocky coasts and the resisting force of coastal rocks in the eastern part of Chiba Prefecture, Japan, *Geogr. Rev. Japan*, **58**, Ser. A, 180–92 (in Japanese with English abstract).
Tsujimoto, H. (1987) Dynamic conditions for shore platform initiation, *Sci. Rept., Inst. Geosci., University of Tsukuba*, Sec. A, **8**, 45–93.
Twenhofel, W. H. (1945) The rounding of sand grains, *Jour. Sedimentary Petrology*, **15**, 59–71.
Twenhofel, W. H. (1950) *Principles of Sedimentation*, 2nd edn, McGraw-Hill, New York, 673pp.
Twidale, C. R. (1976) *Analysis of Landforms*, Wiley, Brisbane, 572pp.
Twidale, C. R. (1982) *Granite Landforms*, Elsevier, Amsterdam, 372pp.
Valentin, H. (1954) Der Landverlust in Holderness, Ostengland, Von 1852 bis 1952, *Die Erde*, **6**, 296–315.
Vaughan, T. W. (1932) Rate of sea cliff recession on the property of the Scripps Institution of Oceanography at La Jolla, California, *Science*, **75**, 250.
Verstappen, H. T. (1960) On the geomorphology of raised coral reefs and its tectonic significance, *Zeit. Geomorphologie N.F.*, **4**, 1–28.
Viggósson, G. and Tryggvason, G. S. (1985) Wave measurement in Iceland, *Proc. Iceland Coastal River Symp.*, Reykjavik, Iceland, pp. 145–64.
Viles, H. A. and Trudgill, S. T. (1984) Long term remeasurements of micro-erosion meter rates, Aldabra Atoll, Indian Ocean, *Earth Surface Processes Landforms*, **9**, 89–94.
Vita-Finzi, C. and Cornelius, P. F. S. (1973) Cliff sapping by molluscs in Oman, *Jour. Sedimentary Petrology*, **43**, 31–2.
Von Engeln, O. D. (1942) *Geomorphology*, MacMillan, New York, 655pp.
Vutukuri, V. S., Lama, R. D., and Saluja, S. S. (1974) *Handbook on Mechanical Properties of Rocks*, Vol. I, Trans. Tech. Publications, Clausthal, 280pp.
Warrick, R. and Oerlemans, J. (1990) Sea level rise, in J. T. Houghton, G. J. Jenkins, and J. J. Ephraums, *Climate Change*, The IPCC Scientific Assessment, Cambridge University Press, Cambridge, pp. 261–81.
Washburn, A. L. (1979) *Geocryology*, Edward Arnold, London, 406pp.
Wefer, G., Flemming, B., and Tauchgruppe, K. (1976) Submarine Abrasion des Geschiebemergels vor Bokniseck (Westl. Ostsee), *Meyniana*, **28**, 87–94.
Weggel, J. R. (1972) Maximum breaker height, *Jour. Waterways Harbors Coastal Eng. Div., Proc. Am. Soc. Civil Engr.*, **98**, 529–48.
Weishar, L. L. and Byrne, R. J. (1978) Field study of breaking wave characteristics, *Proc. 16th Coastal Eng. Conf.*, Am. Soc. Civil Engr., pp. 487–506.

Wellman, H. W. and Wilson, A. T. (1965) Salt weathering, a neglected geological erosion agent in coastal and arid environments, *Nature*, **205**, 1097–8.
Wentworth, C. K. (1938) Marine-bench forming processes: water-level weathering, *Jour. Geomorphology*, **1**, 6–32.
Wentworth, C. K. (1939) Marine bench-forming processes: II, solution benching, *Jour. Geomorphology*, **2**, 3–25.
Wentworth, C. K. (1944) Potholes, pits, and pans: subaerial and marine, *Jour. Geology*, **52**, 117–30.
Wentworth, C. K. and Palmer, H. S. (1925) Eustatic bench of islands of the north Pacific, *Bull. Geol. Soc. Am.*, **36**, 521–44.
Wentworth, C. K. and Hoffmeister, J. E. (1939) Geology of Ulupau Head, Oahu, *Geol. Soc. Am. Bull.*, **50**, 1553–72.
Wheeler, W. H. (1902) *The Sea-Coast*, Longmans Green, London, 361pp.
Wiegel, R. L. (1964) *Oceanographical Engineering*, Prentice-Hall, Englewood Cliffs, New Jersey, 532pp.
Williams, A. T. and Davies, P. (1980) Man as a geological agent: the sea cliffs of Llantwit Major, Wales, U.K., *Zeit. Geomorphologie*, Suppl. Bd. **34**, 129–41.
Williams, A. T. and Davies, P. (1987) Rates and mechanics of coastal cliff erosion in Lower Lias rocks, *Proc. Coastal Sediments '87*, Am. Soc. Civil Engr., pp. 1855–70.
Williams, W. W. (1956) An east coast survey: some recent changes in the coast of East Anglia, *Geogr. Jour.*, **12**, 317–334.
Williams, W. W. (1960) *Coastal Changes*, Routledge and Kegan Paul, London, 220pp.
Wilson, B. W. (1955) Graphical approach to the forecasting of waves in moving fetches, U.S. Army Beach Erosion Board, Tech. Memo., No. 73, 64pp.
Wilson, G. (1952) The influence of rock structures on coastline and cliff development around Tintagel, North Cornwall, *Proc. Geol. Assoc.*, **63**, 20–48.
Winker, E. M. and Wilhelm, E. J. (1970) Salt burst by hydration pressures in architectural stone in an urban atmosphere, *Geol. Soc. Am. Bull.*, **81**, 567–72.
Wood, A. (1968) Beach platforms in the chalk of Kent, England, *Zeit. Geomorphologie N.F.*, **12**, 107–13.
Wood, F. J. (1982) Tides, in M. L. Schwartz, *The Encyclopedia of Beaches and Coastal Environments*, Hutchinson Ross, Stroudsburg, Pennsylvania, pp. 826–37.
Woodroffe, C. D., Stoddart, D. R., Harmon, R. S., and Spencer, T. (1983) Coastal morphology and late Quaternary history, Cayman Islands, West Indies, *Quaternary Res.*, **19**, 64–84.
Wray, J. L. (1977) *Calcareous Algae*, Elsevier, Amsterdam, 185pp.
Wright, L. W. (1967) Some characteristics of the shore platforms of the English Channel Coast and the northern part of the North Island, New Zealand, *Zeit. Geomorphologie N.F.*, **11**, 36–46.
Wright, L. W. (1970) Variation in the level of the cliff/shore platform junction along the south coast of Great Britain, *Marine Geology*, **9**, 347–53.
Yajima, A. (1965) A study on recession of a coastal cliff at Habuseura, Niijima, Japan, Unpub. B.S. Thesis, Dept. Geography, Tokyo Kyoiku University, 17pp. (in Japanese).
Yamaguchi, U. and Nishimatsu, Y. (1977) *Introduction to Rock Mechanics*, 2nd edn, University of Tokyo Press, Tokyo, 266pp (in Japanese).
Yamanouchi, H. (1964) On the retreat of the sea cliffs along the Pacific coast at Haranomachi, Japan, *Geogr. Rev. Japan*, **37**, 138–46 (in Japanese with English abstract).
Yamanouchi, H. (1977) A geomorphological study about the coastal cliff retreat along the southwest coast of the Atsumi peninsula, central Japan, *Sci. Rept., Faculty Educ., Gunma University*, **26**, 95–128.

Yano, K. and Kamata, M. (1963) A vehicle for in situ cone-type penetration testing, *Soil Mechanics Foundation Eng.*, **11**(4), 8–16 (in Japanese).

Yatsu, E. (1988) *The Nature of Weathering—An Introduction*, Sozosha, Tokyo, 624pp.

Yoshikawa, T. (1952) Shore erosion and its bearing on topography of Kurobe alluvial fan, Toyama Prefecture, *Bull. Geogr. Inst., University of Tokyo*, **2**, 92–109.

Yoshikawa, T. (1953) Some consideration on the continental shelves around the Japanese islands, *Natural Sci. Rept., Ochanomizu University*, **4**, 138–50 (in Japanese with English abstract).

Yoshikawa, T. and Saito, M. (1954) Topography of the shallow sea floor in the vicinity of Chikura harbour on the southeastern coast of Boso peninsula, on the Pacific coast of Japan, *Bull. Geogr. Inst., University of Tokyo*, **3**, 40–50 (in Japanese with English abstract).

Zeigler, J. M., Hayes, C. R., and Tuttle, S. D. (1959) Beach changes during storms on Outer Cape Cod, Massachusetts, *Jour. Geology*, **67**, 318–36.

Zeigler, J. M., Tuttle, S. D., Giese, G. S., and Tasha, H. J. (1964), Residence time of sand composing the beaches and bars of Outer Cape Cod, *Proc. 9th Coastal Eng. Conf.*, Am. Soc. Civil Engrs., pp. 403–16.

Zeman, A. J. (1986) Erodibility of Lake Erie undisturbed tills, *Proc. Symp. Cohesive Shores*, National Research Council, Canada, pp. 150–69.

Zeman, A. J. (1989) Numerical modelling of seepage erosion in shore bluffs consisting of glaciolacustrine silts, *Proc. 3rd Int. Symp. Numerical Models in Geomechanics*, Elsevier Applied Science, London, pp. 375–82.

Zenkovich, V. P. (1967) *Processes of Coastal Development*, Oliver & Boyd, Edinburgh, 738pp.

Zenkovich, V. P., Ionin, A. S. and Kaplin, P. A. (1965) Abrasion as the resources of debris in the sea shore zone, *Inst. Oceanogr., Acad. Sci.*, USSR, **76**, 103–25 (in Russian with English abstract).

Zeuner, F. E. (1952) *Dating the Past*, Methuen, London, 495pp.

Appendix One
Conversion factors

LENGTH

kilometre (km)	metre* (m)	centimetre (cm)	millimetre (mm)	inch (in)	foot (ft)	yard (yd)	mile (mi)	fathom (fm)	nautical mile (nm)
1	1×10^3	1×10^5	1×10^6	3.937×10^4	3.281×10^3	1.094×10^3	6.214×10^{-1}	5.468×10^2	5.400×10^{-1}
1×10^{-3}	1	1×10^2	1×10^3	3.937×10	3.281	1.094	6.214×10^{-4}	5.468×10^{-1}	5.400×10^{-4}
1×10^{-5}	1×10^{-2}	1	1×10	3.937×10^{-1}	3.281×10^{-2}	1.094×10^{-2}	6.214×10^{-6}	5.468×10^{-3}	5.400×10^{-6}
1×10^{-6}	1×10^{-3}	1×10^{-1}	1	3.937×10^{-2}	3.281×10^{-3}	1.094×10^{-3}	6.214×10^{-7}	5.468×10^{-4}	5.400×10^{-7}
2.54×10^{-5}	2.54×10^{-2}	2.54	2.54×10	1	8.333×10^{-2}	2.778×10^{-2}	1.578×10^{-5}	1.389×10^{-2}	1.371×10^{-5}
3.048×10^{-4}	3.048×10^{-1}	3.048×10	3.048×10^2	1.2×10	1	3.333×10^{-1}	1.894×10^{-4}	1.667×10^{-1}	1.646×10^{-4}
9.144×10^{-4}	9.144×10^{-1}	9.144×10	9.144×10^2	3.6×10	3	1	5.682×10^{-4}	5.001×10^{-1}	4.938×10^{-4}
1.609	1.609×10^3	1.609×10^5	1.609×10^6	6.336×10^4	5.280×10^3	1.760×10^3	1	5.001×10^{-1}	8.688×10^{-1}
1.829×10^{-3}	1.829	1.829×10^2	1.829×10^3	7.2×10	6	2	1.136×10^{-3}	1	9.872×10^{-4}
1.852	1.852×10^3	1.852×10^5	1.852×10^6	7.291×10^4	6.076×10^3	2.025×10^3	1.151	1.013×10^3	1

*SI base unit

AREA

sq. kilometre (km^2)	sq metre* (m^2)	sq. centimetre (cm^2)	sq. inch (in^2)	sq. foot (ft^2)	sq. yard (yd^2)	sq. mile (mi^2)	hectare (ha)	acre (acre)
1	1×10^6	1×10^{10}	1.550×10^9	1.076×10^7	1.197×10^6	3.861×10^{-1}	1×10^2	2.471×10^2
1×10^{-6}	1	1×10^4	1.550×10^3	1.076×10	1.197	3.861×10^{-7}	1×10^{-4}	2.471×10^{-4}
1×10^{-10}	1×10^{-4}	1	1.550×10^{-1}	1.076×10^{-3}	1.197×10^{-4}	3.861×10^{-11}	1×10^{-8}	2.471×10^{-8}
6.452×10^{-10}	6.452×10^{-4}	6.452	1	6.944×10^{-3}	7.717×10^{-4}	2.490×10^{-10}	6.452×10^{-8}	1.594×10^{-7}
9.290×10^{-8}	9.290×10^{-2}	9.290×10^2	1.44×10^2	1	1.111×10^{-1}	3.587×10^{-8}	9.290×10^{-6}	2.296×10^{-5}
8.361×10^{-7}	8.361×10^{-1}	8.361×10^3	1.296×10^3	9	1	3.229×10^{-7}	8.361×10^{-5}	2.064×10^{-4}
2.590	2.590×10^6	2.590×10^{10}	4.015×10^9	2.789×10^7	3.098×10^6	1	2.59×10^2	6.40×10^2
1×10^{-2}	1×10^4	1×10^8	1.550×10^7	1.076×10^5	1.197×10^4	3.861×10^{-3}	1	2.471
4.047×10^{-3}	4.047×10^3	4.047×10^7	6.273×10^6	4.355×10^4	4.844×10^3	1.560×10^{-3}	4.047×10^{-1}	1

*SI derived unit

VOLUME

cubic metre* (m^3)	cubic centimetre (cm^3)	litre (l)	cubic inch (in^3)	cubic foot (ft^3)	cubic yard (yd^3)	gallon (UK) (gal)	gallon (US) (gal)
1	1×10^6	1×10^3	6.102×10^4	3.532×10	1.308	2.200×10^2	2.642×10^2
1×10^{-6}	1	1×10^{-3}	6.102×10^{-2}	3.532×10^{-5}	1.308×10^{-6}	2.200×10^{-4}	2.642×10^{-4}
1×10^{-3}	1×10^3	1	6.102×10	3.532×10^{-2}	1.308×10^{-3}	2.200×10^{-1}	2.642×10^{-1}
1.639×10^{-5}	1.639×10	1.639×10^{-2}	1	5.787×10^{-4}	2.143×10^{-5}	3.605×10^{-3}	4.329×10^{-3}
2.832×10^{-2}	2.832×10^4	2.832×10	1.728×10^3	1	3.704×10^{-2}	6.229	7.481
7.645×10^{-1}	7.645×10^5	7.645×10^2	4.666×10^4	2.7×10	1	1.682×10^2	2.020×10^2
4.546×10^{-3}	4.546×10^3	4.546	2.774×10^2	1.605×10^{-1}	5.946×10^{-3}	1	1.201
3.785×10^{-3}	3.785×10^3	3.785	2.31×10^2	1.337×10^{-1}	4.950×10^{-3}	8.326×10^{-1}	1

*SI derived unit

Appendix 1

MASS

	tonne* (ton)	kilogramme† (kg)	gramme (g)	milligramme (mg)	ounce (oz)	pound (lb)	slug (slug)	ton (UK) (long ton)	ton (US) (short ton)
	1	1×10^3	1×10^6	1×10^9	3.527×10^4	2.205×10^3	6.852×10	9.843×10^{-1}	1.102
	1×10^{-3}	1	1×10^3	1×10^6	3.527×10	2.205	6.852×10^{-2}	9.843×10^{-4}	1.102×10^{-3}
	1×10^{-6}	1×10^{-3}	1	1×10^3	3.527×10^{-2}	2.205×10^{-3}	6.852×10^{-5}	9.843×10^{-7}	1.102×10^{-6}
	1×10^{-9}	1×10^{-6}	1×10^{-3}	1	3.527×10^{-5}	2.205×10^{-6}	6.852×10^{-8}	9.843×10^{-10}	1.102×10^{-9}
	2.835×10^{-5}	2.835×10^{-2}	2.835×10	2.835×10^4	1	6.25×10^{-2}	1.943×10^{-3}	2.790×10^{-5}	3.125×10^{-5}
	4.536×10^{-4}	4.536×10^{-1}	4.536×10^2	4.536×10^5	1.6×10	1	3.108×10^{-2}	4.454×10^{-4}	5.000×10^{-4}
	1.459×10^{-2}	1.459×10	1.459×10^4	1.459×10^7	5.148×10^2	3.217×10	1	1.436×10^{-2}	1.608×10^{-2}
	1.016	1.016×10^3	1.016×10^6	1.016×10^9	3.584×10^4	2.240×10^3	6.964×10	1	1.120
	9.072×10^{-1}	9.072×10^2	9.072×10^5	9.072×10^8	3.200×10^4	2.000×10^3	6.216×10	8.929×10^{-1}	1

* Metric ton
† SI base unit

DENSITY

	tonne* per cubic metre (ton/m³)	kilogramme per cubic metre† (kg/m³)	killogramme per cubic centimetre (kg/cm³)	gramme per cubic centimetre (g/cm³)	ounce per cubic inch (oz/in³)	ounce per cubic foot (oz/ft³)	pound per cubic inch (lb/in³)	pound per cubic foot (lb/ft³)	slug per cubic foot (slug/ft³)
	1	1×10^3	1×10^{-3}	1	5.780×10^{-1}	9.986×10^2	3.614×10^{-2}	6.243×10	1.940
	1×10^{-3}	1	1×10^{-6}	1×10^{-3}	5.780×10^{-4}	9.986×10^{-1}	3.614×10^{-5}	6.243×10^{-2}	1.940×10^{-3}
	1×10^3	1×10^6	1	1×10^3	5.780×10^2	9.986×10^5	3.614×10	6.243×10^4	1.940×10^3
	1	1×10^3	1×10^{-3}	1	5.780×10^{-1}	9.986×10^2	3.614×10^{-2}	6.243×10	1.940
	1.730	1.730×10^3	1.730×10^{-3}	1.730	1	1.728×10^3	6.25×10^{-2}	1.080×10^2	3.357
	1.001×10^{-3}	1.001	1.001×10^{-6}	1.001×10^{-3}	5.787×10^{-4}	1	3.617×10^{-5}	6.25×10^{-2}	1.943×10^{-3}
	2.768×10	2.768×10^4	2.768×10^{-2}	2.768×10	1.6×10	2.765×10^4	1	1.728×10^3	5.371×10
	1.602×10^{-2}	1.602×10^1	1.602×10^{-5}	1.602×10^{-2}	9.259×10^{-3}	1.6×10	5.787×10^{-4}	1	3.108×10^{-2}
	5.154×10^{-1}	5.154×10^2	5.154×10^{-4}	5.154×10^{-1}	2.979×10^{-1}	5.148×10^2	1.862×10^{-2}	3.217×10	1

* Metric ton
† SI derived unit

TIME

	year (y or yr)	day (d)	hour (h)	minute (min)	second* (s)
	1	3.652×10^2	8.766×10^3	5.259×10^5	3.156×10^7
	2.738×10^{-3}	1	2.4×10	1.44×10^3	8.64×10^4
	1.141×10^{-4}	4.167×10^{-2}	1	6×10	3.6×10^3
	1.901×10^{-6}	6.944×10^{-4}	1.667×10^{-2}	1	6×10
	3.169×10^{-8}	1.157×10^{-5}	2.778×10^{-4}	1.667×10^{-2}	1

* SI base unit

SPEED

kilometre per hour (km/h)	metre per minute (m/min)	metre per second* (m/s)	centimetre per minute (cm/min)	centimetre per second (cm/s)	inch per second (in/s)	inch per minute (in/min)	foot per second (ft/s)	foot per minute (ft/min)	mile per hour (mi/h)	knot (International) (kn or kt)
1	1.667×10	2.778×10^{-1}	1.667×10^3	2.778×10	1.094×10	6.563×10^2	9.116×10^{-1}	5.469×10	6.214×10^{-1}	5.400×10^{-1}
6×10^{-2}	1	1.667×10^{-2}	1×10^2	1.667	6.563×10^{-1}	3.937×10	5.469×10^{-2}	3.281	3.728×10^{-2}	3.240×10^{-2}
3.6	6×10	1	6×10^3	1×10^2	3.937×10	2.362×10^3	3.281	1.968×10^2	2.237	1.944
6×10^{-4}	1×10^{-2}	1.667×10^{-4}	1	1.667×10^{-2}	6.563×10^{-3}	3.937×10^{-1}	5.469×10^{-4}	3.281×10^{-2}	3.728×10^{-4}	3.240×10^{-4}
3.6×10^{-2}	6×10^{-1}	1×10^{-2}	6×10	1	3.937×10^{-1}	2.362×10	3.281×10^{-2}	1.968	2.237×10^{-2}	1.944×10^{-2}
9.144×10^{-2}	1.524	2.54×10^{-2}	1.524×10^2	2.54	1	6×10	8.333×10^{-2}	5	5.682×10^{-2}	4.938×10^{-2}
1.524×10^{-3}	2.540×10^{-2}	4.234×10^{-4}	2.54	4.234×10^{-2}	1.667×10^{-2}	1	1.388×10^{-3}	8.333×10^{-2}	9.470×10^{-4}	8.231×10^{-4}
1.097	1.829×10	3.048×10^{-1}	1.829×10^3	3.048×10	1.2×10	7.2×10^2	1	6×10	6.817×10^{-1}	5.925×10^{-1}
1.829×10^{-2}	3.048×10^{-1}	5.080×10^{-3}	3.048×10	5.080×10^{-1}	2×10^{-1}	1.2×10	1.667×10^{-2}	1	1.136×10^{-2}	9.876×10^{-3}
1.609	2.682×10	4.470×10^{-1}	2.682×10^3	4.470×10	1.76×10	1.056×10^3	1.467	8.8×10	1	8.690×10^{-1}
1.852	3.086×10	5.144×10^{-1}	3.086×10^3	5.144×10	2.025×10	1.215×10^3	1.688	1.012×10^2	1.151	1

* SI derived unit

FORCE

tonne*-force per cubic metre (tonf/m³)‡	kilogramme-force (kgf)‡	gramme-force (gf)‡	dyne (dyn)	newton† (N)	kilonewton (kN)	pound-force (lbf)‡	poundal (pdl)
1	1×10^3	1×10^6	9.807×10^8	9.807×10^3	9.807	2.204×10^3	7.087×10^4
1×10^{-3}	1	1×10^3	9.807×10^5	9.807	9.807×10^{-3}	2.204	7.087×10
1×10^{-6}	1×10^{-3}	1	980.7	9.807×10^{-3}	9.807×10^{-6}	2.204×10^{-3}	7.087×10^{-2}
1.020×10^{-9}	1.020×10^{-6}	1.020×10^{-3}	1	1×10^{-5}	1×10^{-8}	2.248×10^{-6}	7.231×10^{-5}
1.020×10^{-4}	1.020×10^{-1}	1.020×10^2	1×10^5	1	1×10^{-3}	2.248×10^{-1}	7.231
1.020×10^{-1}	1.020×10^2	1.020×10^5	1×10^8	1×10^3	1	2.248×10^2	7.231×10^3
4.536×10^{-4}	4.536×10^{-1}	4.536×10^2	4.448×10^5	4.448	4.448×10^{-3}	1	3.217×10
1.411×10^{-5}	1.411×10^{-2}	1.411×10	1.383×10^4	1.383×10^{-1}	1.383×10^{-4}	3.109×10^{-2}	1

* Metric ton
† SI derived unit
‡ The suffix 'f' is commonly omitted

UNIT WEIGHT

tonne*-force per cubic metre (tonf/m³)‡	kilogramme-force per cubic metre (kgf/m³)‡	kilogramme-force per cubic centimetre (kgf/cm³)‡	gramme-force per cubic centimetre (gf/cm³)‡	newton per cubic metre† (N/m²)	kilonewton per cubic metre (kN/m³)	pound-force per cubic inch (lbf/in³)‡	pound-force per cubic foot (lbf/ft³)‡
1	1×10^3	1×10^{-3}	1	9.807×10^3	9.807	3.612×10^{-2}	6.240×10
1×10^{-3}	1	1×10^{-6}	1×10^{-3}	9.807	9.807×10^{-3}	3.612×10^{-5}	6.240×10^{-2}
1×10^3	1×10^6	1	1×10^3	9.807×10^6	9.807×10^3	3.612×10	6.240×10^4
1	1×10^3	1×10^{-3}	1	9.807×10^3	9.807	3.612×10^{-2}	6.240×10
1.020×10^{-4}	1.020×10^{-1}	1.020×10^{-7}	1.020×10^{-4}	1	1×10^{-3}	3.684×10^{-6}	6.365×10^{-3}
1.020×10^{-1}	1.020×10^2	1.020×10^{-4}	1.020×10^{-1}	1×10^3	1	3.684×10^{-3}	6.365
2.768×10	2.768×10^4	2.768×10^{-2}	2.768×10	2.714×10^5	2.714×10^2	1	1.728×10^3
1.602×10^{-2}	1.602×10	1.602×10^{-5}	1.602×10^{-2}	1.571×10^2	1.571×10^{-1}	5.787×10^{-4}	1

* Metric ton
† SI derived unit
‡ The suffix 'f' is commonly omitted

PRESSURE OR STRESS

tonne*-force per sq. metre (tonf/m^2)‡	kilogramme-force per sq. metre (kgf/m^2)‡	kilogramme-force per sq. centimetre (kgf/cm^2)‡	gramme-force per sq. centimetre (gf/cm^2)‡	pascal† (=N/m^2) (Pa)	kilopascal (=kN/m^2) (kPa)	megapascal (=MN/m^2) (MPa)	pound-force per sq. inch (lbf/in^2 or psi)‡	pound-force per sq. foot (lbf/ft^2)‡	bar (bar)
1	1×10^3	1×10^{-1}	1×10^2	9.807×10^3	9.807	9.807×10^{-3}	1.422	2.048×10^2	9.807×10^{-2}
1×10^{-3}	1	1×10^{-4}	1×10^{-1}	9.807	9.807×10^{-3}	9.807×10^{-6}	1.422×10^{-3}	2.048×10^{-1}	9.807×10^{-5}
1×10	1×10^4	1	1×10^3	9.807×10^4	9.807×10	9.807×10^{-2}	1.422×10	2.048×10^3	9.807×10^{-1}
1×10^{-2}	1×10	1×10^{-3}	1	9.807×10	9.807×10^{-2}	9.807×10^{-5}	1.422×10^{-2}	2.048	9.807×10^{-4}
1.020×10^{-4}	1.020×10^{-1}	1.020×10^{-5}	1.020×10^{-2}	1	1×10^{-3}	1×10^{-6}	1.450×10^{-4}	2.088×10^{-2}	1×10^{-5}
1.020×10^{-1}	1.020×10^2	1.020×10^{-2}	1.020×10	1×10^3	1	1×10^{-3}	1.450×10^{-1}	2.088×10	1×10^{-2}
1.020×10^2	1.020×10^5	1.020×10	1.020×10^4	1×10^6	1×10^3	1	1.450×10^2	2.088×10^4	1×10
7.031×10^{-1}	7.031×10^2	7.031×10^{-2}	7.031×10	6.895×10^3	6.895	6.895×10^{-3}	1	1.44×10^2	6.895×10^{-2}
4.883×10^{-3}	4.883	4.883×10^{-4}	4.883×10^{-1}	4.788×10	4.788×10^{-2}	4.788×10^{-5}	6.944×10^{-3}	1	4.788×10^{-4}
1.020×10	1.020×10^4	1.020	1.020×10^3	1×10^5	1×10^2	1×10^{-1}	1.450×10	2.088×10^3	1

* Metric ton
† SI derived unit
‡ The suffix 'f' is commonly omitted

Appendix 1

ENERGY OR WORK

joule* ($=$m-N) (J)	erg (erg)	metre kilogramme-force (m-kgf)†	foot pound-force (ft-lbf)†	watt hour (W-h)	kilowatt hour (kW-h)	calorie (cal)	kilocalorie (kcal)	British thermal unit (Btu)
1	1×10^7	1.020×10^{-1}	7.376×10^{-1}	2.778×10^{-4}	2.778×10^{-7}	2.389×10^{-1}	2.389×10^{-4}	9.479×10^{-4}
1×10^{-7}	1	1.020×10^{-8}	7.376×10^{-8}	2.778×10^{-11}	2.778×10^{-14}	2.389×10^{-8}	2.389×10^{-11}	9.479×10^{-11}
9.807	9.807×10^7	1	7.231	2.724×10^{-3}	2.724×10^{-6}	2.343	2.343×10^{-3}	9.297×10^{-3}
1.356	1.356×10^7	1.383×10^{-1}	1	3.766×10^{-4}	3.766×10^{-7}	3.239×10^{-1}	3.239×10^{-4}	1.285×10^{-3}
3.60×10^3	3.60×10^{10}	3.671×10^2	2.655×10^3	1	1×10^{-3}	8.598×10^2	8.598×10^{-1}	3.413
3.60×10^6	3.60×10^{13}	3.671×10^5	2.655×10^6	1×10^3	1	8.598×10^5	8.598×10^2	3.413×10^3
4.186	4.186×10^7	4.269×10^{-1}	3.080	1.163×10^{-3}	1.163×10^{-6}	1	1×10^{-3}	3.968×10^{-3}
4.186×10^3	4.186×10^{10}	4.269×10^2	3.087×10^3	1.163	1.163×10^{-3}	1×10^3	1	3.968
1.055×10^3	1.055×10^{10}	1.076×10^2	7.780×10^2	2.930×10^{-1}	2.930×10^{-4}	2.520×10^2	2.520×10^{-1}	1

* SI derived unit
† The suffix 'f' is commonly omitted

POWER

watt* ($=$J/s) (W)	kilowatt ($=$kJ/s) (kW)	erg per second (erg/s)	metre kilogramme-force per second (m-kgf/s)†	foot pound-force per second (ft-lbf/s)†	calorie per second (cal/s)	kilocalorie per second (kcal/s)	horse-power (hp)	Btu per second (Btu/s)
1	1×10^{-3}	1×10^7	1.020×10^{-1}	7.376×10^{-1}	2.389×10^{-1}	2.389×10^{-4}	1.341×10^{-3}	9.479×10^{-4}
1×10^3	1	1×10^{10}	1.020×10^2	7.376×10^2	2.389×10^2	2.389×10^{-1}	1.341	9.479×10^{-1}
1×10^{-7}	1×10^{-10}	1	1.020×10^{-8}	7.376×10^{-8}	2.389×10^{-8}	2.389×10^{-11}	1.341×10^{-10}	9.479×10^{-11}
9.807	9.807×10^{-3}	9.807×10^7	1	7.231	2.343	2.343×10^{-3}	1.315×10^{-2}	9.297×10^{-3}
1.356	1.356×10^{-3}	1.356×10^7	1.383×10^{-1}	1	3.239×10^{-1}	3.239×10^{-4}	1.818×10^{-3}	1.285×10^{-3}
4.186	4.186×10^{-3}	4.186×10^7	4.269×10^{-1}	3.087	1	1×10^{-3}	5.612×10^{-3}	3.968×10^{-3}
4.186×10^3	4.186	4.186×10^{10}	4.269×10^2	3.087×10^3	1×10^3	1	5.612	3.968
7.457×10^2	7.457×10^{-1}	7.457×10^9	7.607×10	5.50×10^2	1.782×10^2	1.782×10^{-1}	1	7.072×10^{-1}
1.055×10^3	1.055	1.055×10^{10}	1.076×10^2	7.780×10^2	2.520×10^2	2.520×10^{-1}	1.414	1

* SI derived unit
† The suffix 'f' is commonly omitted

Appendix Two
Worldwide coastal cliff erosion rates

Location	Lithology	Erosion rate (m/y)	Interval	Method	Source
Australia					
Black Rock Point, near Melbourne, Victoria	Tertiary sediments	0.04–0.81	1973–1986	Surveys	Bird and Rosengren (1986)
Northeast of Point Sturt, Victoria	Siltstone	0.0175	(6000 y)	Surveys	Gill (1973)
Point Sturt, Victoria	Arkose	0.009	(6000 y)	Surveys	Gill (1973)
Warrnambool Victoria	Aeolianite	0.014	(130 y)	Surveys	Gill (1973)
Warrnambool Victoria	Aeolianite	0.04	(6000 y)	Surveys	Gill (1973)
Cape Reamur-Port Fairy, Victoria	Basalt	≈0	—	Surveys	Gill (1973)
Point Peron near Perth	Limestone	0.0002–0.001	1953–1962	Steel pegs	Hodgkin (1964)
Barbados					
Mullins	Coral rock	0.002	(4.5 y)	Steel pegs	Bird *et al.* (1979)
Paynes Bay	Coral rock	0.0005	—	Steel pegs	Bird *et al.* (1979)
Canada					
Headland near Indian Bay, Cape Breton, Nova Scotia	Carboniferous sedimentary rocks	0.75	1902–1908	—	Johnson (1925, p. 320)

Appendix 2

Location	Material	Rate	Period	Method	Reference
North side of Table Head, Cape Breton, Nova Scotia	Carboniferous sedimentary rocks	0.5	1907–1919	Surveys	Johnson (1925, p. 320)
East side of Table Head, Cape Breton, Nova Scotia	Carboniferous sedimentary rocks	1	1907–1913	Surveys	Johnson (1925, p. 320)
Lake Ontario shore, Ontario	Glacial deposits	0–2.04	1955–1973	Air-photogrammetry	Boulden (1975)
Scarborough, Ontario	Glacial deposits	0.76	1952–1976	Surveys, Air photos	Bryan and Price (1980)
Scarborough, Ontario	Glacial deposits	0.4	—	—	Parker *et al.* (1986)
South Marine Drive Sector, Scarborough, Ontario	Glacial deposits	1.1–1.4	—	—	Parker *et al.* (1986)
Stoney Creek, Ontario	Glacial deposits	≲1	—	—	Davidson-Arnott and Keizer (1982)
St Catharines, Ontario	Glacial deposits	0.24–2.8	—	—	Sturm and Kolberg (1986)
Lake Erie shore, Ontario	Glacial deposits	0–3.97	1955–1973	Air-photogrammetry	Boulden (1975)
Central part, Lake Erie, Ontario	Glacial deposits	0.25–2.75	1810–1964	Surveys, Air photos	Gelinas and Quigley (1973)
Port Bruce, Ontario	Glacial deposits	2.2	1964–1977	Maps	Quiglen and Di Nardo (1980)
Western Lake Erie, Ontario	Glacial deposits	<1.5	1931–1970	Air photos	Coakley and Cho (1972)
Lake Huron shore, Ontario	Glacial deposits	0–1.97	1955–1973	Air-photogrammetry	Boulden (1975)
Pt Grey, Vancouver	Sand and silt	0.3	1908–1974	Maps	Lum, K. (vid. Clague and Bornhold, 1980)
Saanich Pen., Vancouver Is.	Unconsolidated sediments	0.3	—	—	Foster, H. D. (vid. Clague and Bornhold, 1980)

(continued)

Table continued

Location	Lithology	Erosion rate (m/y)	Interval	Method	Source
France					
Ault, Somme	Chalk	0.08–0.37	(100 y)	—	Prêcheur (1960, p. 120)
Dieppe, Seine Maritime	Chalk	0.4	—	—	Prêcheur (1960, p. 120)
St. Jouin, Seine Maritime	Chalk	0	1834–1894	Maps	Prêcheur (1960, p. 122)
Germany					
North of Kiel, Baltic Sea	Glacial clay	{0.6 / 0.8}	1873–1934 / 1876–1938	{— / —}	Schütze, H. (vid. Seibold, 1963) / Healy, T. R. and Werner, F. (vid. Healy et al., 1987)
Kiel Bay, Baltic Sea	Glacial clay	0.3	—	—	Zeuner (1952, p. 354)
Heligoland, North Sea	Mesozoic sandstone	1	—	—	
Cape Arkona, Rügen	Chalk	3–4	(100 y)	—	Neumayr, M. and Suess, F. E. (vid. Zeuner, 1952, p. 354)
Iceland					
Surtsey Island	Lava	25–37	1967–1975	Air-Photogrammetric Surveys	Norrman (1980, Figure 4)
Indian Ocean					
Aldabra Atoll	Limestone	0.0010–0.0013	1969–1971	Micro-erosion meter	Trudgill (1976)
Indonesia					
Krakatoa Island	Pyrolyclastic deposits	33	1883–1928	—	Umbgrove, J. H. F. (vid. Guilcher, 1958, p. 70)
Krakatoa Island	Pyroclastic deposits	{3.0–1.4 (m/month) / 5–7}	(Monsoon season of 1981–82) / 1981–1983	{Surveys / Surveys}	Bird and Rosengren (1984)

Appendix 2

Location	Rock type	Rate	Period	Method	Reference
Japan					
Notsuka, Tokachi, Hokkaido	Alluvial deposits	2.3	1964–1974	Surveys	Toyoshima (1974)
Notsuka, Tokachi, Hokkaido	Alluvial deposits	1.3	1920–1975	Maps	Tsujimoto (1987)
Shichiri-Nagahama, Aomori	Quaternary clayey tuff	0.9	1911–1979	Maps	Tsujimoto (1987)
Odosezaki, Aomori	Andesite pyroclastic rocks	0.02	(6000 y)	Surveys	Tsujimoto (1987)
Oga-Kitaura, Akita	Pliocene sandstone	0.5	1911–1978	Maps	Tsujimoto (1987)
Unosaki, Akita	Miocene shale	0.05	(6000 y)	Surveys	Tsujimoto (1987)
Kuji, Iwate	Cretaceous sandstone	0.01	(6000 y)	Surveys	Tsujimoto (1987)
Rikuchu-Noda, Iwate	Oligocene sandy tuff	0.02	(6000 y)	Surveys	Tsujimoto (1987)
Haranomachi, Fukushima	Pliocene sandstone, mudstone	0.02	1912–1959	Maps	Yamanouchi (1964)
Iwaki, Fukushima	Pliocene mudstone	0.13–2.05	1974–1978	Surveys	Aramaki (1978)
Okuma, Fukushima	Pliocene mudstone, sandstone	{0.62, 0.31, 1.08}	{1947–1978, 1961–1963, 1963–1965}	Air-photogrammetric maps, Surveys	Horikawa and Sunamura (1967)
Oragahama, Fukushima	Pliocene siltstone	0.4	1912–1957	Maps	Aramaki (1978)
Suetsugi to Hirono, Fukushima	{Pliocene siltstone, Pliocene sandstone}	{0.80–1.4, 0.74}	{1947–1973, 1947–1973}	Air-photogrammetric maps	Ohshima (1974)
Byobugaura, Chiba	Pliocene mudstone	0.4–1.1	1888–1950	Maps, Surveys	Kawasaki (1954)
Byobugaura, Chiba	Pliocene mudstone	{0.79, 0.91, 1.47}	{1946–1960, 1960–1965, 1965–1967}	Air-photogrammetric maps	Sunamura and Horikawa (1969)
Byobugaura, Chiba		0.73	1884–1969	Maps	Horikawa and Sunamura (1970)
Taito-misaki, Chiba	Pliocene mudstone	{1.11, 0.70}	{1960–1966, 1966–1970}	Air-photogrammetric maps	Sunamura (1973)

(continued)

Table continued

Location	Lithology	Erosion rate (m/y)	Interval	Method	Source
Taito-misaki, Chiba	Pliocene mudstone	0.6	1883–1968	Maps	Sunamura and Tsujimoto (1982)
Ubara, Chiba	Miocene sandstone	0.04	(6000 y)	Surveys	Tsujimoto (1987)
Kominato, Chiba	Miocene mudstone	0.04	(6000 y)	Surveys	Tsujimoto (1987)
Shimoda-Ebisujima, Izu Pen.	Miocene tuffaceous sandstone	0.01	(6000 y)	Surveys	Tsujimoto (1987)
Tatado, Izu Pen.	Miocene volcanic breccia	0.004	(6000 y)	Surveys	Tsujimoto (1987)
Southern coast, Atsumi Pen.	Pleistocene deposits	0.03–0.6	1888–1959	Maps	Yamanouchi (1977)
Toyama coast	Alluvial fan deposits	0.5–1.0	—	—	Yoshikawa (1952)
West coast, Kanazawa city	Alluvial clay	1–2	—	Surveys	Kanazaki (1961)
Toban, Hyogo	Pleistocene deposits	1.0–1.5	1893–1955	Maps	Noda (1971)
Tsushi-Tsunokawa, Hyogo	Quaternary deposits	0.15	1929–1980	Maps	Tsujimoto (1987)
Iwami-Tatamigaura, Shimane	Miocene tuffaceous sandstone	0.03	(6000 y)	Surveys	Tsujimoto (1987)
Hario, Nagasaki	Oligocene sandstone	0.004	(6000 y)	Surveys	Tsujimoto (1987)
Oyano-Kodomari, Kumamoto	Eocene sandstone	0.04	(6000 y)	Surveys	Tsujimoto (1987)
Aoshima, Miyazaki	Miocene sandstone, mudstone	0.03	(6000 y)	Surveys	Kino (1958)
Habuseura, Niijima Island	Volcanic sand (Base-surge deposits)	5.5	1961–1965	Maps, Surveys	Yajima (1965)

Appendix 2

Location	Rock type	Rate	Period	Method	Reference
Habuseura, Niijima Island	Volcanic sand (Base-surge deposits)	1.2	1815–1968	Maps	Sunamura (1987)
Nishinoshima Island	Volcanic ejecta	80	1974–1977	Air-photogrammetric maps	Mogi et al. (1980)
Mexico					
Isla San Benedicto	Tephra	0.9(m/day)	(11 August to 15 November 1952)	Maps	Richards (1965)
New Zealand					
Around the New Zealand coast	Various rock types	0.25–1(mean)	—	Cadastral plans, Air photos, Measurements	Gibb (1978)
Cape Turnagain, south of Napier	Mudstone	2.25	1943–1975	Air photos	Gibb (1978)
Ngapotiki, Wellington	Conglomerate	3.46	1944–1973	Air photos	Gibb (1978)
Point Kean, Kaikoura	Tertiary mudstone	0.24	1942–1974	Air photos	Kirk (1975)
Sharks Tooth Point, Kaikoura	Tertiary mudstone	0.33	1942–1974	Air photos	Kirk (1975)
Fifth Bay, Kaikoura	Tertiary mudstone	0.08–0.10	1942–1974	Air photos	Kirk (1975)
Mudstone Bay, Kaikoura	Tertiary mudstone	0.06–0.13	1942–1974	Air photos	Kirk (1975)
Poland					
Baltic Sea coast	Quaternary deposits	1	—	—	Zenkovich (1967, p. 164)
Sweden					
Hallshuk, Gotland	Silurian limestone	0.018	1899–1955	Photos	Rudberg (1967)
North of Ygne, Gotland	Silurian marl	0.025	1950–1962	Photos	Rudberg (1967)

(continued)

Table continued

Location	Lithology	Erosion rate (m/y)	Interval	Method	Source
UK					
Fourth Bight, Whitby, Yorkshire	Upper Lias shale	0.023	1971–1972	Micro-erosion meter	Robinson (1977b)
Tees estuary to Ravenscar, Yorkshire	Upper ~ Lower Lias shales	0.09	1892–?	Maps, Surveys	Agar (1960)
Tees estuary to Ravenscar, Yorkshire	Glacial deposits	0.28	1892–?	Maps, Surveys	Agar (1960)
Flamborough Head, Humberside	Chalk	0.3	—	—	Matthews (1934, p. 10)
Bridlington to Sewerby, Humberside	Glacial deposits	1.8	—	—	Matthews (1934, p. 12)
Holderness, Humberside	Glacial deposits	3.3	—	—	Matthews (1934, p. 11)
Holderness, Humberside	Glacial deposits	0.29–1.75	1852–1952	Maps, Surveys	Valentin (1954)
Withernsea to Easington, Holderness, Humberside	Glacial deposits	10(max.)	1974–1983	Cliff-top measurements	Pringle (1985)
Norfolk	Glacial deposits	0.9	1880–1967	Maps, Air photos	Cambers (1976)
Sheringham to Happisburgh, Norfolk	Glacial deposits	0.2–1.8	1885–1985	Maps	Clayton (1989)
		0.7	1975–1985	Surveys	
Cromer to Mudesley, Norfolk	Glacial deposits	4.2	1838–1861	—	Matthews (1934, p. 21)
		5.7	1861–1905	—	
West Runton to East Runton, Norfolk	Glacial deposits	0.2	1971–1973	Cliff-top stakes	Cambers (1976)
Suffolk	Glacial deposits	0.8	1880–1950	Maps	Cambers (1976)
Hopton, Suffolk	Glacial deposits	0.8–0.9+	1926–1950	Maps	Steers (1951)
Pakefield, Suffolk	Glacial deposits	0.6–0.9	1926–1950	Maps	Steers (1951)
Covehithe, Suffolk	Glacial deposits	5.1	1925–1950	Maps	Steers (1951)
Southwold, Suffolk	Glacial deposits	3–3.3	1925–1950	Maps	Steers (1951)
Southwold, Suffolk	Glacial deposits	4.5–13.5	—	—	Matthews (1934, p. 21)

Appendix 2

Location	Material	Rate	Dates	Method	Reference
Dunwich, Suffolk	Glacial deposits	1.6	1589–1753	Maps	Robinson (1980a)
		0.9	1753–1824		
		1.5	1824–1884		
		1.1	1884–1925		
		1.2	1589–1977		
North coast, Isle of Sheppey, Kent	London Clay	0.9–2.2	1864(65)–1963(64)	Maps, Surveys	Hutchinson (1973)
Studd Hill, Kent	London Clay	1.5	1872–1898	Maps	So (1967)
		2	1898–1931		
		3.4	1931–1939		
		1.2	1939–1961		
Beltinge, Kent	London Clay	0.7	1872–1907	Maps	So (1967)
		0.9	1907–1933		
		1.1	1933–1939		
		0.9	1939–1959		
E. Minnis Bay, Kent	Chalk	0.30	1872–1938	Maps	May and Heeps (1985)
Grenham Bay, Kent	Chalk	0.08–0.4	1872–1932	Maps	Wood (1968, Figure 1)
E. Epple Bay, Kent	Chalk	0.14	1872–1938	Maps	May and Heeps (1985)
White Ness, Kent	Chalk	0.05	1842–1938	Maps	May and Heeps (1985)
Kingsgate, Kent	Chalk	0.15	1842–1938	Maps	May and Heeps (1985)
Hereson, Kent	Chalk	0.23	1842–1938	Maps	May and Heeps (1985)
Pegwell C. G. Stn., Kent	Chalk	0.05	1839–1938	Maps	May and Heeps (1985)
Lydden Spout, Kent	Chalk	0.51	1873–1933	Maps	May and Heeps (1985)
South Foreland, Kent	Chalk	0.19	1878–1962	Maps	May (1971)
E. Folkestone Warren, Kent	Chalk	0.36	1873–1933	Maps	May and Heeps (1985)
W. Falkestone Warren, Kent	Chalk	0.09	1873–1933	Maps	May and Heeps (1985)
Seven Sisters, Sussex	Chalk	0.51	1873–1962	Maps	May (1971)
Birling Gap, Seven Sisters, Sussex	Chalk	0.91	1875–1916	Maps, Surveys	May (1971)
Birling Gap, Seven Sisters, Sussex	Chalk	0.99	1950–1962		
		0.28–0.98	1951–1962	Maps	May and Heeps (1985)

(continued)

Table continued

Location	Lithology	Erosion rate (m/y)	Interval	Method	Source
Rottingdean, Sussex	Chalk	0.66 0.13	1873–1929 1929–1951	Maps	May and Heeps (1985)
Roedean, Sussex	Chalk	0.76 0.76	1826–1884 1884–1897	Maps	May and Heeps (1985)
Middle Bottom, Dorset	Chalk	0.05	1882–1962	Maps	May (1971); May and Heeps (1985)
Scratchy Bottom, Dorset	Chalk	0.46	1882–1962	Maps	May and Heeps (1985)
White Nothe to Hamburg Tout, Dorset	Chalk	0.22	1882–1962	Maps	May (1971)
Ballard Down, Dorset	Chalk	0.23	1882–1962	Maps	May (1971)
Fairy Dell, Dorset	Marls	0.4–0.5	1887–1969	Maps, Air photos	Brunsden and Jones (1980)
Vale of Glamorgan, Wales	Lower Lias limestone, shale	0.008–0.099	1967–1969	Cliff-top stakes	Trenhaile (1972)
Ogmore-by-Sea to Barry, Wales	Lower Lias limestone, mudstone	0.068	1977–1985	Stakes	Williams and Davies (1987)
South of Aberaeron, West Wales	Aberystwyth Grits (Greywacke and mudstone)	0.06	1880–1970	Maps	Jones and Williams (1991)
Near Aberarth, West Wales	Glacial clay	0.65	1880–1970	Maps	Jones and Williams (1991)
Southeast County Down, N. Ireland	Glacial deposits	0.21–0.84	1834–1962	Maps, Air photos	McGreal (1979c)
USA Martha's Vineyard, Mass.	Glacial deposits	1.7	1846–1886	Surveys	Johnson (1925, p. 403)

Appendix 2

Location	Deposit	Rate	Period	Method	Reference
Boston Harbour, Mass.	Glacial deposits	0.23 / 0.31	1860–1908 / 1903–1905	Surveys / Maps	Johnson (1925, p. 400)
Boston Harbour, Mass.	Glacial deposits	0.1	1958–1962	Steel spikes	Kaye (1967)
Thompson Island, Boston Harbour, Mass.	Glacial deposits	0.2	1938–1977	Air photos	Jones et al. (1985)
Outer Cape Cod, Mass.	Glacial deposits	1	1848–1888	—	Marindin, H. L. (vid. Davis, 1909, p. 694)
Outer Cape Cod, Mass.	Glacial deposits	0.96	1848–1888	Surveys	Johnson (1925, p. 400)
Outer Cape Cod, Mass.	Glacial deposits	$0.9 - 1.5^+$	(Long continued observations)		Schalk (1938)
Outer Cape Cod, Mass.	Glacial deposits	0.8	1879–1959	Surveys	Zeigler et al. (1964)
Outer Cape Cod, Mass.	Glacial deposits	0.12–2.2	1938–1974	Air photos	Gatto (1978)
Outer Cape Cod, Mass.	Glacial deposits	0.3	(4y)	Surveys	Giese and Aubrey (1987)
Outer Cape Cod, Mass.	Glacial deposits	0.3	1953–1958	Surveys	Zeigler et al. (1959)
Highland, Outer Cape Cod, Mass.	Glacial deposits	0–0.5	1933–1966	—	Davies et al. (1972)
North Shore, Long Island, N.Y.	Glacial deposits	0.5	(80 y)	Charts, Air photos	Bokuniewicz and Tanski (1980)
North Shore, Long Island, N.Y.	Glacial deposits	0.9^+	1833–1883	—	Johnson (1925, p. 393)
East Fort Point, Long Island, N.Y.	Glacial deposits				
Eastern shore of Chesapeake Bay, Md.	Miocene-Pleistocene deposits	1–2.5	—	—	Dalrymple et al. (1986)
Western shore of Chesapeake Bay, Md.	Miocene-Pleistocene deposits	0.8–0.9	—	—	Dalrymple et al. (1986)

(continued)

Table continued

Location	Lithology	Erosion rate (m/y)	Interval	Method	Source
Western Chesapeake Bay, Md.	Miocene sediments	0.3	1847–1934	Maps	Leatherman (1986)
Randle Cliff Beach, Western Chesapeake Bay, Md.	Miocene sediments	0.15	1951–1978	Surveys	Leatherman (1986)
Colonial Beach, Virginia	Pleistocene deposits	0.5	—	—	Dean et al. (1986)
Lake County, Ohio	Glacial deposits	0.3–0.9 0.3(max.)	1876–1937 1937–1973	Maps, Air photos Air photos	Carter (1976)
Berrien County, Mich.	Glacial deposits	3.8	1971–1974	Air photos	Birkemeier (1980, Table 10)
Eastern shore, Lake Michigan, Mich.	Glacial deposits	2.5	1970–1974	Surveys	Birkemeier (1981)
Lake Michigan shore, Mich. and Wisc.	Glacial deposits	0.4	(122 to 147 y)	Surveys and measurements	Buckler and Winters (1983)
Shannon Point, Wash.	Glacial deposits	0.2	1893–1969	Maps	Schwartz (1971)
North of Yaquina Bay, Ore.	Miocene sandstone	0.6	1880–1963	Photos	Byrne (1963)
Taft, Lincoln City, Ore.	Pleistocene terrace sands	<0.02	(50 y)	Photos	Komar and Shih (1991)
Rockaway Beach, Calif.	Franciscan volcanic and metavolcanic rocks	0.15	—	Maps	Sorensen (1968)
Montara, Calif.	Miocene conglomerate	0.26–0.29	1912–1965	Maps	Sorensen (1968)
Santa Cruz, Calif.	Mio-Pliocene mudstone	0–0.2	1943–1963	Air photos	Griggs and Johnson (1979)
Santa Cruz, Calif.	Miocene mudstone with shale	0.4	(80 y)	Maps, Surveys	US Army Corps. Eng. (vid. Sorensen, 1968)
Santa Barbara, Calif.	Miocene shale Mio-Pliocene shale Pliocene siltstone	0.2 0.08 0.3	1951–1965 1923–1967 1927–1947	Surveys	Norris (1968)

Appendix 2

Location	Rock type	Rate	Period	Method	Reference
Isla Vista, Calif.	Mio-Pliocene shale	0.05–0.4	1966–1973	Maps, Surveys	Cottonaro (1975)
Leucadia to Pacific Beach, Calif.	Sandstone, claystone	0–0.5	1970–1975	Nails	Lee et al. (1976)
La Jolla, Calif.	Cretaceous-Eocene sandstone	0.0003–0.0006	—	Dated inscriptions and graffiti	Emery (1941), Emery and Kuhn (1980)
La Jolla, Calif.	Cretaceous sandstone, Eocene shale	0.01–0.2	1940–1979	Photos	Emery and Kuhn (1980)
La Jolla, Calif.	Alluvium	0.5	(11 y)	Photos	Shepard and Grant (1947)
La Jolla, Calif.	Alluvium	0.28–0.50	1918–1930	Measurements	Vaughan (1932)
La Jolla, Calif.	Alluvium	0.09–0.3	1912–1975	Maps	Hannan, D. L. (vid. Emery and Kuhn, 1980)
Sunset Cliffs, San Diego, Calif.	Upper Cretaceous sandstone, siltstone	0.012	(75 y)	Photos, Observations	Kennedy (1973)
Sunset Cliffs, San Diego, Calif.	Pre-Miocene sandstone, shale	0.01–0.3	—	Photos, Maps	City of San Diego (vid. Lee, 1980)
USSR (former)					
Barrents Sea coast	Granitic rocks	0.001–0.002	—	—	Zenkovich, V. P. (vid. Zenkovich et al., 1965)
Laptev Sea coast	Glacial clay	4–6	—	—	Hmuijuikov, P. K. (vid. Zenkovich et al., 1965)
Laptev Sea coast	Glacial clay	4	—	—	Gakeli, J. J. (vid. Zenkovich et al., 1965)
Laptev Sea coast	Glacial clay	50(max.)	—	—	Gregorieb, E. P. (vid. Zenkovich et al., 1965)
Laptev Sea coast	Glacial clay	3.5–4	—	—	Vilynerr, B. A. (vid. Zenkovich et al., 1965)

(continued)

Table continued

Location	Lithology	Erosion rate (m/y)	Interval	Method	Source
Bering Sea coast	Moraine deposits	1–2(max.)	—	—	Iomin, A. S. (vid. Zenkovich et al., 1965)
Okhotsk Sea coast	Volcanic ash	50(max.)	—	—	Luimarev, B. I. (vid. Zenkovich et al., 1965)
Okhotsk Sea coast	Quaternary brown loam and clay	40(max.)	—	—	Zenkovich, V. P. (vid. Zenkovich et al., 1965)
Baltic Sea coast	Granitic rocks	0.001–0.002 (max.)	—	—	Ljungner, E. (vid. Zenkovich et al., 1965)
Baltic Sea coast	Sandstone	3(max.)	—	—	Zenkovich, V. P. (vid. Zenkovich et al., 1965)
Baltic Sea coast	Sandy deposits	20(max.)	—	—	Zenkovich, V. P. (vid. Zenkovich et al., 1965)
Baltic Sea coast	Sandy deposits	2	—	—	Zenkovich, V. P. (vid. Zenkovich et al., 1965)
Baltic Sea coast	Sandy deposits	1–1.5	—	—	Kaplin, P. A. and Bolduirev, V. L. (vid. Zenkovich et al., 1965)
Azov Sea coast	Quaternary brown loam and clay	3	—	—	Mamuikina, V. A. (vid. Zenkovich et al., 1965)
Primorsko-Atchtarsk, Sea of Azov	Clay	12	—	—	Zenkovich (1967, p. 165)
Black Sea coast	Flysh	0.02–0.03	(20 y)	Surveys, Photos	Zenkovich et al. (1965)

Appendix 2 277

Black Sea coast	Flysh, shale	0.01–0.02	—	Zenkovich, V. P. (vid. Zenkovich et al., 1965)
Black Sea coast	Coquinite	0.002–0.005	—	Aksentiev, G. N. (vid. Zenkovich et al., 1965)
Black Sea coast	Crystallized limestone	0.003	—	Aksentiev, G. N. (vid. Zenkovich et al., 1965)
Black Sea coast	Massive limestone	0.3–0.5 (max.)	—	Zenkovich, V. P. (vid. Zenkovich et al., 1965)
Black Sea coast	Limestone with loess	2–3	—	Jachiko, I. J. (vid. Zenkovich et al., 1965)
Black Sea coast	Limestone with loess	0.61	—	Aksentiev, G. N. (vid. Zenkovich et al., 1965)
Black Sea coast	Quaternary loess	0.5–1.0	—	Zamiry, P. K. (vid. Zenkovich et al., 1965)
Black Sea coast	Quaternary conglomerate	12	—	Zenkovich, V. P. (vid. Zenkovich et al., 1965)
Black Sea coast	Quaternary brown loam and clay	1(max.)	—	Zenkovich, V. P. (vid. Zenkovich et al., 1965)
Black Sea coast	Quaternary clay	2–3	—	Zenkovich, V. P. (vid. Zenkovich et al., 1965)
Black Sea coast	Diluvial deposits	0.11	—	Zenkovich, V. P. (vid. Zenkovich et al., 1965)

Author index

Abbott, A. T., 190
Abrahams, A. D., 179
Adey, W. H., 80
Agar, R., 124, 270
Ahr, W. M., 121
Aida, T., 55
Airy, G. B., 8
Alexander, H. S., 197, 199
Allen, J. R. L., 132
Allison, R. J., 79
Andersen, K. H., 69
Ängeby, O., 197
Apted, J. P., 70
Aramaki, M., 79, 97(Fig.), 114, 149(Fig.), 184, 185, 219, 267
Askin, R. W., 123
Aubrey, D. G., 92, 273

Bagnold, R. A., 33
Baker, G., 78, 190, 191, 193
Baker, R. F., 212
Barnes, H. L., 78
Barrell, J., 133, 139, 153
Bartrum, J. A., 141, 164, 165, 167, 176, 179, 193, 204, 206, 208
Bascom, W., 18
Battistini, R., 190
Battjes, J. A., 21, 27
Baulig, H., 144
Belikov, B. P., 70, 71
Bell, F. G., 55, 56(Fig.), 61
Bernaix, J., 53
Bieniawski, Z. T., 55
Biésel, F., 31
Bird, E. C. F., 2, 72, 78, 80, 88, 105, 107, 132, 139, 141, 143, 144, 148, 150, 151, 163, 167, 177, 190, 191, 193, 194, 196, 197, 199, 200, 202, 204, 217, 219, 220, 221, 223, 225, 264, 266
Bird, J. B., 88, 187, 188, 264
Birkeland, P. W., 78, 132
Birkemeier, W. A., 90, 226, 274
Bishop, A. W., 67(Fig.)
Blackmore, P. A., 33, 35
Bokuniewicz, H., 86, 87, 273
Bornhold, B. D., 143, 265
Bosence, D. W. J., 80
Boulden, R. S., 265
Bourcart, J., 205, 208
Bowen, A. J., 26
Boyd, G. L., 130
Bradley, W. C., 119, 132, 133, 134, 139, 153, 158, 164, 165
Bray, J. W., 62, 114
Bretschneider, C. L., 7, 45
Brighenti, G., 69
Broch, E., 55
Brodeur, D., 122, 176
Bromhead, E. N., 114, 224
Brunsden, D., 70, 72, 109, 272
Bryan, R. B., 107, 109, 265
Buckler, W. R., 106, 107, 274
Burshtein, L. S., 53
Buttle, J. M., 107
Byrne, J. V., 79, 88, 190, 191, 274
Byrne, R. J., 22

Cailleux, A., 208
Cambers, G., 86, 90, 103, 107, 270
Camfield, F. E., 24, 36, 37
Carr, A. P., 43, 44
Carson, M. A., 109
Carter, C. H., 46, 47, 79, 80, 88, 90, 91, 102, 103, 149, 216, 217, 226, 274

Carter, R. W., 2, 132, 219
Cernia, J. N., 51
Challinor, J., 153
Chan, C. K., 69
Chandler, R. J., 70
Chappell, J., 2
Chien, N., 82
Chieruzzi, R., 212
Cho, H. K., 265
Clagne, J. J., 143, 265
Clayton, K. M., 211, 270
Coakley, J. P., 120, 122, 123, 125, 128, 130, 156, 265
Coastal Engineering Research Center, 8, 10, 12, 20, 24, 35, 36, 37, 41, 44, 46, 209
Colback, P. S. B., 53
Coleman, J. M., 80
Cooke, R. U., 72
Cornelius, P. F. S., 80
Cotton, C. A., 78, 141, 143, 144, 164, 165, 167, 169, 172, 177, 193
Cottonaro, W. F., 225, 275
Crabtree, R. W., 121, 123
Craig, A. K., 121
Craig, D., 107
Croad, R. N., 129
Curtis, C. D., 70

Dally, W. R., 26
Dalrymple, R. A., 273
Daly, R. A., 143
Dana, J. D., 164, 165
Davidson, C. F., 141, 164
Davidson-Arnott, R. G. D., 122, 123, 125, 128, 130, 139, 162, 218, 265
Davies, D. S., 87, 273
Davies, J. L., 2, 42, 78, 128
Davies, P., 107, 189, 223
Davis, R. A., Jr., 90
Davis, W. M., 139, 143, 144, 153, 273
De Boer, G., 103
Dean, J. L., 213, 274
Dean, R. G., 8, 12, 13, 17, 38, 39, 42, 155

Dearman, W. R., 55, 56, 61
Deere, D. U., 61, 68
Denny, D. F., 33
Dent, O. F., 139, 141, 150, 163
Di Nardo, L. R., 88, 265
Dick, T. M., 86, 105
Dietz, R. S., 132, 133
Dionne, J. C., 122, 176, 197(Fig.), 198(Fig.), 199
Doornkamp, J. C., 106
Dubois, R. N., 225
Duckmanton, N. K., 179
Duncan, N., 61, 65

Eagleson, P. S., 8, 12, 13, 17
Edil, T. B., 107, 116
Edwards, A. B., 141, 150, 163, 165, 166, 167, 177, 183, 193, 204
Elston, E. D., 199
Emanuel, K. A., 228
Emery, K. O., 2, 80, 88, 122, 123, 125, 141, 176, 185, 186, 188, 189, 199, 200, 201, 207, 275
Emmons, W. H., 78
Evans, I., 53, 60
Evans, J. W., 80, 123
Everts, C. H., 92, 109
Eyles, N., 107

Fairbridge, R. W., 78, 132, 133, 141, 164, 170, 188, 196
Farmer, I., 67, 69
Feda, J., 53
Fenneman, N. M., 132
Filloux, J., 120
Flaxman, E. M., 79
Fleming, C. A., 98, 99, 216, 217
Flemming, N. C., 153, 158, 188
Focht, J. A., Jr., 69
Focke, J. W., 80, 186, 187, 188
Foster, H. L., 185, 186
Fox, F., 211
Franklin, J. A., 55
Friedman, G. M., 132

Fryde, W. T., 216
Fulton-Bennett, K., 213, 218

Galvin, C. J., Jr., 20, 21, 22, 23, 24
Gardner, R., 193
Garner, H. F., 132
Garrels, R. M., 153
Gary, M., 132
Gatto, L. W., 88, 273
Gaughan, M. K., 21, 22, 119
Gelinas, P. J., 88, 100, 101, 116, 265
Gibb, J. G., 269
Giese, G. S., 92, 273
Gill, E. D., 125, 126, 139, 141, 148, 150, 151, 163, 164, 165, 167, 169, 171, 172, 173, 176, 177, 179, 180, 193, 195, 199, 203, 204, 205, 208, 264
Gilluly, J., 78
Goda, Y., 5, 7, 22, 23, 24
Goudie, A., 193, 219
Graff, J., 43, 44
Grainger, P., 109
Grant, U. S., 105, 191, 219, 275
Greenwood, B., 119
Griggs, G. B., 86, 106, 119, 134, 139, 158, 164, 165, 213, 218, 223, 274
Grisez, L., 203, 205
Groen, P., 45, 46
Grosvenor, N. E., 53
Groves, G. W., 45, 46
Guidicini, G., 55
Guilcher, A., 2, 78, 141, 143, 144, 164, 188, 189, 200, 201, 202, 204, 266
Gulliver, F., 132
Guy, D. E., Jr., 79, 80, 90, 91, 103, 149, 226
Guza, R. T., 27

Haas, B. J., 116
Hails, J. R., 93
Hale, D. B., 119

Hall, K. R., 79
Hanna, F. K., 88, 123
Hansen, J., 228
Harris, D. L., 45
Harris, W. B., 122, 125, 127, 216, 223
Hatai, K., 197(Fig.), 198(Fig.), 199
Hattori, M., 120
Hawkes, I., 53, 55
Hayami, S., 21
Hayashi, M., 65, 66
Hayes, M. O., 120
Healy, T. R., 74, 80, 121, 139, 141, 156, 164, 166, 176, 196, 266
Heeps, C., 88, 109, 193, 271, 272
Henkel, I., 199
Herrmann, H. G., 69
Hewson, P. J., 33, 35
Higgins, C. G., 184, 186, 187, 188
High, C., 88, 123
Hills, E. S., 141, 150, 164, 165, 166, 167, 172, 193, 196, 200, 204
Hinds, N. E. A., 143
Hobbs, D W, 53
Hobbs, N. B., 67
Hodgkin, E. P., 80, 88, 124, 186, 187, 188, 264
Hoek, E., 62, 114
Hoffman, J. S., 225
Hoffmeister, J. E., 141
Höllermann, P., 204, 208
Holmes, A., 78, 132, 193
Holmes, D. L., 78, 132, 193
Hom-ma, M., 30, 31, 32, 35, 36, 37
Hondros, G., 55
Horibe, T., 69
Horikawa, K., 7, 8, 23, 29, 32, 35, 36, 37, 46, 81, 82, 88, 122, 124, 126, 132, 137, 138(Fig.), 210, 212, 213, 214, 215, 218, 267
Hoshino, M., 133
Houpert, R., 53
Houston, W. N., 69
Hudson, R. Y., 31
Hume, W. F., 208
Hutchinson, J. N., 107, 108(Fig.), 109, 113, 114, 115, 116, 125, 126, 127,

Hutchinson (*continued*)
  129, 185, 189, 199, 200, 221, 222, 223, 224, 271

Ijima, T., 7, 25, 26
Ikeda, K., 68
Imanaga, I., 164
Imbrie, J., 133
Inman, D. L., 28, 119, 120
Inoue, M., 64
Inozemtsev, Yu. P., 79
International Bureau for Rock Mechanics, 54
Ippen, A. T., 23
Irfan, T. Y., 55, 56, 61
Ito, R., 197(Fig.), 198(Fig.)
Iversen, H. W., 20, 21, 22

Jaeger, C., 51
Janssen, J. P. F. M., 27
Jehu, T. J., 121
Jennings, J. N., 202, 205
Jerwood, L. C., 121
Johannessen, C. L., 80, 164, 207
Johnson, D. W., 132, 133, 139, 141, 153, 164, 165, 172, 190, 191, 264, 265, 272, 273
Johnson, R. E., 86, 106, 223, 274
Jones, D. G., 79, 88, 103, 107, 111, 272
Jones, D. K. C., 109, 272
Jones, J. R., 273
Jonsson, I. G., 119
Judson, S., 141
Jutson, J. T., 141, 163, 165, 167, 193, 196

Kaizuka, S., 133
Kalaugher, P. G., 109
Kamata, M., 63(Fig.)
Kamphuis, J. W., 79, 100, 101, 119, 155, 156, 161
Kanazaki, H., 268
Kawana, T., 186

Kawasaki, I., 105, 267
Kayanne, H., 165
Kaye, C. A., 88, 107, 186, 188, 273
Keizer, H. I., 218, 265
Kelletat, D., 206
Kennedy, M. P., 275
Kézdi, Á., 51
King, C. A. M., 2, 78, 106, 119, 132, 133, 139, 158
Kino, Y., 268
Kirk, R. M., 88, 119, 124, 126, 128, 164, 166, 172, 176, 269
Kirkgoz, M. S., 33, 35
Kobayashi, M., 149(Fig.)
Kobayashi, R., 53, 58
Koh, R. C. Y., 22
Kohno, F., 151
Kolberg, M., 265
Kolberg, T. O., 90
Komar, P. D., 3, 6, 8, 21, 22, 23, 29, 40, 78, 103, 119, 274
Kraus, N. C., 29, 119
Kuenen, Ph. H., 78, 133, 153
Kuhn, G. G., 2, 86, 88, 107, 122, 125, 176, 190, 191, 200, 201, 275
Kulin, G., 23

Lama, R. D., 63, 65
Lang, J. G., 125, 172, 176, 177, 179, 180
Larson, E. E., 78, 132
Layzell, M. G. J., 167, 176, 177, 180
Le Bourdiec, P., 190
Le Méhauté, B., 8, 22
Leatherman, S. P., 107, 225, 274
Lee, K. L., 69
Lee, L. J., 88, 139, 225, 275
Leet, L. D., 141
Longuet-Higgins, M. S., 20, 26, 27
Longwell, C. R., 78, 132, 133, 139
Lorriman, N. R., 103
Louis, H., 153
Lugt, H. J., 129
Lundborg, N., 53
Lupini, J. F., 224

Author Index

McCowan, J., 23
Macdonald, G. A., 190
MacFadyen, W. A., 186
McDougal, W. G., 103
McGreal, W. S., 79, 87, 90, 92, 103, 107, 111, 272
McGreevy, J. P., 206, 208
Machatschek, F. 153
MacIntyre, I. G., 80
McLean, R. F., 121, 123, 141, 164
Magoon, O. T., 213
Manabe, S., 228
Marsland, A., 67
Martini, I. P., 202, 205
Mather, K. F., 190, 191
Matsukura, Y., 73, 203, 205, 208
Matsuoka, N., 203, 205, 208
Matthews, E. R., 270
May, V. J., 88, 107, 109, 193, 225, 271, 272
Mellor, M., 53, 55
Menard, H., 132, 133
Mercan, D. W., 80, 121
Miche, M., 31
Michell, J. H., 22
Mii, H., 122, 133, 165, 176, 194, 197(Fig.), 198, 199, 200, 201, 202, 203, 205
Miller, A. A., 133
Miller, R. P., 61
Minikin, R. R., 32
Mitsuyasu, H., 31, 32, 33
Mizuguchi, M., 23, 25, 26
Mogi, A., 269
Mogi, K., 53
Munk, W. H., 7, 21
Muronachi, T., 63(Fig.)
Mustoe, G. E., 202, 203, 207, 208

Nagai, S., 33, 35
Nairn, R. B., 82, 129, 155, 161, 162, 221
Nash, D., 114
Neumann, A. C., 80, 121, 186, 188
Newell, N. D., 133

Nicholls, N., 228
Nishimatsu, Y., 58
Noda, H., 268
Nohara, M., 197(Fig.), 198(Fig.), 199
Norris, R. M., 274
Norrman, J. O., 88, 136, 266
North, W. J., 123
Nunn, P. D., 188

Oak, H. L., 179
Oerlemans, J., 225, 226
Ohshima, H., 79, 97(Fig.), 149(Fig.), 267
Okamoto, T., 55
Okazaki, S., 21, 22
Okumura, K., 58
Ollier, C., 70, 72, 74
Omi, M., 64
Ongley, M., 164
Onodera T. F., 67
Otsubo, M., 35

Palmer, H. S., 164
Parker, G. F., 265
Partenscky, H. W., 33, 35
Patrick, D. A., 21
Pethick, J., 2, 29, 141, 166
Phillips, B. A. M., 164, 172
Philpott, K. L., 2, 3, 131, 161, 219, 221
Pierson, W. J., 7
Pinchin, B. M., 221
Pirazzoli, P. A., 186
Plakida, M. E., 35
Pomeroy, C. D., 53, 60
Porter, S. C., 132
Prêcheur, C., 191, 193, 266
Price, A. G., 107, 109, 265
Pringle, A. W., 88, 93, 94, 103, 270
Prior, D. B., 109
Protodyakonov, M. M., 53, 54, 59
Pugh, D. T., 38

Quigley, R. M., 88, 90, 100, 101, 116, 226, 265

Raichlen, F., 47
Ralph, K. J., 122, 125, 127, 216, 223
Raudkivi, A. J., 120
Reffell, G., 165, 166, 171, 172
Revelle, R., 123
Rich, J. L., 132
Richards, A. F., 269
Richards, K. S., 103
Ritter, D. F., 1, 78
Robinson, A. H. W., 88, 92, 93, 96, 98, 103, 221, 271
Robinson, D. A., 121
Robinson, L. A., 87, 88, 104, 119, 120, 121, 123, 124, 126, 127, 128, 139, 158, 159, 160, 163, 184, 185, 196, 270
Rode, K., 133, 134
Rosengren, N. J., 88, 105, 107, 190, 191, 217, 223, 225, 264, 266
Ross, C. W., 33
Rudberg, S., 269
Rundgren, L., 31
Russell, R. J., 133, 165, 187, 188

Saeki, H., 25, 27
Sainflou, M., 31
Saito, M., 133
Sakamoto, N., 7
Sanders, J. E., 132
Sanders, N. K., 81, 82, 141, 163, 165, 170, 171, 172
Sanglerat, G., 63
Sasaki, N., 25, 27
Sato, S., 197
Sato, Sh., 134
Schalk, M., 273
Schlichting, H., 129
Schneider, J., 121
Schwartz, M. L., 219, 274
Seed, H. B., 69
Seibold, E., 266
Shackleton, N. J., 2
Shepard, F. P., 28, 86, 105, 107, 190, 191, 219, 275
Shih, S-M., 103, 274

Short, A. D., 126, 141, 196
Shuisky, Y. D., 219
Shumway, G., 133
Silvester, R., 7, 45, 216
Sims, P., 223
Singh, D. P., 59, 60
Skinner, B. J., 132
Smalley, I. J., 72
Snead, R. E., 132, 190, 191
So, C. L., 88, 103, 124, 139, 271
Sorensen, R. M., 139, 155, 157, 213, 274
Sparks, B. W., 78, 105, 141, 143, 151, 153, 202
Spencer, T., 74, 80, 123, 125, 188
Stanton, R. J., Jr., 121
Stearns, H. T., 141
Steers, J. A., 191, 193, 270
Stewart, R. W., 26
Stouffer, R. J., 228
Strahler, A. N., 78
Street, R. L., 24
Strum, M., 265
Summers, D., 90
Summers, L., 216
Sunamura, T., 1, 21, 22, 23, 25, 34, 35, 55, 56(Fig.), 59(Fig.), 68, 76, 77, 79, 81, 82, 84, 85, 86, 88, 89, 90, 93, 95, 96, 97(Fig.), 102, 103, 104, 105, 111, 119, 120, 122, 124, 126, 132, 133, 134, 135, 136, 137, 138(Fig.), 141, 149(Fig.), 151, 153, 155, 157, 162, 163, 164, 165, 167, 168, 169, 172, 173, 174, 175, 176, 178, 186, 192, 200, 210, 212, 213, 214, 215, 218, 226, 227, 228, 267, 268, 269
Suwardi, A., 105
Suzuki, T., 68, 70, 80, 121, 127, 176, 194
Svendsen, I. A., 26
Sverdrup, H. U., 7
Swinchatt, J. P., 121
Swinerton, A. C., 176, 199

Takahashi, K., 55, 56(Fig.), 59(Fig.), 73, 80, 121, 127, 149(Fig.), 176

Author Index

Takahashi, T., 141, 164, 172
Takeda, I., 120
Takenaga, K., 186
Tanabe, K., 197(Fig.), 198(Fig.), 199
Tanimoto, I., 198(Fig.), 199
Tanski, J., 86, 87, 273
Tayama, R., 133
Taylor, D. W., 109
Ternan, L., 223
Thomas, R. H., 225, 226(Fig.)
Thornbury, W. D., 78
Thornton, E. B., 27
Titus, J. G., 225
Tjia, H. D., 187, 188
Toyoshima, O., 267
Toyoshima, Y., 133, 144, 164, 165, 189, 198(Fig.), 199
Trenhaile, A. S., 2, 74, 80, 121, 123, 124, 126, 139, 141, 142, 150, 151, 164, 166, 167, 171, 172, 173, 176, 177, 180, 196, 197, 202, 205, 206, 208, 272
Tricart, J., 164, 202, 206
Trudgill, S. T., 72, 74, 80, 88, 121, 123, 124, 125, 126, 186, 187, 200, 202, 266
Tryggvason, G. S., 136
Tsujimoto, H., 55, 56(Fig.), 59(Fig.), 79, 142, 143, 144, 146, 147, 148, 149(Fig.), 151, 152, 164, 165, 166, 267, 268
Twenhofel, W. H., 132, 134, 153
Twidale, C. R., 193, 202, 207, 208

Valentin, H., 88, 98, 99, 100, 270
Vallejo, L. E., 107, 116
Vaughan, T. W., 275
Verstappen, H. T., 186, 187, 188
Viggósson, G., 136
Viles, H. A., 125, 126
Vita-Finzi, C., 80
Von Bulow, P., 107
Von Engeln, O. D., 133, 139
Vutukuri, V. S., 55, 57, 63, 65

Warrick, R., 225, 226
Washburn, A. L., 72
Wafer, G., 123, 124, 139, 156
Weggel, J. R., 21, 24
Weishar, L. L., 22
Wellman, H. W., 72
Wentworth, C. K., 141, 164, 176, 184, 186, 187, 188, 190, 193, 196, 199, 200
Wheeler, W. H., 210
Wiegel, R. L., 7, 8, 10, 12, 21
Wild, B. L., 53
Wilhelm, E. J., 73
Williams, A. T., 79, 88, 103, 107, 111, 189, 223, 272
Williams, W. W., 87, 91, 105
Wilson, A. T., 72
Wilson, B. W., 7
Wilson, G., 79, 191
Winker, E. M., 73
Winters, H. A., 106, 107, 274
Wood, A., 119, 139, 160, 271
Wood, F. J., 38
Woodroffe, C. D., 186, 188
Wray, J. L., 80
Wright, L. W., 139, 141, 160, 163, 164

Yajima, A., 268
Yamaguchi, U., 58
Yamanouchi, H., 79, 97, 98, 99, 267, 268
Yano, K., 63(Fig.)
Yatsu, E., 69, 70, 72, 73, 74
Yoshikawa, T., 133, 165, 268

Zeigler, J. M., 273
Zeman, A. J., 79, 86, 90, 101, 105, 107, 122, 226
Zenkovich, V. P., 2, 78, 80, 88, 123, 124, 126, 128, 132, 133, 134, 139, 141, 144, 148, 153, 158, 163, 189, 190, 218, 269, 275, 276, 277
Zeuner, F. E., 266

Location index

Aberaeron, 272
Aberarth, 111, 272
Agay, 205
Aichi, 155
Akabane, 155
Akane-Jima I., 143(Fig.), 144
Akita, 267
Alaska, 44
Aldabra Atoll, 124, 125, 126, 186, 187, 266
Angel de la Guarda I., 144
Angola, 204
Antrim, County, 219
Aomori, 267
Aoshima I., 127, 149(Fig.), 268
Arecibo, 190
Arran, I. of, 206
Artillery Rocks, 205
Atlantic City, 45
Atsumi Peninsula, 97, 99, 268
Auckland, 121, 179, 206
Auckland Is., 143
Ault, 266
Australia, 163, 165, 172, 204
Azov Sea, 276

Baikal, L., 133
Baku, 207
Ballard Down, 193, 272
Ballycastle, 206
Baltic Sea, 123, 154, 190, 266, 269, 276
Banks Peninsula, 143, 144
Barbados, 187, 188
Barrents Sea, 275
Barry, 107, 272
Barton-on-Sea, 216
Batanta, 187
Bay of Islands, 164

Bean Gulch, 207
Beirut, 190
Belle Ile, 143
Bellingham, 207, 208
Beltinge, 103, 271
Benouville, 193
Benten-Jima, 143, 144, 148
Bering Sea, 276
Bermuda, 186, 188
Berrien County, 274
Birling Gap, 271
Black Rock Point, 191, 223, 264
Black Sea, 128, 133, 158, 219, 276, 277
Bokniseck, 123, 124, 156
Boso Peninsula, 205
Boston Harbour, 273
Brazil, 164
Bridlington, 270
Burnie, 43(Fig.)
Byobugaura, 89, 90, 102, 106, 124, 126, 132, 133, 137, 138, 155, 163, 212, 213, 214, 215, 218, 267

Cabo Rojo, 187
California, 164, 190, 191, 200, 213, 218, 274, 275
California, Gulf of, 144
California, Southern, 92, 133, 219, 225
Campbell I., 143
Canary Is., 204
Cape Arago, 207
Cape Arkona, 266
Cape Breton I., 190, 191, 264, 265
Cape Doob, 158
Cape Gilyanly, 189
Cape Paterson, 204
Cape Reamur, 264
Cape St Vincent, 143

Cape Turnagain, 269
Castlepoint, 164
Cayman Is., 186
Chesapeake Bay, 273, 274
Chiba, 142, 149(Fig.), 155, 166, 267, 268
Chikura, 133
Christchurch, 143
Clacton-on-Sea, 121, 125, 127, 210, 211, 216, 223
Clare, County, 121, 125
Cleveland, 216
Clion-sur-Mer, 205
Colonial Beach, 213, 274
Covehithe, 91, 105, 270
Craig Ddu, 107
Crimea, 144
Cromer, 270
Curacao, 186, 187, 188

Dainoyama, 97(Fig.), 149(Fig.)
Devon, North, 164
Devon, South, 93
Dieppe, 266
Dorset, 272
Down, County, 272
Downderry, 223
Dunwich, 92, 96, 103

Easington, 93, 103, 270
East Fort Point, 273
Elba, I. of, 205
England, 191
English Channel, 44, 190
Eno-shima, 199
Eire, 200
Epple Bay, 160, 271
Erie, L., 47, 90, 99, 101, 102, 103, 149(Fig.), 216, 219, 265
Etretat, 193

Fairy Dell, 272
Fifth Bay, 269

Fifty Mile Point, 218
Fiji, 188
Flamborough Head, 98, 270
Folkestone Harbour, 222, 223, 271
Folkestone Warren, 221, 224
Fourth Bight, 270
Fukushima, 97, 155, 184, 185, 219, 267
Fundy, Bay of, 44

Gaspé, 166, 196
Gelendzhik, 128
George R., 195
Glace Bay, 191
Glamorgan, 43(Fig.)
Gotland, 269
Gourmalon, 203, 205
Grand Cayman I., 188
Great Lakes, 46, 90, 116, 144, 154, 225
Greece, 186, 187
Grenham Bay, 160, 163, 271
Grimsby, 128, 129, 156
Gunwalloe, 221

Habuseura, 268, 269
Hallsands, 93, 221
Hallshuk, 269
Hamburg Tout, 272
Happisburgh, 270
Haranomachi, 267
Hario, 268
Harrington Sound, 186
Hatake I., 205
Hawaii, 144, 164, 187, 190, 199, 200
Helen Drive, 90
Heligoland, 266
Helston, 221
Hengistbury Head, 225
Hereson, 271
Highland, 273
Hirono, 267
Hobart, 191
Holderness, 93, 94, 98, 100, 103, 129, 189, 200, 216, 270

Location Index

Hopton, 270
Houtman's Abrolhos Is., 188
Humberside, 270
Huron, L., 265
Hyogo, 268

Iceland, 136
Idehama, 184, 185
Indian Bay, 264
Isla San Benedicto, 269
Isla Vista, 275
Isle of Man, 164
Isle of Wight, *see* Wight, I. of
Iwaki, 219, 267
Iwami-Tatamigaura, 268
Iwate, 166, 267
Izu Peninsula, 172, 178, 179, 268

Japan, 164, 165, 200
Jólnir, 136
Joss Bay, 113
Jubilee Point, 72
Jump-off-Joe, 191

Kaikoura, 126, 127, 128, 172, 176, 179, 269
Kamaran I., 186
Kamogawa, 143(Fig.)
Kanazawa, 268
Kanto, Southern, 133
Kauai, 190
Kent, 271
Key Minor, 186, 188
Kiel Bay, 124, 266
Kieler Bucht, 156
Kii Peninsula, 144
Kilkeel, 92, 103
Kilkeel Bay, 111, 116
Kingsgate, 271
Kohriyama, 114
Kominato, 166, 268
Krakatau, 105, 106, 266
Kuji, 267

Kumamoto, 268
Kurokawa R., 197

La Jolla, 125, 176, 191, 200, 201, 207, 275
Lake County, 274
Langkawi I., 187, 188
Laptev Sea, 275
Le Havre, 193
Leucadia, 275
Lincoln City, 103, 274
Lingrow, 184, 185
Llantwit Major, 223
Loire-Atlantique, 205
Long I., 87, 273
Long Point, 100
Lorne, 196, 204, 205
Los Angeles, 40, 42
Lydden Spout, 271
Lyme Regis, 211, 223

Malta, Island of, 190
Mamba Point, 206
Manukau Harbour, 206
Margate, 43(Fig.)
Martha's Vineyard, 272
Maryland, 273, 274
Massachusetts, 272, 273
Matsushima, 184, 185
Michigan, L., 90, 106, 274
Middle Bottom, 272
Minnesota, 197
Minnis Bay, 160, 271
Miura Peninsula, 127
Miyagi, 184, 185
Miyazaki, 149(Fig.), 268
Moluccas, 186
Montara, 155, 157, 274
Moonlight Head, 193
Moresby I., 143
Morocco, 200
Moss Beach, 199
Mudesley, 270
Mudstone Bay, 269

Mullins, 264
Muriwai, 206
Murmansk, 144

Nagao-bana, 143(Fig.), 144, 148
Nagasaki, 268
Namforsen, 197
Napier, 269
Netherlands Antilles, 186
Newcastle, 204
Newfoundland, 166
New Guinea, 187
Newhaven, 219
Newhaven Harbour, 219, 220
Newport, 191
Newport Head, 166
New South Wales, 163
New York, 273
New Zealand, 164, 165, 193, 269
Ngapotiki, 269
Niijima I., 103, 133, 149(Fig.), 227, 268, 269
Nishinoshima I., 269
Nojima-zaki, 203, 205, 208
Norfolk, 103, 211, 270
Norfolk I., 124
North Sea, 44
North Shore, 273
Notsuka, 267
Nova Scotia, 190, 264, 265

Oahu, 186, 188, 190, 196
Odosezaki, 166, 267
Oga-Kitaura, 267
Ogmore-by-Sea, 107, 272
Ohio, 274
Okhotsk Sea, 276
Okuma, 89, 90, 97(Fig.), 124, 126, 132, 267
Old Hive, 93
Omika, 155
Oniga-jyo, 144
Ontario, L., 122, 123, 125, 130, 221, 265

Oragahama, 97(Fig.), 149(Fig.), 267
Oregon, 164, 190, 207, 274
Orikigawa, 97(Fig.), 149(Fig.)
Orkney I., 193
Otway, 125, 177, 179, 180, 193, 194, 195, 203, 205
Outer Cape Cod, 92, 273
Oyano-Kodomari, 268

Pacific Beach, 275
Painesville, 212
Painesville-on-the-Lake, 216, 217(Fig.)
Pakefield, 270
Paynes Bay, 264
Pegwell, 271
Plymouth, 93, 221
Point Kean, 269
Point Peron, 186, 187, 188, 196, 264
Point Sturt, 264
Pointe du Raz, 143
Port Bruce, 116, 156, 265
Port Burwell, 131, 219
Port Burwell Harbour, 219
Port Campbell, 191, 193
Port Fairy, 264
Port Jackson, 164
Port Nicholson, 144
Port Phillip Bay, 196
Portballintrae, 219
Potomac R., 213
Provence, 205
Pt Grey, 205, 265
Puerto Rico, 186, 188
Puget Sound, 207

Québec, 163, 196

Randle Cliff Beach, 274
Ravenscar, 270
Reculver, 103
Red Sea, 186
Rikuchu-Noda, 166, 267
Rockaway Beach, 274

Location Index

Roedean, 272
Rondeau, 100
Rottingdean, 272
Rügen, 266
Runton, 270
Russell, 164
Ryukyu Is., 186

Saanich Peninsula, 265
St Catherines, 265
St George R., 205
St Helena I., 143
St Jouin, 266
St Michael 40, 42
St Sturt, 204, 205
San Diego, 275
San Remo, 204, 205
San-In, 133, 165, 189
Sand le Mere, 100
Santa Barbara, 274
Santa Cruz, 106, 133, 134, 155, 157, 213, 223, 274
Santa Cruz I., 189
Scarborough, 265
Scratchy Bottom, 272
Seaford Head, 219, 220
Seaton, 223
Seine Maritime, 266
Serra do Sombreiro, 204
Sertung, 105
Seven Sisters, 271
Sewerby, 270
Shannon Point, 274
Shareham, 87
Sharks Tooth Point, 269
Sheppey, I. of, 107, 116, 184, 185, 271
Sheringham, 270
Shichiri-Nagahama, 267
Shikoku, 133
Shimane, 268
Shimoda, 143(Fig.), 144, 268
Shinyashiki, 14, 148
Shionomisaki, 133
Shirahama, 202, 205
Sizewell Bank, 92

Skye, I. of, 206
Sochi, 218
Somme, 266
South Foreland, 271
Southwold, 270
Spurn Head, 98
Stoney Creek, 128, 218, 265
Suetsugi, 267
Suffolk, 270, 271
Sunset Cliffs, 275
Surtla, 136
Surtsey, 136, 266
Sussex, 271, 272
Swansea, 43(Fig.)
Sydney, 165, 171, 172

Table Head, 191, 265
Taft, 274
Tahiti, 143
Taito-misaki, 111, 192, 267, 268
Tanabe Bay, 133
Tasmania, 43(Fig.), 141, 163, 165, 172
Tatado, 268
Tateyama, 149(Fig.)
Taylor's Falls, 197
Tees, R., 270
Tenerife, 204
Thanet, I. of, 43(Fig.), 113, 124, 160
Thompson I., 273
Thorpeness, 92
Tintagel, 191
Toban, 268
Tokachi, 267
Tokyo, 149(Fig.)
Toledo, 46, 47
Toronto, 221
Torrey Pine Beach, 27
Tottori, 143(Fig.)
Toyama, 268
Tsemes Bay, 128
Tsushi-Tsunokawa, 268
Turimetta Head, 166
Tuscany, 205

Ubara, 268

Unosaki, 267

Vale of Glamorgan, 107, 124, 272
Vancouver I., 199, 265
Vatulele, 188
Victoria, 141, 163, 190, 193, 199, 200
Virginia, 274

Waiheke I., 206
Wakayama, 119, 194, 201, 205
Wales, 107, 189, 191, 272
Warden Point, 107, 116, 127, 184, 185
Warren, The, 221, 222, 223
Warrnambool, 167, 264
Wellington, 269

West Bay, 219, 221
West Indies, 186
Whangaparaoa, 196
Whangarei Heads, 206
Whitby, 104, 270
White Ness, 271
White Nothe, 272
Wight, I. of, 193
Wilsons Promontory, 143
Withernsea, 93, 103, 270

Yakama, 197
Yaquina Bay, 274
Yellow Sea, 44
Ygne, 269
Yorkshire, 124, 127, 128, 196
Yusuji, 97(Fig.), 149(Fig.)

Subject index

abrasion, 78
   pothole formation, 198
   ramp formation, 196
   sea cave development, 190
   sea floor lowering, 120, 127, 128
   Type-B platform lowering, 176
abrasion benches, 141
abration platforms, 139
abrasion ramps, 196
accretion
   beach, 2
acoustic transducer, 123
aerial photographs, 88
age of specimen, 82(Fig.)
algae, 121
   encrusting, 194
alveoli or alveoles, see honeycombs
American Society of Testing and
   Materials (ASTM), 53
amplitude, 10
angle
   of effective slope, 109
   of internal friction, 57, 109
   of potential failure plane, 109
angular frequency, 10
assailing force of waves, 75–9
   associated with Type-B platform
     formation, 167, 173
   factors affecting, 90
   on sea floor bedrocks, 117–201
arches, see sea arches
artificial headlands, 216

barnacles, 121
basal erosion
   wave-induced, 76(Fig.)
beach dumping (or fill), 3, 213
beach mining, 219, 221

beach nourishment
   artificial, 2, 212, 213, 219, 221
beach platforms, 139
beach sediment controlling cliff erosion,
   92–5
bedrock erosion, underwater, 117–38
bedrock (sea floor) lowering rates
   exponential decay function, 131
   measurements, 123–6
   rapid lowering, 136
   spatial variations, 128
   temporal variations, 126
   till coasts. 129–32
benches, 139
   abrasion, 141
   denuded, 141
   marine, 141
   shore, 141
   wave-cut, 139, 141
Bernoulli equation, 9
bidirectional flow on Type-B platforms,
   194
biological erosion (bioerosion), 74,
   121
   associated with notch geometry, 186,
     188
bivalves, 74, 121
blowholes, 190
bore, 105, 177
boring organisms, 74, 121
   rock strength reduction, 121
Brazilian test, 55
breaker types, 21
breaking waves, 20–6, 29–30, 182
   classification, 21–2
   height, 21–3
   pressure, 31–5
   type of, 21
   water depth of, 23–4

breakwaters, 219
  cliff erosion induced by breakwater construction, 219–21
brittleness index, 55
broken waves, 25–6, 29–30, 182
  height, 25–6
  pressure, 35–7
bulkheads, 209, see also sea walls

carving of graffiti on cliff face, 225
Cathedral Arch, 191
cavernous weathering, 202
caves, see also sea caves
  artificially excavated, 225
cavitation, 78
cell-like structure, see honeycombs
centrifugal motion of water in potholes, 198
chemical erosion, 72
chitons, 74, 121
cliff erosion by waves, 80–106
  associated with rip current embayments, 103
  associated with sea level rise, 225–8
  controlled by beach width, 102
  controlled by marine stacks, 98
  induced by sediment removal, 93
cliff erosion rates, 86–8, 264–77
  average, 87
  instantaneous, 86
  measurements, 87–8
  measuring techniques, 88, 264-77
  spatial variations, 96–106
  temporal variations, 88–96
  vs. cliff height, 105–6
cliff instability induced by human activities, 223–5
cliff line recession
  parallel, 102–3
cliff materials, 79–80
  mechanical strengths, 79
  representative strength, 79
  resisting force, 79–80
cliff profile changes, 107–16
  mode of mass movement. 107–9

parallel recession, 111–16
cliffed coasts, 2
clinoid sponges, 74
cnoidal-wave theory, 8
coalescence
  of potholes, 197
  of solution pools, 201
  of tafoni, 203
coastal lapies, 202
coastal platforms, 139, 141
cohesive shores, 219
cohesive strength, 57
  cliff erosion studies, 79
collapsing breakers, 21
compressive strength, 53–4
  cliff erosion studies, 79
  platform studies, 141–2, 146–52
  representative strength parameter, 65
  ridge and furrow formation, 194
compressive wave velocity, 63–5, 67
conservation of natural environments, 217
conversion factors, 257–63
corrasion, 78
cost-benefit relationship, 216
Couette flow, 129
cracks, 65
  formation of, 127, 129
Culmann theory, 109
cyclones, 44

dam construction of rivers, 219
debris at cliff base, 75, 76(Fig.)
  removal of, 105–6
debris falls, 107
deep-seated (rotational) landslide, 221, 223
deep water condition, 12
deformation, 51
  deformation modulus, 52
density, 49–50
denuded benches, 141
detached breakwaters, 213
direct shear test, 57
direct test, see uniaxial tensile test

Subject Index

discontinuities, 61, 65, 79, 84
   sea cave development, 188–9
   strength reduction, 65–8. 121
dispersion relation, 10
diurnal tides, 40, 41(Fig.)
drowning associated with plunging cliff development, 145

earth falls, 107
Earth–Moon–Sun system, 38–9
echinoids, 74, 121
echo soundings, 165
effective slope angle, 109
elasticity, modulus of, 52
Elephant Rock, 191
encrusting organisms, 80
energy flux, 16
engineering structures, 209
   breakwaters, 219
   bulkheads, 209
   detached breakwaters, 209, 213
   groynes, 209, 216
   jetties, 219
   revetments, 209
   sea walls, 209, 211–13
engineering works, 209–17
Environmental Protection Agency (EPA), 225
ephemeral landforms, 183
   sea arches, 191
   sea caves, 189
equilibrium time, 136, 138
erosion, 2
erosion of cohesive soils, 129
erosion ramps, 196
erosive force of waves, 85
erratics, 129
eustatic sea level changes, 1
experiments, 81
   field, 81
   laboratory, 81
   prototype-(or full-) scale, 81

falls, 107, 108(Fig.)

fatigue, 68–70, 80
   fatigue limit, 69
   fatigue strength, 69
   sea floor bedrocks, 122
faults, 65
   blowholes, 189
   sea caves, 189
   sea tunnels, 191
   stacks, 191–3
Federal Court of Canada, 219
feedback, 84, 85
   negative, 129
   sea cave development, 190
field experiments, 81
field measurements, 81
flakes, 194
   formation of, 127
flat-bottomed pools, 200
flows, 107, 108(Fig.)
Folkestone Warren landslides, 221–4
foot protection works, 218
fracture frequency, 67
freeze-thaw action, 72
friction factor, 151
front depth, 175–6, 178–9
frontal water depth, 30, 145
frost action, 72, *see also* frost weathering
frost weathering, 72
   intertidal bed rocks, 121–2, 127
   tafoni and honeycombs, 208
   Type-B platforms, 176

gastropods, 74, 121
geomorphology, 1
   of rocky coasts, 1–2
global warming, 225
   sea level rise, 225–6
grazing organisms, 74, 121
   rock strength reduction, 121
   solution pool formation, 200
group velocity, 17
groynes, 209, 216, 218, 223

harbour works, 219

headlands-and-bay concept, 216
Holocene (marine) transgression, 1, 139, 143, 171, 194, 226
honeycombs, 202–8
　arkose 202–5, 207, 208
　coalescence, 203
　geological structures, 203
　origin, 208
　size, 202–7
human impacts on rocky coasts, 209–28
　artificial beach nourishment, 221
　beach mining, 221
　cliff slope cutting, 223
　engineering works 209–20, 223
　sea level rise due to global warming, 225–8
　trampling, 225
hurricanes, 44
　bluff erosion by, 87
　sea level fluctuation by, 45
hydration, 73
　notch formation, 186
hydraulic action of waves, 78
　sea cave development, 190
　on plunging cliffs, 144
hydrolysis, 70
hyperbolic functions, 11

ice action, *see* frost action
ice effect, protective, 90
impact strength, 59–60
　impact strength index (ISI), 60
impulsive pressure, 30, 31, 33
induration
　due to encrusting by algae, 194
　ferruginous, 194
Intergovernmental Panel on Climate Change (IPCC), 225
intermediate-depth water condition, 12
internal friction
　angle of, 57, 109
IPCC, *see* Intergovernmental Panel on Climate Change
irregular-shaped specimens, 54

jetties, 219
　cliff erosion induced by jetty construction, 219
joints, 65
　sea cave development, 188–9
　stack development, 191–3

laboratory experiments, 81
　cliff erosion, 81–5
　cohesive shores, 129, 161–2
　sand effects on cliff erosion, 93, 105
　shore platforms, 167–71
laboratory tests, 81
Laplace equation, 9
lithification, 188, 194
London Bridge, 191
longshore currents, 28
long-term erosion rate, 96
low rock platforms, 141
low rock terraces, 141
low-tidal sea cliffs, 166
low-tide cliffs, 166
low-water cliffs, 166
lunar day, 40

macrotides, 44
maps, 88
marine benches, 141
marine potholes, *see* potholes
marine stacks, *see* stacks
marine terraces, 2
mass movement, 75, 76(Fig.), 107–16
　associated with human activities, 221–5
　factors affecting, 107
　type, 107
mean high water (MHW), 44
mean high water neap (MHWN), 44
mean high water spring (MHWS), 44
mean higher high water (MHHW), 44
mean low water (MLW), 44
mean low water neap (MLWN), 44
mean low water spring (MLWS), 44
mean lower low water (MLLW), 44
mean sea level (MSL), 44

Subject Index

mean tide level (MTL), 44
mean water level (MWL), 26
mechanical action of waves, 78
 on plunging cliffs, 145
mechanical strength, 48
 rocks, 50–65
 soils, 50–8, 63
MEM, see Micro-erosion meter
Merlin's Cave, 191
mesotides, 44
MHHW, see mean higher high water
MHW, see mean high water
MHWN, see mean high water neap
MHWS, see mean high water spring
microcaliper, 123
micro-erosion meter (MEM), 88
 measurements, 104, 123, 124, 125, 126, 127, 128, 158, 176, 184, 187, 196
microtides, 42
mixed tides, 40, 41(Fig.)
MLLW, see mean lower low water
MLW, see mean low water
MLWN, see mean low water neap
MLWS, see mean low water spring
model cliffs, 81–4, 170–1
 mechanical properties, 83
 models for Type-A platform development, 161-3
 horizontal-erosion antecedent, 162
 simultaneous erosion, 163
 vertical-erosion antecedent, 161
Mohr's circle, 51, 56–7
Mohr's envelope, 57(Fig.)
moisture content, 50
mound-type sea walls, 211, 217
moving layer
 thickness of, 119
MSL, see mean sea level
MTL, see mean tide level
mudflows, 108
mudslides, 108
MWL, see mean water level

nails, 88

neap tides, 42
nearshore currents, 28–9, 76(Fig.)
Needles, The, 193
Newton's law of gravitation, 38
nips, 184
notch development
 abrasion, 187
 biological activities, 187
 solution, 187
 speed, 184, 187
notches, 184–8
 double, 188
 height, 188
 intertidal, 186, 187
 level, 188
 measurements, 184
 in plunging cliffs, 184
 profile, 185
 single, 188
 subtidal, 169–70, 186, 188
NX core, 55, 68

Old Hat type platform, 164
 elevation, 164
 level of permanent saturation, 164
Old Man of Hoy, 193
onshore currents, 28
ord, 93
 associated with cliff erosion, 93–4
organic accretion, 188
orthogonals, 19
oxidation, 71

pans, 200
parallel recession
 of cliff profile, 111–16
 of cliff line, 102–3
pegs, 88, 187
penetration strength, 63
 cliff erosion studies, 79, 98, 99
penetrometers, 63, 98
permanent saturation
 level of, 164

phase velocity, 9
pinnacles, 202
planar slides, 107, 108(Fig.), 111
plaster, 82
platforms
  abrasion, 139
  beach, 139
  coastal, 139, 141
  horizontal, 139, 141
  low rock, 141
  Old Hat type, 164
  rock, 141
  shore, 139, 141
  sloping, 139
  storm-wave, 141
  submarine, 139
  wave-cut, 139, 141
plucking, 78
  cohesive shores, 129
  sea floor bedrocks, 123, 126
  Type-A platforms, 163
plunging breakers, 21, 119
plunging cliffs, 139–50, 182
  frontal depth, 144
  lithology of, 143–4
  origin, 143–4
  profile, 142–3
  slope, 144
  strength of, 148–50
  survival of, 144–5
pneumatic action in sea cave
    development, 190
P–N–J method, 7
point-load test, 55
Poisson's ratio, 51
pools, *see* solution pools
porosity, 50
Portland cement, 83
potential failure plane
  angle of, 109
potholes, 196–200
  abrasion, 198–9
  abrasive material, 198
  coalescence, 197
  concretions, 199
  dimensions, 197(Fig.), 198(Fig.)

  erratics, 199
  joints, 199
  rock types, 199
  solution, 199
  till platforms, 200
  vortex, 197
pressure, 9, 16
  dynamic, 16
  static, 16
projection of sea level rise
  by EPA, 225
  by IPCC, 225

Quarrying, 78, 123, 126, *see also* plucking

Ramparts, 193
  arkose platforms, 194–5
  wet/dry weathering, 193
  wet-edge hypothesis, 193, 194
ramps, 196
reduction process in weathering, 71
reflected waves, *see* standing waves
refraction coefficient, wave, 19
relative rock-strength parameter, 183
representative strength parameter, 65
resisting force
  of cliff material, 79–80
  of sea floor bedrocks, 120–2
respiration, 200
resurgences, 45
revetments, 209, *see also* sea walls
rigid-type sea walls, 211, 217
rimmed pools, 200
  size, 201
rip currents, 28
riprap sea walls, 211, 213
rips, *see* rip currents
rock falls, 107
rock mass factor, 67
rock platforms, 141
rock quality designation (RQD), 68
rock strength
  deterioration of, 68–74

Subject Index

by biological factors, 74, 121
by fatigue, 68–70
by weathering, 70–4
rocks, 48
  compressive strength, 52–4
  compressive wave velocity, 63–5
  impact strength, 59–60
  mechanical strength, 50–65
  physical properties, 48–50
  Schmidt hummer rebound number, 60–2
  shear strength, 56–9
  tensile strength, 54–6
rocky coasts, 2–3, 48
  evolution model, 180–3
rotational slides, 107, 108(Fig.)
  deep-seated, 107, 108(Fig.), 114
  shallow-seated, 107, 108(Fig.), 116
rubble-mound sea walls, 211, 213

salt crystallization, 72
salt weathering, 72–3
  tafoni and honeycomb formation, 208
saturation, degree of, 50
scanning electron microscopy in tafoni studies, 208
Schmidt hammer rebound number, 60–2
  cliff erosion studies, 79
scouring at the toe of structures, 218
scuba, 128, 165
sea arches, 190–3
  ephemeral, 191
  double, 191
  rapid change, 191
sea caves, 188–90
  blowholes, 189, 190
  ephemeral, 189
  formative processes, 190
  in plunging cliffs, 189
sea level fluctuations
  Quaternary, 2, *see also* Holocene (marine) transgression
sea level rise due to global warming, 225

cliff erosion associated with, 225–8
sea tunnel, 191
sea urchins, 200
sea walls, 209, 211–13, 216, 217, 218, 223
  failure of, 216
  riprap, 211, 213
  rubble-mound, 211
  steel sheet pilings, 211
  timber, 211
  type of sea walls, 209
seaward cliffs (drops or scarps), 151, 166–72, 177
sediment, depth of, 119, 120
sediment budget, 3
sediment transport
  longshore, 3
  onshore/offshore, 3
  by nearshore currents, 28–9
sediment volume controlling cliff erosion, 3
Seiches, 47
semidiurnal tides, 40, 41(Fig.)
shallow water condition, 12
shear box, 57
shear force
  wave-induced, 151
shear strength, 56–9
  cliff erosion studies, 79
  platform studies, 151
  ridge and furrow formation, 194
  slope failure, 109
shear stress, 51, 109
  wave-induced, 119
  of sediment-laden flow, 120
Shields parameter, 120
shoaling coefficient, 18
shock pressure, 32, *see also* impulsive pressure
shore benches, 141
shore platforms, 139–80
  horizontal, 139–41
  Old Hat type, 164
  sloping, 139–40
  Type-A, 141–2, 150–63
  Type-B, 141–2, 150–2, 163–80

shore platforms (*continued*)
  vs. plunging cliffs, 144–50
shore potholes, *see* potholes
shore protection structures, 211
short-term erosion rate, 90
SI units, 52, 257–63
skittle-like boulder in potholes 199
slaking, 74, 121, 127, *see also* wet/dry weathering
slides, 107
slope failure, 223
slope instability, 107
S–M–B method, 7
S–N curves, 69
snails, 200
soils, 48
  compressive strength, 52–3
  mechanical strength, 50–8, 63
  penetration strength, 63
  physical properties of, 48–50
  shear strength, 56–9
solitary-wave theory, 8
solution, 71–2
  notch formation, 186
  pothole initiation, 199
  recession of seaward drop, 167
  Type-B platform lowering, 176
solution basins, 200
solution bench, 188
solution pits, 201
solution pools, 200–2
  biochemical processes, 200
  with elevated rims, 200
  cross-shore distribution, 201
sonic velocity, 67, 147, *see also* compressive wave velocity
spectral analysis, 7
spilling breakers, 21
spiral grooves in potholes, 198
sponges, 121
spring tides, 41
stacks, 190–3
  development of, 193
standing waves, 29, 182
  plunging cliff development, 148
  pressure intensity, 29–31

scouring in front of sea walls, 218
still-water level (SWL), 8(Fig,), 9
still-water line, 26(Fig.)
Stokes-wave theory, 8
storm surges, 38, 44–7
  1953 North Sea storm surge, 91, 105, 211
  Type-A platform development, 162
storm terraces, 141
storm-wave platforms, 141
strain, 50–2
strength reduction
  by biological factors, 74
  by repeated (cyclic) loading, 69
  by weathering, 70–4
stress, 50–2
  cyclic, 69
  hydration, 73
  normal, 50
  peak, 52
  shear, 51
  thermal, 72
sructural ramps, 196
surf base, 132
  Type-A platform development, 157
surf similarity parameter, 21
surges, *see* storm surges
surging breakers, 21
surveying, 88
submarine platforms, 139
subtidal notches, 169–70
SWL, *see* still-water level

tafoni, 202–8
  growth rate, 203, 208
  origin, 208
tensile strength, 54–6
  cliff erosion studies, 79
  ridge and furrow formation, 194
tension crack, 112(Fig.), 113
terraces
  low rock, 141
  marine, 2
  storm, 141
  wave-cut, 141

terrestrial photographs, 88
thermal weathering, 72
  tafoni and honeycomb formation, 208
tides, 38–44
  astronomical, 38
  diurnal, 40, 41(Fig.)
  mixed, 40, 41(Fig.)
  neap, 42
  semidiurnal, 40, 41(Fig.)
  spring, 41
  tidal bulge, 39
  tidal curve, 40
  tidal day, 40
  tidal range, 42, 44
  tide generating force, 39, 40
toe weighting structures, 223
topples, 107, 108(Fig.)
trampling, 225
triaxial compression test, 56
Twelve Apostles, 193
Type-A platforms, 153–63, 182, 183
  cliff-platform junction, 163
  development, 158–63
  elevation, 160
  gradient, 154, 156, 158
  models, 153, 161–3
  profile, 154–7
  ramp, 159(Fig.)
  vs. Type-B platforms, 150–3
  width, 142(Fig.), 154, 158
Type-B platforms, 141, 142(Fig.), 163–80, 182, 183
  elevation, 172–7
  front depth, 175–6
  initiation, 166–72
  laboratory experiments, 167–71, 173
  level of permanent saturation, 164
  Old Hat type, 164
  profile, 165–6
  rate of lowering, 176–7
  seaward cliffs (drops or scarps), 166–72
  secondary lowering processes, 176–7
  vs. Type-A platforms, 150–3
  width, 177–80

Typhoons, 44
unconfined (or uniaxial) compressive strength, 52–4, 65, *see also* compressive strength
uniaxial tensile test, 54–5
unibidirectional flow on Type-B platforms, 194
unit weight, 49–50
unloading, 74, 223

velocity index, 67
velocity potential, 9
visor, 184
void ratio, 50
vortex
  associated with cliff erosion, 105
  notch formation, 186
  pothole development, 197
  three-dimensional, 129

water content, 50
water curing, 82(Fig.)
water level rise due to meteorological factors, 45–6
water particle motions, 13–15
  to-and-fro, 14
wave base, 129, 132–8
  definitions, 132
  prediction, 134
  Type-A platform development, 153, 157
  values, 133
wave-cut benches, 139, 141
wave-cut platforms, 139, 141
wave-cut ramps, 196
wave-cut terraces, 141
wave energy, 16–18
  rate of transmission, 16
wave forecasting, 7–8
  P–N–J method, 7
  S–M–B method, 7
wave furrows, 196
wave height, 5, 8
  change in, 17–18

wave number, 10
wave period, 5, 9
wave power, 17
   cliff erosion studies, 100–1
wave prediction, *see* wave forecasting
wave ramps, 141, 196
wave refraction, 18–20
wave set-up, 26–7
wave theories, 8
   Airy, 8
   finite-amplitude, 8
   linear, 8–20
   small-amplitude, 8
wave velocity, 9
wavelength, 8, 10
waves, 5–37
   development of, 5–6
   hydraulic action, 78
   irregular, 7
   mechanical action, 78
   pressure of, 16, 30–7
   regular, 8
   significant, 5
   swell, 5
   type of, 29–30
   wind-generated, 5
weathering, 70–4
   chemical, 70–2

frost, 72
granite, 70
limestone, 72
mechanical, 72–4
muricate, 72
notch formation, 186
salt, 72–3
thermal, 72
wet/dry, 74
weathering joints, 80, 127
wedge action, 78
wet/dry weathering, 74
   rampart formation, 193
   ridge and furrow formation, 194
   Type-B platform lowering, 127, 176
wet-edge hypothesis, 193, 194
wind erosion in tafoni and honeycomb
   formation, 208
wind set-up, 45
wind waves, 5

X-ray diffraction analyses in tafoni
   studies, 208

zero-upcrossing method, 5